Communications and Control Engineering

Springer
London
Berlin
Heidelberg
New York
Barcelona
Hong Kong
Milan
Paris
Singapore
Tokyo

Published titles include:

Constructive Nonlinear Control
R. Sepulchre, M. Janković and P.V. Kokotović

A Theory of Learning and Generalization
M. Vidyasagar

Adaptive Control
I.D. Landau, R. Lozano and M.M' Saad

Stabilization of Nonlinear Uncertain Systems
Miroslav Krstić and Hua Deng

Passivity-based Control of Euler-Lagrange Systems
Romeo Ortega, Antonio Loría, Per Johan Nicklasson and Hebertt Sira-Ramírez

Stability and Stabilization of Infinite Dimensional Systems with Applications
Zheng-Hua Luo, Bao-Zhu Guo and Omer Morgul

Nonsmooth Mechanics (2nd edition)
Bernard Brogliato

Nonlinear Control Systems II
Alberto Isidori

L_2-Gain and Passivity Techniques in nonlinear Control
Arjan van der Schaft

Control of Linear Systems with Regulation and Input Constraints
Ali Saberi, Anton A. Stoorvogel and Peddapullaiah Sannuti

Robust and H_∞ Control
Ben M. Chen

Computer Controlled Systems
Efim N. Rosenwasser and Bernhard P. Lampe

Dissipative Systems Analysis and Control
Rogelio Lozano, Bernard Brogliato, Olav Egeland and Bernhard Maschke

Control of Complex and Uncertain Systems
Stanislav V. Emelyanov and Sergey K. Korovin

Robust Control Design Using H^∞ Methods
Ian R.Petersen, Valery A. Ugrinovski and Andrey V.Savkin

Model Reduction for Control System Design
Goro Obinata and Brian D. O. Anderson

Control Theory for Linear Systems
Harry L. Trentelman, Anton Stoorvogel and Malo Hautus

Simon G. Fabri and Visakan Kadirkamanathan

Functional Adaptive Control

An Intelligent Systems Approach

With 74 Figures

 Springer

Simon G. Fabri, PhD
Department of Electrical Power and Control Engineering, University of Malta,
Msida MSD 06, Malta

Visakan Kadirkamanathan, PhD
Department of Automatic Control and Systems Engineering,
The University of Sheffield, Sheffield, S1 3JD

Series Editors
E.D. Sontag • M. Thoma

ISSN 0178-5354

ISBN 1-85233-438-X   Springer-Verlag London Berlin Heidelberg

British Library Cataloguing in Publication Data
Fabri, Simon G.
  Functional adaptive control : an intelligent systems
  Approach. - (communications and control engineering)
  1. Adaptive control systems 2. Intelligent control systems
  I. Title II. Kadirkamanathan, Visakan
  629.8
ISBN 185233438X

Library of Congress Cataloging-in-Publication Data
Fabri, Simon G., 1965-
  Functional adaptive control : an intelligent systems approach. .Simon
G. Fabri and Visakan Kadirkamanathan.
      p. cm. -- (Communications and control engineering)
  Includes bibliographical references and index.
  ISBN 1-85233-438-X (alk. paper)
  1. Adaptive control systems. 2. Intelligent control systems. I.
Kadirkamanathan, Visakan, 1962- II. Title. III. Series.
  TJ217 .F33 2001
  629.8' 36-dc21
                                                    2001016083

Apart from any fair dealing for the purposes of research or private study, or criticism or review, as permitted under the Copyright, Designs and Patents Act 1988, this publication may only be reproduced, stored or transmitted, in any form or by any means, with the prior permission in writing of the publishers, or in the case of reprographic reproduction in accordance with the terms of licences issued by the Copyright Licensing Agency. Enquiries concerning reproduction outside those terms should be sent to the publishers.

© Springer-Verlag London Limited 2001
Printed in Great Britain

The use of registered names, trademarks, etc. in this publication does not imply, even in the absence of a specific statement, that such names are exempt from the relevant laws and regulations and therefore free for general use.

The publisher makes no representation, express or implied, with regard to the accuracy of the information contained in this book and cannot accept any legal responsibility or liability for any errors or omissions that may be made.

Typesetting: Camera ready by authors
Printed and bound by Athenæum Press Ltd., Gateshead, Tyne & Wear
69/3830-543210   Printed on acid-free paper   SPIN 10779960

To Miriam, Ġemima and Jean-Marc (S.F)

To my mother (V.K)

Preface

The field of intelligent control has recently emerged as a response to the challenge of controlling highly complex and uncertain nonlinear systems. It attempts to endow the controller with the key properties of adaptation, learning and autonomy. The field is still immature and there exists a wide scope for the development of new methods that enhance the key properties of intelligent systems and improve the performance in the face of increasingly complex or uncertain conditions.

The work reported in this book represents a step in this direction. A number of original neural network-based adaptive control designs are introduced for dealing with plants characterized by unknown functions, nonlinearity, multimodal behaviour, randomness and disturbances. The proposed schemes achieve high levels of performance by enhancing the controller's capability for adaptation, stabilization, management of uncertainty, and learning. Both deterministic and stochastic plants are considered.

In the deterministic case, implementation, stability and convergence issues are addressed from the perspective of Lyapunov theory. When compared with other schemes, the methods presented lead to more efficient use of computational storage and improved adaptation for continuous-time systems, and more global stability results with less prior knowledge in discrete-time systems.

Stochastic systems are handled by using concepts from stochastic adaptive optimal control, leading to novel dual adaptive control schemes that take into consideration the extent of the uncertainty when generating the control signal. This yields a highly improved transient response, particularly in the absence of prior network training. Multiple model techniques are used to address the complexity arising from multimodal behaviour. These lead to a reconfigurable control scheme for uncertain jump-nonlinear systems, and also an adaptive form of gain scheduling for plants characterized by unknown complex nonlinearities.

A range of simulation experiments are included, confirming that these novel designs are truly effective for dealing with the stringent conditions usually associated with intelligent control.

November 2000 *Simon G. Fabri and Visakan Kadirkamanathan*

Acknowledgements

The book arose out of the first author's research towards his degree of *Ph.D.* at the Department of Automatic Control & Systems Engineering, University of Sheffield, United Kingdom, under the supervision of the second author. During the period of three years from 1996 – 1999, we have had numerous interesting conversations on various aspects of control and estimation and have been a source of inspiration to each other.

The first author's *Ph.D.* research programme was financially supported by a CVCP Overseas Research Students (ORS) award of the U.K., a staff development scholarship from the University of Malta and funds from the Department of Automatic Control & Systems Engineering of the University of Sheffield. These funds were made possible by the support of Professor Joseph A. Agius of the University of Malta and Professor Peter J. Fleming of The University of Sheffield. The second author acknowledges the support of his academic colleagues who have had to share the teaching load and the support of the University of Sheffield which granted a study leave to enable this book to be written.

A very special thank you goes to our wives Miriam and Manjula for their interest, encouragement, patience and moral support during these particularly busy years. Thanks also to our parents and siblings who have always been an example to follow.

Finally we would like to thank the ACSE Department's general office and programming staff for secretarial and technical assistance. We also acknowledge all those who helped indirectly, perhaps by a word of encouragement or a sincere expression of interest. In particular we would like to mention our colleagues Mr M.H. Jaward, Mr P. Li, Dr Z.-Q. Lang and Dr M.-Y. Chen, and the first author's colleagues in the Faculty of Engineering at the University of Malta.

Contents

List of Figures xv

Acronyms xix

Mathematical Notation xxi

Part I. Introduction

1. **Introduction** 3
   1.1 Intelligent Control Systems 3
   1.2 Approaches to Intelligent Control 5
       1.2.1 Contribution of Adaptive Control 5
       1.2.2 Contribution of Artificial Intelligence 6
       1.2.3 Confluence of Adaptive Control and AI: Intelligent Control 8
   1.3 Enhancing the Performance of Intelligent Control 10
       1.3.1 Multiple Model Schemes: Dealing with Complexity ... 10
       1.3.2 Stochastic Adaptive Control: Dealing with Uncertainty 14
   1.4 The Objectives and their Rationale 16

Part II. Deterministic Systems

2. **Adaptive Control of Nonlinear Systems** 23
   2.1 Introduction 23
   2.2 Continuous-time Systems 23
       2.2.1 Control by Feedback Linearization 24
       2.2.2 Control by Backstepping 27
       2.2.3 Adaptive Control 29
   2.3 Discrete-time Systems 36
       2.3.1 Affine Approximations and Feedback Linearization ... 42
       2.3.2 Adaptive Control 42
   2.4 Summary 46

3. Dynamic Structure Networks for Stable Adaptive Control — 47
- 3.1 Introduction — 47
- 3.2 Problem Formulation — 49
- 3.3 Fixed-structure Network Solutions — 50
- 3.4 Dynamic Network Structure — 54
- 3.5 The Control Law and Error Dynamics — 56
- 3.6 The Adaptive System — 58
- 3.7 Stability Analysis — 59
- 3.8 Evaluation of Control Parameters and Implementation — 62
  - 3.8.1 The Disturbance Bound — 62
  - 3.8.2 Choice of the Boundary Layer — 65
  - 3.8.3 Comments — 67
  - 3.8.4 Implementation — 68
- 3.9 Simulation Examples — 70
  - 3.9.1 Example 1 — 70
  - 3.9.2 Example 2 — 73
- 3.10 Summary — 77

4. Composite Adaptive Control of Continuous-Time Systems — 79
- 4.1 Introduction — 79
- 4.2 Problem Formulation — 81
- 4.3 The Neural Networks — 81
- 4.4 The Control Law — 83
- 4.5 Composite Adaptation — 84
  - 4.5.1 The Identification Model — 84
  - 4.5.2 The Adaptation Law — 86
- 4.6 Stability Analysis — 88
- 4.7 Determination of the Disturbance Bounds — 91
- 4.8 Simulation Examples — 92
  - 4.8.1 Example 1 — 92
  - 4.8.2 Example 2 — 98
- 4.9 Summary — 100

5. Functional Adaptive Control of Discrete-Time Systems — 101
- 5.1 Introduction — 101
- 5.2 Problem Formulation — 102
- 5.3 The Neural Network — 103
- 5.4 The Control Law — 104
- 5.5 The Adaptive System — 104
- 5.6 Stability Analysis — 105
- 5.7 Tracking Error Convergence — 111
- 5.8 Simulation Examples — 112
  - 5.8.1 Example 1 — 113
  - 5.8.2 Example 2 — 115
- 5.9 Extension to Adaptive Sliding Mode Control — 117

    5.9.1 Definitions of a Discrete-time Sliding Mode 117
    5.9.2 Adaptive Sliding Mode Control 120
    5.9.3 Problem Formulation 121
    5.9.4 The Control Law 122
    5.9.5 The Adaptive System 122
    5.9.6 Stability Analysis 123
    5.9.7 Sliding and Tracking Error Convergence 123
    5.9.8 Simulation Example 124
  5.10 Summary... 126

Part III. Stochastic Systems

6. **Stochastic Control** 131
   6.1 Introduction 131
   6.2 Fundamental Principles 132
   6.3 Classes of Stochastic Control Problems 134
   6.4 Dual Control... 136
       6.4.1 Degrees of Interaction............................ 137
       6.4.2 Solutions to the Implementation Problem............ 137
   6.5 Conclusions.. 145

7. **Dual Adaptive Control of Nonlinear Systems** 147
   7.1 Introduction 147
   7.2 Problem Formulation 148
   7.3 Dual Controller Design 149
       7.3.1 GaRBF Dual Controller 149
       7.3.2 Sigmoidal MLP Dual Controller 153
       7.3.3 Analysis of the Control Laws 157
   7.4 Simulation Examples and Performance Evaluation 158
       7.4.1 Example 1....................................... 158
       7.4.2 Example 2....................................... 161
   7.5 Summary... 163

8. **Multiple Model Approaches** 165
   8.1 Introduction 165
   8.2 Basic Formulation 165
       8.2.1 Multiple Model Adaptive Control................... 168
       8.2.2 Jump Systems 170
   8.3 Adaptive IO Models 178
       8.3.1 Scheduled Mode Transitions 179
   8.4 Summary... 185

9. Multiple Model Dual Adaptive Control of Jump Nonlinear Systems ... 187
- 9.1 Introduction ... 187
- 9.2 Problem Formulation ... 189
- 9.3 The Estimation Problem ... 191
  - 9.3.1 Known Mode Case ... 191
  - 9.3.2 Unknown Mode Case ... 193
- 9.4 Self-organized Allocation of Local Models ... 197
- 9.5 The Control Law ... 200
  - 9.5.1 Known Mode Case ... 200
  - 9.5.2 Unknown Mode Case ... 201
- 9.6 Simulation Examples and Performance Evaluation ... 204
  - 9.6.1 Example 1 ... 205
  - 9.6.2 Example 2 ... 210
- 9.7 Summary ... 212

10. Multiple Model Dual Adaptive Control of Spatial Multimodal Systems ... 213
- 10.1 Introduction ... 213
- 10.2 Problem Formulation ... 215
- 10.3 The Modular Network ... 216
- 10.4 The Estimation Problem ... 217
  - 10.4.1 Local Model Parameter Estimation ... 217
  - 10.4.2 Validity Function Estimation ... 220
- 10.5 The Control Law ... 224
  - 10.5.1 Known System Case ... 224
  - 10.5.2 Unknown System Case ... 227
- 10.6 Simulation Examples and Performance Evaluation ... 231
  - 10.6.1 Example 1 ... 231
  - 10.6.2 Example 2 ... 236
  - 10.6.3 Performance Evaluation ... 239
- 10.7 Summary ... 240

Part IV. Conclusions

11. Conclusions ... 245

References ... 250

Index ... 265

List of Figures

2.1	Generic 2-layer feedforward neural network.	32
3.1	Neural network based adaptive control scheme.	51
3.2	Selective node activation technique: using a hypersphere or a hypercube.	56
3.3	Initial hypercube of activated units.	64
3.4	Φ-based activation technique.	70
3.5	Simulation 1: Tracking error.	72
3.6	Simulation 1: Activated nodes.	72
3.7	Two degrees of freedom robotic manipulator.	73
3.8	Simulation 2: Tracking error.	76
4.1	Generation of the estimation error.	86
4.2	Trial 1 of Example 1; (a) System output (b) Tracking error with direct adaptation.	93
4.3	Trial 1 of Example 1; (a) Function approximation (b) Approximation error with direct adaptation.	93
4.4	Trial 2 of Example 1; (a) System output (b) Tracking error with composite adaptation.	94
4.5	Trial 2 of Example 1; (a) Function approximation (b) Approximation error with composite adaptation.	94
4.6	Trial 3 of Example 1; (a) System output (b) Tracking error with *high gain* direct adaptation.	95
4.7	Trial 3 of Example 1; (a) Function approximation (b) Approximation error with *high gain* direct adaptation.	95
4.8	Trial 3 of Example 1; time variation of an arbitrarily chosen network parameter with *high gain* direct adaptation.	96
4.9	Trial 4 of Example 1; (a) System output (b) Tracking error with *high gain* composite adaptation.	96
4.10	Trial 4 of Example 1; (a) Function approximation (b) Approximation error with *high gain* composite adaptation.	97
4.11	Trial 4 of Example 1; time variation of an arbitrarily chosen network parameter with *high gain* composite adaptation.	97
4.12	Results of Example 2; (a) Direct adaptation (b) Composite adaptation	99

4.13 Steady-state tracking errors of Example 2; (a) Direct adaptation (b) Composite adaptation 99

5.1 The $(e_1, \|\tilde{\mathbf{w}}\|)$ plane: the semi-ellipses represent the level curves of $V(k)$ with the outer ones corresponding to larger values; the vertical dashed lines denote the boundaries of region I; and the solid line reflects the constraints of a typical trajectory........... 107
5.2 Results of Example 1; (a) Output (solid) and reference (dashed) (b) Control input 114
5.3 Results of Example 1; (a) Initial tracking error (b) Steady-state tracking error 114
5.4 Results of Example 1; Function $f(\mathbf{x})$ (solid) and the network approximation (dashed) 114
5.5 Results of Example 2; Tracking error 115
5.6 Results of Example 2; Steady-state convergence............... 116
5.7 Results of Example 2; Function approximation 116
5.8 The trajectory of an ideal discrete-time sliding regime in a second order system. Points (A, B, C, D) represent the reaching phase and points (D, E, F, G), the sliding phase. 118
5.9 Sliding error $s(k)$ 125
5.10 Steady-state convergence of $s(k)$ to the boundary layer 125
5.11 Tracking error $e(k)$... 126
5.12 Steady-state convergence of $e(k)$ to the error bound 126

7.1 Output and Tracking error; (a) HCE (b) Cautious (c) Dual. N.B: In plot a(i), the y-axis is truncated to enable clear visualization of the steady-state tracking. The actual amplitude during the initial period of the response could be seen in plot a(ii), which is purposely drawn at a different scale from the rest. 159
7.2 Accumulated cost; (a) HCE (b) Cautious (c) Dual 160
7.3 Effect of q; (a) HCE ($q = 1$) (b) Cautious ($q = 0.0001$) (c) Dual ($q = 0.0001$) 160
7.4 Output and Tracking error; (a) HCE (b) Cautious (c) Dual 161
7.5 Accumulated cost; (a) HCE (b) Cautious (c) Dual 162
7.6 Output and Tracking error; (a) HCE (b) Cautious (c) Dual 163
7.7 Accumulated cost; (a) HCE (b) Cautious (c) Dual 163

8.1 The DUL multiple model control scheme..................... 169
8.2 Evolution diagram for the optimal scheme with $H = 2$. Each block represents one Kalman filter, the event $M_j(k)$ to which it is matched and the corresponding state estimate, covariance and probability updates. The arrow links show how information is propagated from one time instant to the next along a particular sequence. 173
8.3 Evolution diagram for GPB with $d = 2$ 174

8.4	Evolution diagram for GPB with $d = 1$	175
8.5	Evolution diagram for IMM algorithm	176
8.6	Partitioning of the scheduling space. Note that in general a partition need not be continuous, like P_1 and P_2 in the diagram.	181
8.7	A modular network representation.	182
9.1	The curves represent $p(y\|\hat{M}, \hat{S}, \tilde{I})$, the probability density functions (pdf) for two models i and j. p_j represents the pdf for a model that has not yet been subjected to mode learning, and p_i is the pdf of model i at some time k after it was subjected to learning. The shape of the pdfs shows how the variance decreases as a consequence of learning. During activity of the mode captured by model i, the plant output will be well within the support of p_i, as shown by point $y(k_1)$. By contrast, if at some time k_2 a different mode is active, the innovations ϵ_i of model i will increase rapidly, leading to a small value of $p(y(k_2)\|\hat{M}_i(k_2), \hat{S}(k_2 - 1), \tilde{I}^{k_2})$.	198
9.2	MAPIDC control law; (a) reference input (b) plant output	206
9.3	MAPIDC control law; (a) tracking error (b) local model allocation (c) control	206
9.4	MMIDC control law; (a) output and reference input (b) tracking error	207
9.5	MMIDC control law; (a) local model allocation (b) control	208
9.6	Probability density function at $25s$ instant	208
9.7	Probability density function at $29s$ instant	209
9.8	Probability density function at $40s$ instant	209
9.9	(a) Mode sequence (b) MMIDC performance (y and y_d)	210
9.10	Various single model adaptive controllers (a) Kalman filter estimator (b) Exponential forgetting (c) Random walk (d) Covariance resetting (e) Modified random walk (f) Modified covariance resetting	211
10.1	Illustration of the proof of Proposition 10.5.1 for a simple three mode system. Note that only u_3 "satisfies" its own partition.	227
10.2	Example 1 with softmax gating and control law (10.42).	232
10.3	Local model parameter estimation for Example 1.	232
10.4	Example 1 with softmax gate; (a) Validity functions (b)Actual and estimated (dashed) plant dynamics	233
10.5	Example 1 with softmax gating and control law (10.39)	233
10.6	Example 1 with GMK gating and control law (10.42)	234
10.7	Local model parameter estimates for example 1 with GMK gate	234
10.8	Example 1 with GMK gate; (a) Validity functions and Kernels (dashed) (b)Actual and estimated (dashed) dynamics	235
10.9	Example 1 with GMK gate and control law (10.39)	236
10.10	Example 2 with softmax gate	237
10.11	Example 2 with GMK gate	237
10.12	Local model estimates for Example 2 with softmax gate	238

10.13 Local model estimates for Example 2 with GMK gate 238
10.14 Partition boundaries with softmax: actual (solid) and estimated (dashed) 239
10.15 Partition boundaries with GMK; actual (solid) and estimated (dashed) 239
10.16 System output of Example 2 with control law (10.39); (a) Softmax gate (b) GMK gate 240

Acronyms

AAC	Actively adaptive controller
AI	Artificial intelligence
ARMAX	Auto-regressive moving average with exogenous input
ARX	Auto-regressive with exogenous input
ASOD	Actively suboptimal dual controller
CE	Certainty equivalence
CL	Closed-loop
CMAC	Cerebellar model articulation controller
DEA	Detection-estimation algorithm
EKF	Extended Kalman filter
ELS	Extended least squares
EM	Expectation-maximization
GaRBF	Gaussian radial basis function
GMK	Gaussian mixture kernel
GMV	Generalized minimum variance
GPB	Generalized pseudo-Bayes
HCE	Heuristic certainty equivalence
IDC	Innovations dual control
iid	Identically, indipendently distributed
IMC	Internal model control
IMM	Interacting multiple model
IO	Input-output
ISI	Incomplete state information
LQG	Linear quadratic Gaussian
MAD	Model adaptive control
MMAE/C	Multiple model adaptive estimation/control
MAP	Maximum *a posteriori*
MAPIDC	Maximum *a posteriori* innovations dual control
ME	Mixture of experts
MIMO	Multiple-input/multiple-output
MLP	Multi-layer perceptron
MM	Multiple model
MMIDC	Multiple model innovations dual control
MRAC	Model reference adaptive control

NARMA	Nonlinear auto-regressive moving average
NN	Neural network
OLOF	Open-loop optimal feedback
PAF/C	Partitioned adaptive filtering/control
RBF	Radial basis function
RNN	Recurrent neural network
RSA	Random sampling algorithm
SISO	Single-input/single-output
SOFLIC	Self-organizing fuzzy logic controller
SPR	Strictly positive real
UUB	Uniform ultimate boundedness
VSC	Variable structure control

Mathematical Notation

Notation	Description
$\mathbf{a}, \mathbf{A}$	Boldface characters denote vectors and matrices
$\mathbf{A}^T$	Transpose of a matrix
$|x|$	Absolute value of scalar x
$\|\cdot\|$	Euclidean vector norm
$\dot{y}, y^{(r)}$	First and rth derivative of y with respect to time
$\Re$	The set of real numbers
$\Re^n$	n-dimensional Euclidean space
$\Re \mapsto \Re^n$	A mapping from $\Re$ to $\Re^n$
$\nabla h(\mathbf{x})$	Gradient of $h(\mathbf{x})$, i.e., $[\partial h/\partial x_1 \cdots \partial h/\partial x_n]$
$L_\mathbf{a} h(\mathbf{x})$	Lie derivative of $h(\mathbf{x})$ with respect to $\mathbf{a}(\mathbf{x})$
$f_1 \circ f_2$	Composition of functions f_1, f_2, i.e., $f_1(f_2(\cdot))$
$x \in X$	Element x belongs to set X
$x \notin X$	Element x does not belong to set X
$X \subset Y$	Set X is contained in set Y
$:=$	Defined as
$\equiv$	Equivalence
$\approx$	Approximately equal
$x \gg y$	x is much greater than y
$\mathbf{x} \sim N(\mathbf{a}, \mathbf{B})$	$\mathbf{x}$ is normally distributed with mean $\mathbf{a}$ and covariance $\mathbf{B}$
$E\{x\}$	The mathematical expectation of x
$E\{x|y\}$	The expectation of x conditioned on y
$\text{cov}\{\mathbf{x}\}$	The covariance matrix of vector $\mathbf{x}$
$p(x|y)$	The probability density of x given y
$\Pr(x)$	The probability of x
$\min_a$	The minimum with respect to a
$\arg\min_a$	The value of a that minimizes
$\exp(\cdot)$	Exponential

Part I

Introduction

1. Introduction

1.1 Intelligent Control Systems

The main goal of control engineering is to ensure that some system of interest performs according to a given specification, often under conditions of uncertainty and with as little human intervention as possible. In general, the uncertainty arises because of insufficient knowledge about the system itself or the environment in which it operates. This could be due to a highly complex plant, components that change their characteristics because of failure or drift, and the presence of unpredictable external disturbances. The specification, which need not be constant so as to accommodate for possible changes in the control objectives, defines how the variables of interest within the system are required to behave. This goal demands the system to effect some form of self-regulation. Control theory shows that in general, the most reliable way of achieving this, is by connecting a suitably designed controller in a feedback configuration with the system.

A large body of control techniques, including *classical control*, *optimal control* and *robust control*, are based on designing fixed controllers that can only meet this goal under very restricted conditions. For example, the specification is assumed constant and both the plant and the disturbances must be known very accurately. In situations where the plant parameters are uncertain, the technique of *adaptive control* [23, 144, 186] offers a definite improvement because it is able to maintain adequate performance in the presence of unknown or time-varying parameters. However, conventional adaptive control techniques can hardly claim to be reliable when drastic and sudden changes take place, say like those brought about by the onset of plant, sensor or actuator faults. Such radical and discontinuous changes in the structure of the system are not usually accommodated for in these schemes. Combining this with the fact that the uncertainty permitted is often limited only to linear plant parameters, one is led to the conclusion that even conventional adaptive control is far off from fulfilling the above-mentioned goal in its generality.

Modern advances in technology have led to highly complex processes and plants that must be controlled within very tight specifications and under higher levels of autonomy. In such situations, strict adherence to the above goal is becoming ever more crucial. For example, satellites and spacecraft are expected to operate in a variety of different conditions with high levels of performance and with as little human intervention as possible [213]. Similarly, the modern powerful and highly populated microprocessors that have contributed to the revolution in information technology, depend on complex manufacturing processes whose success centres on the ability to control process parameters with very high levels of accuracy.

Considerations of this kind have recently led to the notion of *intelligent control* [12, 100, 159, 198, 214, 268]. Indeed, a controller that is able to meet such an elaborate specification, must exhibit some of the features normally attributed to human intelligence, like adaptation, learning and planning. Quoting from the report of the Task Force on Intelligent Control, appointed by the IEEE Control Systems Society [12]:

> an intelligent system must be highly **adaptable** to significant unanticipated changes and so **learning** is essential. It must exhibit a high degree of **autonomy** in dealing with changes. It must be able to deal with significant **complexity**, and this leads to certain sparse types of functional architectures such as hierarchies.

This definition indicates that the key features of intelligent systems are:

- *Adaptation*: the ability to handle uncertainty by *continuously* estimating the relevant unknown knowledge. When conditions change, adaptation treats each distinct situation as novel, even if it had appeared before [25].
- *Learning*: the ability to modify behaviour when conditions change, by utilizing knowledge *memorized* from previous estimations.
Adaptation endows the system with the flexibility to face new and unknown situations, whilst learning provides the ability to utilize previous knowledge so that behaviour could be modified quickly and without the need of re-estimation when facing previously encountered situations.
- *Autonomy*: the ability to appropriately deal with uncertainty, without external (human) intervention.
- *Complexity*: the ability to deal with complex systems typically characterized by nonlinear dynamics, high dimensional decision spaces, distributed sensors and actuators, high (possibly non-Gaussian) noise levels and multiple modes of operation [190].

Narendra [183] broadly classifies intelligent controllers as those having to deal with

- complexity
- nonlinearity
- uncertainty.

Plant complexity inevitably leads to poor models that exhibit a high degree of uncertainty and are characterized by nonlinear dynamics involving unknown functionals, rather than merely unknown parameters. The situation becomes even more complex if the plant arbitrarily switches to different modes of operation that exhibit very different dynamic behaviour, or if it is subject to external stochastic disturbances.

Within the framework of these definitions, intelligent control theory should aim to develop techniques that enhance the degree of adaptation, learning and autonomy of a control system operating on ever more complex and uncertain plants.

1.2 Approaches to Intelligent Control

Control engineers and theorists have shown interest in automatic control systems that exhibit learning and planning characteristics normally associated with the faculties of human intelligence since the 1960's [82, 251]. During the last 15 years or so there emerged a new interest in the field, under the research banner of *intelligent control*. This was driven by developments in adaptive control theory and by advances in concepts of artificial intelligence (AI), such as neural networks, fuzzy logic, expert systems and genetic algorithms.

1.2.1 Contribution of Adaptive Control

Adaptive control theory [23] started in the 1950's and evolved as a solution to the problem of maintaining adequate control performance in the presence of parametric uncertainty. By the early 1980's, great progress had been reported on the convergence, stability and robustness properties of adaptive control for linear systems, both in discrete and continuous time [94, 186, 192, 223]. Meanwhile, during the mid-1980's, the theory of nonlinear control for known feedback-linearizable systems also reached its maturity [113]. This stimulated research into stable nonlinear adaptive control and by the late 1980's the first results on nonlinear adaptive control, applied specifically to robotic systems, began to appear [60, 232]. This was subsequently expanded to cover more general classes of nonlinear systems (such as those found in references [137, 141, 206, 224] among others), culminating in the work of Kokotović and colleagues on adaptive backstepping and tuning functions in the 1990's [144]. Most of the results on nonlinear adaptive control are concerned with continuous-time systems and unfortunately, the corresponding advances in discrete-time systems have been slower and less general in scope [138, 275, 281].

Despite this progress, the theory of *nonlinear adaptive control* is not usually categorized as an intelligent control scheme. The reason is that it deals only with *parametric uncertain* nonlinear systems *i.e.*, systems characterized by *known* nonlinear functions and *unknown* constant parameters that additionally, are almost always required to appear linearly in the system equations. This usually implies a relatively detailed prior knowledge of the system to be controlled, which limits the scope of application of this method.

On the other hand, intelligent control schemes are meant to deal with higher levels of uncertainty, as for example nonlinear systems characterized by *functional uncertainty* i.e., unknown nonlinear functions and not just parameters [183, 220]. Moreover, although adaptive systems are flexible enough to follow parametric changes in time, any estimates calculated prior to a recent change of parameters are not stored, and adaptation has to start afresh whenever the system reverts to an "old" situation. Hence, adaptive systems suffer from limitations of "short-term memory" and as such, lack the faculty of *learning* through experience [13]; an essential ingredient of intelligent control. To sum up, one could say that in complex situations that demand the use of an intelligent controller, adaptive control theory has much to offer, but it is not enough on its own.

1.2.2 Contribution of Artificial Intelligence

In a highly complex plant, it is not uncommon to have a situation where it is practically impossible to identify an accurate mathematical model of the system. This is not simply due to the uncertainty arising from the presence of unknown parameters or functions. It concerns also the basic form that the model equations should take, particularly if the system is nonlinear and time varying. Unfortunately, all control system design techniques depend upon the availability of a mathematical model, even if it includes some degree of parametric or functional uncertainty. So how could we proceed in such a situation? The solution to this problem has been aided by ideas from the field of artificial intelligence. For such complex plants, it may be the case that human experts have gathered enough knowledge to be able to control the system through hands-on experience. In this case, the idea is to encode the relevant knowledge of human experts into an automatic controller. This approach is called *supervised control* [108, 267].

Supervised control can be implemented by the use of *expert systems* [16]. These encode human knowledge as a set of *IF...THEN* production rules and make use of an automated inference mechanism that utilizes this knowledge to provide adequate control actions for the conditions active in the system at any given time. Unfortunately, the description of human experts for controlling such systems tends to be vague and imprecise, utilizing phrases such as "If the temperature is rising very fast then reduce the power quickly". This brings about the problem of how to systematically represent vague concepts like "very fast" or "reduce quickly". The solution is to use the technique of *fuzzy logic* representation [277]. Fuzzy logic is a system that solves the general problem of representing and operating upon vague or *fuzzy* concepts like those typically appearing in the production rules of a supervised controller. In a control system however, the data coming from the sensors and the output which the controller applies to the actuators is a "crisp" number possessing no vagueness. Hence, when using fuzzy logic for control, techniques have been developed for mapping the crisp data from sensors onto a fuzzy domain (fuzzification process). Inversely, the fuzzy decision from the rule base is mapped back onto a crisp domain (defuzzification process) for output to the actuators [199].

Another approach for implementing supervised control involves the use of Artificial Neural Networks [108, 268]. Originally inspired by the neural structure in the brain of biological organisms, artificial neural networks consist of a network of interconnected basic processing elements (neurons) operating in parallel. The neurons are characterized by a parameterized nonlinear function linking the inputs and output of a neuron. Use of different types of nonlinear functions and interconnection structures has resulted in various classes of neural networks such as the Multi-Layer Perceptron (MLP), Radial Basis Function (RBF) and CMAC networks [4, 101, 174, 203]. The interconnections between neurons are also parameterized by a gain. The parameters (sometimes called "weights") thus affect the overall characteristics of the network and under some specific mathematical conditions, a suitable choice of parameters enables the network to approximate any nonlinear mapping between its inputs and outputs. The parameter values that are required to approximate a desired function could be estimated from a set of exemplary input/output data extracted from the actual function. This is called a *training set* and the parameter estimation process is called *training* or *learning*. Following training, the network has the powerful property of generalizing the approximation even for those points in the input/output space that did not appear in the training set. In supervised control, the network is made to capture the knowledge of a human expert by training it with input/output data from the actual plant when operated by the expert. The data consists of sensory information that the expert operator has received from the plant and the expert's reaction to it, *i.e.*, what inputs the operator applies to the plant. In contrast to expert/fuzzy systems, neural networks encode the human's knowledge via input/output data derived from the human's actions, and not from linguistic instructions.

Although supervised control has helped to solve a number of problems, it cannot be really described as intelligent control. The reason is that basically, it maps the knowledge of humans onto an "artificial" system - the expert, neural or fuzzy controller - and the system itself is fixed. It lacks the crucial elements of intelligent systems, *adaptation* and *learning*, that are required to continuously estimate and track any uncertain or changing dynamics. Unfortunately, a common misconception is that any controller which uses techniques from AI is, by default, inherently intelligent [13]! However, there is still scope to use supervised control at a higher level of abstraction within an intelligent control framework. Consider an intelligent control system consisting of a set of various estimation and control algorithms, purposely included so as to be able to handle a large number of different and uncertain situations. In this case, a higher-level decision mechanism is required to co-ordinate the different algorithms. It has been suggested that this supervisory task could be handled by a system trained from the knowledge of human experts, as in supervised control [20].

Another class of neural network control schemes that also lack adaptation, is the *inverse control* methodology [108, 183]. This class includes techniques such as Direct Inverse Control or neural network-based Internal Model Control (IMC). In contrast to supervised control, the network training does not rely on the availability of a human expert; instead it is trained to model the inverse of the plant to be controlled from an off-line system identification exercise. In direct inverse control [210, 267], the plant inverse model network is connected directly in series with the plant in an open-loop configuration so as to approximate a unity operator between the reference input and the output. Due to the absence of feedback, this scheme is not very robust. The IMC approach [107] is more robust and based on a theoretically sounder framework [67]. It utilizes a second neural network that was previously trained to model the dynamics of the plant and which is connected together with the inverse model network and the plant inside a closed-loop configuration. A suboptimal implementation of this scheme, which does not require a plant inverse model network, has also been proposed [158]. The IMC approach is limited to plants that are open-loop stable.

Recently, a new design technique for non-adaptive neural control has been developed. The approach, sometimes called *implicit function emulation* [89, 155], is based upon more accurate and theoretically rigorous considerations than the inverse or supervised control classes. This method will be discussed in more detail in Section 2.3.2.

1.2.3 Confluence of Adaptive Control and AI: Intelligent Control

Recent developments in both neural and fuzzy controllers have led to the possibility of adaptation in these systems as well. For example, because the neural network approach approximates nonlinear functions by a parameterised model, it is a natural (though not straightforward!) progression to extend concepts from conventional parameter adaptive control theory to the neural control case [193]. A similar approach has also been successful in fuzzy control [152, 260]. Hence, by combining neural and fuzzy representation concepts with adaptive control techniques, the parametric-uncertainty barrier of conventional adaptive control theory was broken. This led to the possibility of introducing adaptation for controlling complex nonlinear systems having unknown *functions* rather than parameters, which is an essential aspect of intelligent control. In this book, adaptive systems that deal with functional uncertainty will be referred to by the term *functional adaptive*.

Conventional adaptive control offers two main methods for handling adaptation: the *direct* (also called Model Reference Adaptive Control MRAC) and *indirect* (also called Self-Tuning) methods [23, 186]. In the direct method, the controller parameters are adjusted directly according to some measure representing the tracking error, *i.e.*, the difference between the system output and that of a reference model that specifies how the output should respond. In the indirect method, an identification model is set up and its parameters are varied, so that it captures the dynamics of the plant. Parameter adjustment is driven by a measure of the estimation error, *i.e.*, the difference between the output of the plant and the identification model. The estimated parameters are then plugged into a control law that generates a control signal based upon these estimates.

Both these methods have been successfully extended to adaptive control by neural networks, though only for particular classes of nonlinear plants. In some cases, the stability of the system is also ensured [53, 54, 70, 120, 204, 219, 255, 274]. This approach is very promising because it borrows heavily from the well-established results of stable adaptive control theory. The adaptive version of the fuzzy controller is called the Self-Organizing Fuzzy Logic Controller (SOFLIC) [100, 152, 208]. Adaptation is introduced to vary the rules or the parameters of the fuzzy representation from the human expert, thus taking into account any process variation or inaccuracy in the prior expert information. The adaptation mechanism is often itself another fuzzy logic system, which acts upon a performance measure that assesses system operation. Recently, adaptation mechanisms that are more closely based on conventional direct and indirect adaptive control, ensuring system stability for particular classes of nonlinear systems, have also been proposed within the fuzzy framework [239, 260]. Yet another approach involves the combination of neural networks and fuzzy logic: the fuzzy-neural approach [152]. This method aims to reconcile the advantages of both systems: the ability of fuzzy systems to encode *a priori* human expert knowledge from linguistic information together with the techniques of parameter adaptation in neural networks for refining and updating the knowledge in the fuzzy representation.

Despite the advances in control of complex systems by the fuzzy/neural-adaptive paradigms, a large number of open problems remain to be addressed. Research efforts are aimed towards handling more general nonlinear plants (including Multi-Input/Multi-Output systems), improving the speed of convergence of parameter estimates, dealing with higher levels of uncertainty and complexity, and taking into consideration stability and robustness issues.

Recently, another paradigm is gaining popularity within the context of intelligent control. This is the *Evolutionary Computational* approach, which tries to mimic the functions of evolution, natural selection and genetics in biological systems [102]. One type of evolutionary computational methods is the *Genetic Algorithm*. The genetic algorithm is a method of performing a parallel, stochastic and directed search within the knowledge space of a system, to seek the "fittest" solution. In the control context, a population of controllers is operated upon by special functions corresponding to reproduction, crossover and mutation. This forms a new generation of controllers whose features are a mixture of those from the previous generation. A figure of merit, called the fitness function, is attributed to each member of a generation, determining which controllers meet the desired performance objectives most closely. Those that have better figures of merit are given priority when forming the next generation. This process is repeated so that ideally, after a number of generations, one ends up with the fittest controller that satisfies the desired objectives most accurately. This is typically done off-line, but recently an on-line version (called the Genetic Model Reference Adaptive Controller) has been developed, where the fitness function is evaluated from measured process data [198]. Hence, it is able to adapt to changes in process parameters in real time.

1.3 Enhancing the Performance of Intelligent Control

Functional adaptive control and AI methodologies are the basic ingredients of intelligent control design. Nevertheless, in order to be able to deal with higher levels of uncertainty and complexity, and also introduce features such as learning and autonomy, the contribution of a number of complementary techniques should be given serious consideration. In the following sections we discuss some approaches that appear to be promising for enhancing the performance of intelligent control systems in the face of higher levels of complexity and uncertainty. These techniques represent the essential framework behind some of the innovative designs that are presented in the subsequent chapters of this book.

1.3.1 Multiple Model Schemes: Dealing with Complexity

Most control design techniques rely on the availability of a model that characterizes some aspect of the system being controlled. This applies equally well to adaptive control where, by definition, an accurate plant model is unavailable. In fact indirect adaptive schemes require an identification model to capture the dynamics of the plant during control operation, and direct adaptive schemes depend on a reference model to specify how the closed-loop system should respond. Hence the availability of a model that characterizes the plant or the closed-loop system, whether obtained *a priori* or during control action, is a key feature of control design.

In practice there exist some systems that typically exhibit a number of distinct modes of behaviour during their operation. This type of plant complexity is called *multimodality*. In this book we distinguish between two types of multimodal complexity:

- *Temporal multimodality*: When the plant dynamics are subject to switch mode *abruptly* and *arbitrarily* in time. In particular, the onset of a mode switch and the currently active mode do not depend on any of the variables that are accessible for measurement or estimation. This situation can arise when a plant is operating in suddenly-changing environments or when a fault condition occurs.

- *Spatial multimodality*: When the plant is nonlinear and characterized by a highly complex function that exhibits very different features in different zones of its input, state or operating space. In this case, the plant dynamics change mode according to the operating conditions, and the currently active mode depends on some variables that are measurable.

One way of handling systems that are subject to multimodal complexity is to extend standard control design techniques by utilizing *several* models for the different modes of operation. This leads to *Multiple Model* techniques, whose use within both temporal and spatial multimodal scenarios is reviewed next.

Temporal Multimodality. Due to the nature of the environment in which it is expected to operate, an intelligent controller must ensure good performance even in the presence of temporal multimodality. On its own, standard adaptive control is usually not equipped to handle such situations because of both the abrupt timing and the extent of the changes. Even if parameter tracking estimators (such as exponential forgetting) were utilized within an adaptive control scheme, the abrupt nature of temporal multimodality makes it difficult for the "new" parameters to be tracked fast enough to ensure an acceptably small transient degradation in control performance. The use of multiple models offers one possible solution to this problem. This issue is very much related to the field of *fault detection and control reconfiguration* [213]. Control reconfiguration refers to the technique of first detecting and isolating component failures, and then switching onto a controller that had been designed for that particular failure mode. This way, the time taken to recover from a fault is usually much less than when using adaptive control.

The Multiple Model control approach, which was developed from the Partitioning Theory of adaptive control attributed to Lainiotis [151], has been successfully used in real applications for control reconfiguration [24, 99, 172]. This method is usually applied to control linear state-space stochastic systems having either unknown or switching parameters [65, 153, 221, 261, 264]. Parameter space is discretized into a finite number of partitions such that the nominal value of the parameters in each partition effectively corresponds to a *possible* plant model. This leads to a set of candidate models, one of which represents the dynamics of the actual plant. A probabilistic criterion derived from Bayes' rule, is then used to calculate the probability that a candidate model represents the actual plant, given only the input/output data. This involves setting up a Kalman filter for each candidate model and using the statistical estimates from each filter to calculate the probability measure. The appeal of the method for control reconfiguration follows because if the plant parameters suddenly change, a "new" plant model is quickly determined from the set of candidate models. This assumes that the new plant mode corresponds to (or at least is close to) one of the candidate models. The parameters of the candidate model having the highest probability of representing the actual mode, and the corresponding state estimate from its Kalman filter, could then be used to generate the new control signal. Alternatively it is more common to generate the control as a probability-weighted sum of all the control signals calculated from the parameters and state estimate of every candidate model.

This approach differs from our earlier definition of adaptation because no parameters are being estimated as such. In effect, the parameter estimation process has been transformed into a parameter *selection* process by detecting which of the candidate models best represents the current plant dynamics. However this scheme does not suffer from the short-term memory limitation of conventional adaptive systems, because each candidate model is effectively a "memory" of a particular plant mode, leading to the desirable feature of *learning* described before. The disadvantage is that it is much more memory intensive, especially when parameter space is large and partitioning has to be done at a high resolution. The latter may be important to ensure that, once a change in plant parameters occurs, there is always a candidate model in the vicinity of the new plant mode. This situation is relieved if it can be safely assumed that when a fault or change in dynamics takes place, the parameters can only take values from a finite, known set. This condition is satisfied in the case of systems that are subject to sensor/actuator faults [171].

As a more sophisticated alternative, Narendra *et al.* [188] suggest combining the multiple model and standard adaptive approaches, thereby exploiting the advantages of both schemes. In this case, in addition to switching to the particular model which best represents the current plant mode in order to generate the control; its parameters may also be adapted if the mode is not accurately represented by the parameters of the chosen model. This way, good performance is maintained even when the partitioning is done at a high granularity, because adaptation compensates for any parameter errors. Hence, the multiple models endow the system with memory, which leads to fast reconfiguration and learning, whilst adaptation refines the quality of the control in the face of any initial parameter error during the new configuration. Similar ideas were applied to improve the transient response of linear adaptive control systems, for which there also exists a stability proof [187]. A natural extension of this technique is to estimate models for previously unknown dynamics as well, leading to a candidate model set that grows in time. Multiple model techniques have not been used solely for control; they have also been applied to perform state estimation and system identification of temporal multimodal systems. These applications are reviewed in some more detail in Chapter 8.

Spatial Multimodality. In an intelligent control scenario, it is not uncommon for the plant to be characterized by a complex nonlinear function that exhibits spatial multimodality. Under these circumstances it is usually very difficult to accurately identify the plant dynamics by one higher-order, homogeneous, nonlinear model that is valid over the full range of operating space, such as a conventional neural network. It is usually more straightforward and accurate to use the principle of *divide and conquer* whereby the operating space is segmented into a number of partitions, each corresponding to a zone over which the dynamics exhibit similar features, and identifying a different *local model* for each partition. In effect, every local model is a representation of one particular mode. The local model could be a standard neural network or even a linear model, depending on the problem at hand. This modelling approach also requires a set of *validity functions* (one for each local model) that specify the partition in state space over which the corresponding model is valid. A global model is then formed by interpolating over the local models, as indicated by the validity functions. Unfortunately there is no standard name for such modelling techniques and different proponents have come up with their own terminology, some of which include: Regime Decomposition [123], Local Model Networks [106], Mixture of Experts [125] and Modular Neural Networks [118]. The approach is also related to fuzzy modelling techniques for nonlinear systems [242, 278]. In this book we make use of the term *modular networks* when referring to multiple modelling techniques for representing spatial multimodality.

In contrast to temporal multimodality, for the spatial multimodal case we assume the availability of information that specifies which mode is active at any given time. This is justified because the mode activity depends on some operating conditions that are measurable. When using a modular network to represent the system, this information is provided by the validity functions.

If applied to control, the modular network technique closely relates to the well-known *Gain Scheduling* methodology of control engineering [217, 226, 227]. In gain scheduling, the control signal is calculated by interpolating the gains of a set of linear local controllers, each of which is valid only over a localized range of operating conditions as indicated by a set of scheduling variables. The local controllers are designed on the basis of a linearized approximation of the plant's nonlinear dynamics about a set of different operating points chosen by the designer. If a modular network model for the plant is available, a similar technique could be used to effect control. The set of local controllers is designed according to the dynamics of the network's local models and the scheduling is specified by the validity functions. Several different standard methods could be used to design the local controllers. Brown *et al.* [45] utilize internal model control design and show simulation results for a modular network-based scheme applied to a process control problem. Hunt and Johansen [106] use pole placement design and the technique is applied to control an electrically stimulated muscle [90], while Jacobs and Jordan [118] use feedforward inverse control and apply the system on a robotics task. Further details on modular networks are given in Chapter 8, particularly those approaches that are based on probabilistic considerations.

1.3.2 Stochastic Adaptive Control: Dealing with Uncertainty

The theory of *stochastic adaptive control* handles uncertain systems by treating the unknown parameters as random variables (stochastic models) whose statistical properties (*e.g.*, mean, variance, probability density) reflect the extent of our knowledge about their value and are taken into consideration in the control law. The control law is derived on the basis of optimality *i.e.*, by minimizing a performance index that usually takes the form of the expected value of some cost function [23, 270]. Stochastic adaptive systems differ from the more common indirect or direct adaptive methods because the latter simply substitute the unknown parameter estimates inside a control law that strictly speaking, is optimal only if the parameters were known. For this reason, the uncertainty of the estimates is not taken into consideration in the control algorithm and, in almost all cases, this leads to a suboptimal solution. The exception is a rather restricted class of systems that are said to be *certainty equivalent* [29]. Nevertheless, most adaptive controllers are designed on the assumption of certainty equivalence, even if it does not hold. For this reason such controllers are also known as *heuristic* or *forced* certainty equivalent controllers. In the 1960's, Fel'dbaum [76, 77, 78] discovered that control laws derived on the basis of stochastic adaptive theory and which utilize measured information to its full extent[1], generate a control signal that possesses two important and attractive properties:

1. *Direction*: it directs the system output to follow some desired value, with due consideration given to the uncertainty of the parameter estimates.
2. *Probing*: it elicits further information from the system so as to reduce future parameter uncertainty, thereby improving the estimation.

Fel'dbaum coined the term *dual control* to categorize control laws that possess these two properties, both of which are important. A control law that lacks probing might be too *cautious* because whilst taking into consideration the uncertainty of the parameter estimates, it does nothing to reduce it. This often gives rise to a sluggish response. On the other hand, whilst efficient parameter estimation is important to effect good quality tracking, overemphasizing the role of probing at the expense of direction would defeat the main objective of effecting good tracking action. Hence it is desirable that both probing and direction coexist in a co-ordinated manner.

The problem is that these two actions are somewhat conflicting: the control input required for efficient parameter estimation (probing) demands a rich (persistently exciting) signal, whilst that required for good quality tracking (direction) typically does not. Hence a compromise must somehow be reached. The distinctive feature of dual controllers designed from stochastic adaptive theory is that they inherently manage to strike a balance between these two conflicting demands. This follows because when minimizing the

---
[1] so-called Closed-Loop Policies in the jargon of optimal control

performance index, the expectation is taken with respect to all random variables, which also includes the parameter estimates. Hence the tracking error, which usually is the main argument of the cost function, is minimized with the quality of the identification taken into account. This implies that if, say, a reduction in tracking error critically depends on obtaining better parameter estimates, then the plant is probed to reach this aim as quickly as possible. In the meantime caution is applied so that the parameter error does not seriously degrade the tracking performance.

For this reason the performance of dual controllers is generally superior to that of non-dual systems (*e.g.*, heuristic certainty equivalent controllers) especially when the control horizon is short, the parameter uncertainty is large or the parameters are changing rapidly. A non-dual controller that does not probe, simply makes use of the input/output signals that happen to occur naturally in the system and so parameter estimation will not necessarily be of good quality. Such a system is called *passively adaptive*. On the other hand, dual control seeks an *actively adaptive* policy because it generates a *planned* control signal in such a way as to minimize parameter uncertainty [29].

Unfortunately dual control has not been widely used in practice because of computational complexity issues associated with solving the equations required to find an optimal solution. The reason is that although, in principle, the optimal control sequence is given by solving the well-known *Bellman Equation* from dynamic programming [37], this necessitates a computer with unrealistically large computational and storage capability, except perhaps for the simplest of systems, which are practically irrelevant [19, 116]. For this reason most practical stochastic adaptive control schemes seek a suboptimal solution that retains, to a certain extent, the desirable properties of ideal dual control whilst permitting implementation. [51, 80, 167, 176, 209, 248, 249].

It is interesting to note that, once again, the ideas from stochastic adaptive control theory could be supplemented by other ideas from the field of AI, notably the *reinforcement learning* paradigm [32, 33]. Inspired by studies on animal learning in experimental psychology, reinforcement learning refers to a class of algorithms whereby the system only receives information about the current value of a performance measure (cost function). This is different from supervised learning algorithms where the system is typically furnished with information in the form of an error between the actual and a desired output. The latter inherently embodies information on the extent to which the parameters must be adjusted (via the magnitude of the error) and in what direction, *i.e.*, if the adjustment is to be an increase or decrease (via the sign of the error). By contrast, the performance measure in reinforcement learning does not convey any directional information. Consequently the adaptive system must actively probe the plant to estimate this information. As in dual control, this often conflicts with the input required to force the output to behave in some desired way. This is often called the *conflict between exploration and exploitation* and is nothing else but the analogue of the conflict between identification and control in stochastic adaptive control theory. The most sophisticated technique of reinforcement learning is called *Sequential Reinforcement Learning*. It makes use of an *adaptive critic* system whose aim is similar to that of dual control, *i.e.*, to improve the ability of the system in striking a balance between exploration and exploitation. In fact, sequential reinforcement learning methods could be interpreted as schemes that try to approximate dynamic programming for finding computationally feasible optimal solutions.

Dual control offers attractive features that should not be ignored in the quest for more intelligent systems, because it seeks to adapt to the current situation as fast as possibly permitted by striking an optimal compromise between active learning and control performance. Its philosophy of characterizing the plant, the uncertainty and the environment within a *unified* stochastic framework is appealing for the typical scenario in which an intelligent controller is expected to operate, where complexity and uncertainty co-exist and interact. The intelligent controller is expected to overcome these obstacles in a co-ordinated manner to guarantee the best possible performance as defined by a sensible performance index. Without doubt, the philosophy of dual control aims to proceed along such a direction.

1.4 The Objectives and their Rationale

From the previous discussion it is clear that the most promising approach for realizing an intelligent control system is one where ideas from AI are used to *complement* the techniques of control theory, rather than compete with them. Control theory builds upon results that are firmly established, well investigated, tried and tested. It would be foolish to ignore a field having such a robust tradition. On the other hand, one should be flexible to exploit emerging ideas that could enrich the results of control theory, even if they originate from a diverse field like AI. This reflects the philosophy underlining the work reported in this book: whilst most of the proposed designs are firmly rooted in control and systems theory, unknown nonlinear functionals are handled by neural networks. These are seen as a very effective tool for obtaining parameterized models of nonlinear mappings. Of course, no methodology is perfect and neural networks bring with them some problems of their own *e.g.*, issues related to choice of network structure, handling of a large amount of parameters and processing elements, nonlinear parameterization and finite approximation accuracy. These limitations should, however, not detract us from using neural networks because in the appropriate scenario, their advantages far outweigh the disadvantages and in any case, there are ways of relieving the effects of these problems. Indeed, Narendra [183] explicitly states that

>...in some cases they (neural networks) are emerging as the only viable alternatives.

1.4 The Objectives and their Rationale

The types of plants considered in this book are deterministic and stochastic. The systems are nonlinear and characterized by functional uncertainty. This is further complicated by the presence of additive random disturbances and multimodal complexity. Hence the contribution of this work to intelligent control stems from the nature of the plant and the environment being considered, which covers functional uncertainty, nonlinearity, multimodality and randomness. These are the typical conditions under which an intelligent controller is expected to operate so as to meet the challenge of extending the performance and autonomy of conventional control schemes.

In this book, the above challenge is met by introducing novel algorithms that enhance the controller's capacity for adaptation, stabilization, management of uncertainty, and learning. Uncertainty is managed by the determination of novel control laws that utilize more fully the probabilistic information provided by the estimator. This is achieved by extending techniques from stochastic adaptive control to the functional adaptive case, leading to novel dual control laws for intelligent control schemes. These enhance the performance of the system and obviate the need for a separate system identification phase prior to applying the control. In addition, these dual control principles are combined with multiple model techniques so as to derive dual control laws for handling both temporal and spatial multimodal complexity.

More specifically, the new results presented in this book address the following issues:

1. The introduction of dynamic structure Gaussian Radial Basis Function (GaRBF) neural-networks for adaptive control of a class of functional uncertain, nonlinear continuous-time systems. The proposed dynamic structure network starts with no basis functions at all. These are then selectively incremented sequentially, according to the evolution of the system's state in space, which in most cases leads to a much smaller number of basis functions had a conventional, fixed-structure neural network been used. This feature has very important practical implications because the number of basis functions required in fixed-structure GaRBF networks increases exponentially with the order of the system, often leading to prohibitively large memory requirements. The dynamic structure scheme leads to a network that is very economic in terms of network size by exploiting the fact that in most realistic situations, the system dynamics span only a small subset of state-space. In addition, the adaptation and control laws proposed are designed so as to ensure closed-loop stability in a Lyapunov sense and robustness to unmodelled dynamics.
2. The introduction of a composite adaptation law for neural network-based adaptive control of a class of functional uncertain, nonlinear continuous-time systems. The adaptation law utilizes both tracking and estimation errors to sequentially update the GaRBF network parameters. This enhances the convergence rate of the parameter estimates and makes the controller react faster to functional uncertainty, leading to superior track-

ing error convergence. At the same time, closed-loop stability and robustness are guaranteed via Lyapunov techniques.
3. The derivation of a new adaptive control scheme for a class of functional uncertain, nonlinear discrete-time systems. The scheme is based on augmented error techniques and GaRBF neural networks, and it guarantees stability and robustness via Lyapunov techniques. The proposed method is also modified to cover the class of discrete-time, adaptive sliding mode control. These novel designs lead to more global stability results, and require less *a priori* information on the unknown optimal network parameters than other discrete-time, functional adaptive control schemes, both within the sliding mode class or otherwise.
4. The introduction of dual adaptive control schemes for a class of functional uncertain, stochastic, nonlinear systems in discrete-time. Two different controllers are proposed, based on GaRBF and MLP neural networks respectively. The system exhibits great improvement over non-dual adaptive schemes that have been utilized so far for handling functional uncertain systems, particularly during the transient response. This follows because the uncertainty is taken into consideration when generating the control signal, leading to the introduction of caution and probing-like effects for better management of uncertainty.
5. The derivation of a novel multiple model, neural network, dual adaptive control scheme for handling temporal multimodality in a class of nonlinear stochastic systems subject to unknown dynamics. This leads to a controller that is able to handle simultaneously functional uncertainty and temporal multimodal complexity. The adaptation laws ensure the estimation of the uncertain mode dynamics, whilst the multiple models introduce learning features, in the sense of memorization. This leads to fast control reconfiguration in the presence of mode jumps. The dual nature of the control law is particularly important in this case because, in addition to functional uncertainty, the system is further complicated by temporal multimodality.
6. The design of a novel dual adaptive controller based on modular networks for handling a class of unknown, nonlinear stochastic systems characterized by spatial multimodality. The uncertainty covers both the mode dynamics and their scheduling, and the scheme could be interpreted as an adaptive form of gain scheduling. Hence it represents an alternative and original approach towards handling functional uncertainty, by partitioning the nonlinearity into a number of simpler local problems so as to facilitate the design of a global adaptive control method. As in the previous cases, the dual nature of the control law helps to improve the general performance of the system in the presence of higher levels of uncertainty.

The book is organized as follows: Chapters 3, 4, 5, 7, 9 and 10 respectively address the new results listed above. Chapters 2, 6 and 8 present an overview of some important theoretical results and published contributions on nonlinear adaptive control, stochastic adaptive control and multiple model estimation techniques, respectively. These provide both the background and the motivation for the new results presented in the book. Finally Chapter 11 summarizes the contributions of this book and their relevance to intelligent control.

Part II

Deterministic Systems

2. Adaptive Control of Nonlinear Systems

2.1 Introduction

An important attribute of intelligent control is the ability to deal with nonlinear systems that are subject to uncertainty. Hence when designing intelligent control schemes, the theoretical background that already exists in the fields of adaptive and nonlinear control should be given its due importance. In this chapter we therefore present a brief review of several aspects of nonlinear systems theory and adaptive control of nonlinear plants.

A number of important structural properties of nonlinear systems that represent a generalization of familiar concepts in linear systems theory (such as minimum phase systems and input/output models) will be discussed. Both continuous and discrete-time cases are considered. A review of the important classes of nonlinear systems for which stable adaptive control has been established, is also included. These are categorized into two main problem classes: (i) parametric uncertain and (ii) functional uncertain systems. Clearly, methods for functional uncertain systems are more relevant to this book because they address a higher level of uncertainty and are usually classified as intelligent control schemes. Nevertheless, parametric uncertain adaptive techniques are also discussed so as to clarify further the differences between the two problems and their solutions.

As far as adaptive control of functional uncertain systems is concerned, we concentrate exclusively on neural network-based solutions. Hence the chapter also includes a brief overview of neural networks and the issues that are particularly relevant within a control scenario. To keep the exposition simple, only single-input/single-output (SISO) systems are considered.

2.2 Continuous-time Systems

Consider a general autonomous SISO, continuous-time, nonlinear system characterized by the state-space equations

$$\begin{aligned} \dot{\mathbf{x}} &= \mathbf{f}(\mathbf{x}, u) \\ y &= h(\mathbf{x}) \end{aligned} \quad (2.1)$$

where $\mathbf{x} \in \Re^n$ is the state vector, u and y represent the input and output respectively, $\mathbf{f} : \Re^n \times \Re \mapsto \Re^n$ is a nonlinear vector function and $h : \Re^n \mapsto \Re$ is a scalar nonlinear function. Control design techniques for such a general class of systems do not exist, even when the nonlinear functions $\mathbf{f}$ and h are known. Consequently most of the results in nonlinear control theory are restricted to particular classes of nonlinear dynamics.

In the mid-1980's, the use of differential geometry led to the development of *feedback linearization* as a method for stably controlling particular types of nonlinear systems that were previously intractable [113]. This gave a strong impetus to nonlinear control theory, which continues to develop up to the present day, albeit along different and more advanced directions.

2.2.1 Control by Feedback Linearization

The method of feedback linearization centres on cancelling nonlinear dynamics by a proper choice of nonlinear state feedback. This transforms the system into a linear form, on which standard linear control techniques could be applied to obtain a stable closed loop system. Naturally, this could only be used on systems that satisfy particular conditions and is not a general technique. Very often, feedback linearization is applied to systems having dynamics that are affine (linear) in the control input, as shown below for the SISO case:

$$\begin{aligned} \dot{\mathbf{x}} &= \mathbf{a}(\mathbf{x}) + \mathbf{b}(\mathbf{x})u \\ y &= h(\mathbf{x}) \end{aligned} \qquad (2.2)$$

where $\mathbf{a}, \mathbf{b}, h$ are smooth, nonlinear functions. Under certain conditions (to be explained in the sequel) it is possible to obtain a direct input-output relation between u and y, by successive differentiation of y with respect to time

At this point we will digress for a while to briefly explain the concept of Lie derivatives, because it serves as a useful tool for handling such problems [234]. In general, the Lie derivative of a scalar function $h(\mathbf{x})$ with respect to a vector function $\mathbf{a}(\mathbf{x})$, denoted by $L_\mathbf{a} h(\mathbf{x})$, is defined as:

$$L_\mathbf{a} h(\mathbf{x}) := \nabla h(\mathbf{x}) \mathbf{a}(\mathbf{x})$$

where ∇h denotes the gradient of $h(\mathbf{x})$, i.e. $[\partial h/\partial x_1 \cdots \partial h/\partial x_n]$. Notice that the Lie derivative is a scalar, so that the process of taking Lie derivatives could be chained and is denoted as follows:

$$\begin{aligned} L_\mathbf{a}^i h(\mathbf{x}) &:= \nabla (L_\mathbf{a}^{i-1} h(\mathbf{x})) \mathbf{a}(\mathbf{x}) \\ L_\mathbf{b} L_\mathbf{a}^i h(\mathbf{x}) &:= \nabla (L_\mathbf{a}^i h(\mathbf{x})) \mathbf{b}(\mathbf{x}). \end{aligned}$$

Hence, differentiating y in Equation (2.2) with respect to time and using Lie derivatives we get:

$$y^{(1)} = \frac{\partial y}{\partial \mathbf{x}} \dot{\mathbf{x}} = L_\mathbf{a} h(\mathbf{x}) + L_\mathbf{b} h(\mathbf{x}) u, \qquad (2.3)$$

where $y^{(i)}$ denotes the ith derivative of y with respect to time.

- If $L_bh(\mathbf{x}) \neq 0$ for all $\mathbf{x}$ inside a region $\Omega \subseteq \Re^n$, the above input-output relation between y and u is well-defined, since the term containing the control never vanishes inside Ω. In this case, the system is said to have a *relative degree one* in region Ω, because the relation was obtained after one differentiation.
- If $L_bh(\mathbf{x}) = 0$ for *all* points inside Ω, we might still get a chance of obtaining a well-defined input-output (IO) relation by differentiating y repeatedly. The rth derivative of y could be easily worked out as

$$y^{(r)} = L_a^r h(\mathbf{x}) + L_b L_a^{r-1} h(\mathbf{x}) u.$$

If the repeated differentiations satisfy the conditions that $\forall \mathbf{x} \in \Omega$

$$\begin{aligned} L_b L_a^{r-1} h(\mathbf{x}) &\neq 0 \\ L_b L_a^{i-1} h(\mathbf{x}) &= 0, \quad \forall i = 1, \cdots, (r-1), \end{aligned} \quad (2.4)$$

then we would have obtained a well-defined IO relation after r differentiations. Such a system is said to have a relative degree r in region Ω and its dynamics could be expressed in the following input-output affine form

$$y^{(r)} = f(\mathbf{x}) + g(\mathbf{x}) u, \qquad (2.5)$$

where $f(\mathbf{x}) = L_a^r h(\mathbf{x})$ and $g(\mathbf{x}) = L_b L_a^{r-1} h(\mathbf{x})$. If conditions (2.4) are not restricted to Ω, but are satisfied for all $\mathbf{x} \in \Re^n$, the relative degree is said to be *globally defined*.

From Equation (2.5) and the fact that inside Ω, $g(\mathbf{x}) \neq 0$ by definition of the relative degree, it is straightforward to see that the state feedback control law:

$$u = \frac{-f(\mathbf{x}) + v(t)}{g(\mathbf{x})} \qquad (2.6)$$

cancels the nonlinearities and reduces the closed loop dynamics to the following linear form with respect to an auxiliary input $v(t)$:

$$y^{(r)} = v(t).$$

This is the reason for calling such systems feedback linearizable. If v is chosen as

$$v(t) = y_d^{(r)} - \alpha_r e^{(r-1)} - \cdots - \alpha_1 e$$

where y_d denotes the reference input which y is required to track, $e := y - y_d$ denotes the tracking error and the coefficients α_i are chosen such that $\Gamma(s) = s^r + \alpha_r s^{r-1} + \cdots + \alpha_1$ is a Hurwitz polynomial in the Laplace variable s, then the tracking error and its $r-1$ derivatives converge to zero asymptotically because the closed-loop dynamics reduce to the equation

$$e^{(r)} + \alpha_r e^{(r-1)} + \cdots + \alpha_1 e = 0$$

which, by virtue of the choice of coefficients α_i, is stable.

However this represents an rth order system, when in fact the original open loop plant was nth order. Hence the use equation (2.6) must have rendered $(n-r)$ states unobservable. This is exactly analogous to the concept of pole-zero cancellation in linear systems theory. For overall stability we must ensure that these unobservable, *internal states* remain bounded as well, even though they do not affect the output directly. But how could these $(n-r)$ unobservable states be characterized? Using differential geometry, it is relatively easy to show [113] that for a system having a relative degree $r \leq n$ in region Ω, at each $\mathbf{x}_o \in \Omega$ there is a neighbourhood $U_{\mathbf{x}_o}$ where it is always possible to find a diffeomorphism[1] $\mathbf{T}(\cdot) : U_{\mathbf{x}_o} \mapsto \Re^n$ transforming the states $\mathbf{x}$ to a set of states $\mathbf{z} = \mathbf{T}(\mathbf{x})$ called normal states, such that the system representation is transformed into a canonical form called *the normal form*, as follows:

$$\dot{z}_{1_1} = z_{1_2}$$
$$\vdots$$
$$\dot{z}_{1_{r-1}} = z_{1_r}$$
$$\dot{z}_{1_r} = \phi(\mathbf{z}_1, \mathbf{z}_2) + \gamma(\mathbf{z}_1, \mathbf{z}_2)u \qquad (2.7)$$
$$\dot{\mathbf{z}}_2 = \mathbf{I}(\mathbf{z}_1, \mathbf{z}_2)$$
$$y = z_{1_1}.$$

Note that the normal state vector $\mathbf{z}$ has been partitioned as $\mathbf{z} = [\mathbf{z}_1^T \ \mathbf{z}_2^T]^T$. Vector $\mathbf{z}_1 \in \Re^r = [h \ L_a h \cdots L_a^{r-1} h]^T$ represents the output and its $(r-1)$ derivatives, and z_{1_i} denotes the ith element of $\mathbf{z}_1$. Vector $\mathbf{z}_2 \in \Re^{n-r}$ represents the *internal states* that do not influence neither the output nor its $(r-1)$ derivatives. The functions $\phi(\mathbf{z}_1, \mathbf{z}_2) = f(T^{-1}(\mathbf{z})) \equiv f(\mathbf{x})$, and $\gamma(\mathbf{z}_1, \mathbf{z}_2) = g(T^{-1}(\mathbf{z})) \equiv g(\mathbf{x})$. $\mathbf{l}(\mathbf{z}_1, \mathbf{z}_2)$ is a vector function that characterizes the internal states $\mathbf{z}_2$.

The stability of the internal states is analysed by the concept of *zero dynamics*. These are defined as the special case when the system dynamics are subjected to the constraint that the output and its $(r-1)$ derivatives are always zero. This is equivalent to saying that $\mathbf{z}_1 = \mathbf{0}$, which can be obtained by starting with initial conditions $\mathbf{z}_1(0) = \mathbf{0}$ and setting $u(t) = -f(\mathbf{x})/g(\mathbf{x})$ to keep $\dot{z}_{1_r} = 0$ continuously and with it, its r integrals $z_{1_r}, \cdots, z_{1_1}$. Hence under this condition, the dynamics of the system in normal form reduce to

$$\dot{\mathbf{z}}_1 = \mathbf{0}$$
$$\dot{\mathbf{z}}_2 = \mathbf{I}(\mathbf{0}, \mathbf{z}_2)$$
$$y = 0.$$

Note that these dynamics characterize exclusively the internal states. Hence, the equation $\dot{\mathbf{z}}_2 = \mathbf{I}(\mathbf{0}, \mathbf{z}_2)$ defines the *zero dynamics* and it determines the stability properties of the internal states.

---
[1] A diffeomorphism is a function from $\Re^n \mapsto \Re^n$ that is continuously differentiable and whose inverse exists and is also continuously differentiable

The following theorem explicitly defines the conditions under which system (2.2) can be controlled to track a reference input with the whole state vector remaining bounded [224].

Theorem 2.2.1. *To make system (2.2) asymptotically track a reference input $y_d(t)$ and all the states* **x** *remain bounded, the following conditions must be satisfied:*

1. *The system has a globally defined relative degree r.*
2. *The zero dynamics are globally exponentially stable. This is sometimes called the Minimum Phase condition.*
3. *The internal dynamics* $\mathbf{I}(\mathbf{z}_1, \mathbf{z}_2)$ *are Lipschitz* [2] *in* $\mathbf{z}_1, \mathbf{z}_2$.
4. *The control maintains the output vector* $\mathbf{z}_1$ *bounded and forces the output $y(t)$ to asymptotically track $y_d(t)$. The feedback linearization control law (2.6) ensures this condition.*
5. *The reference input must be such that $y_d, y_d^{(1)}, y_d^{(2)}, \ldots, y_d^{(r)}$ are bounded.*

Proof. See reference [224]. □

2.2.2 Control by Backstepping

To ensure a stable closed loop system, the technique of feedback linearization imposes rather stringent conditions on the plant nonlinearities. Recently, a more flexible design procedure called *backstepping* [144], based on Lyapunov stability techniques has been developed. In backstepping design, it is possible to avoid cancellation of some nonlinearities which might be useful and also obtain better transient response by adding extra nonlinearities. In this technique however, the system has to be represented or transformed into one of the following, rather specific canonical forms:

(a) Strict-feedback systems. These have the form

$$\dot{\mathbf{x}} = \mathbf{a}(\mathbf{x}) + \mathbf{b}(\mathbf{x})\xi_1$$
$$\dot{\xi}_1 = a_1(\mathbf{x}, \xi_1) + b_1(\mathbf{x}, \xi_1)\xi_2$$
$$\dot{\xi}_2 = a_2(\mathbf{x}, \xi_1, \xi_2) + b_2(\mathbf{x}, \xi_1, \xi_2)\xi_3$$
$$\vdots$$
$$\dot{\xi}_{k-1} = a_{k-1}(\mathbf{x}, \xi_1, \cdots, \xi_{k-1}) + b_{k-1}(\mathbf{x}, \xi_1, \cdots, \xi_{k-1})\xi_k$$
$$\dot{\xi}_k = a_k(\mathbf{x}, \xi_1, \cdots, \xi_k) + b_k(\mathbf{x}, \xi_1, \cdots, \xi_k)u$$

---
[2] A function $\mathbf{f}(t, \mathbf{x})$ that is continuous in t and $\mathbf{x}$, is said to be Lipschitz if it satisfies the condition

$$\|\mathbf{f}(t, \mathbf{x}) - \mathbf{f}(t, \mathbf{y})\| \leq k\|\mathbf{x} - \mathbf{y}\|, \quad \forall \mathbf{x}, \mathbf{y} \in B, \quad \forall t \in [0, T],$$

where B is a neighbourhood of the origin and k, T are positive, finite constants.

where $\mathbf{x} \in \Re^n$, $u \in \Re$, $\mathbf{a}$ and $\mathbf{b} : \Re^n \mapsto \Re^n$, a_i and $b_i : \Re^{n+i} \mapsto \Re$, $\xi_i \in \Re$. Note that strict-feedback systems have k affine subsystems (in terms of ξ) separating the input u from the dynamics of $\mathbf{x}$. Additionally, the ξ subsystem has a triangular structure, i.e. $\dot{\xi}_i$ depends nonlinearly on $\xi_1 \cdots \xi_i$, $\forall i = 1, \cdots, k$.

(b) *Pure-feedback systems.* These are more general than strict-feedback systems and have the form

$$\dot{\mathbf{x}} = \mathbf{a}(\mathbf{x}) + \mathbf{b}(\mathbf{x})\xi_1$$
$$\dot{\xi}_1 = a_1(\mathbf{x}, \xi_1, \xi_2)$$
$$\dot{\xi}_2 = a_2(\mathbf{x}, \xi_1, \xi_2, \xi_3)$$
$$\vdots$$
$$\dot{\xi}_{k-1} = a_{k-1}(\mathbf{x}, \xi_1, \cdots, \xi_k)$$
$$\dot{\xi}_k = a_k(\mathbf{x}, \xi_1, \cdots, \xi_k, u)$$

where $\mathbf{x} \in \Re^n$, $u \in \Re$, $\mathbf{a}$ and $\mathbf{b} : \Re^n \mapsto \Re^n$, a_i and $b_i : \Re^{n+i+1} \mapsto \Re$, $\xi_i \in \Re$. Pure-feedback systems are not restricted to the affine structure of the strict-feedback class in the ξ subsystem, although they still have a triangular structure.

(c) *Block-strict-feedback systems.* If each of the equations in the ξ subsystem of the strict-feedback class is replaced by a full state-space equation (having **both** a measurement and state equation), we obtain the block-strict-feedback class

$$\dot{\mathbf{x}} = \mathbf{a}(\mathbf{x}) + \mathbf{b}(\mathbf{x})y_1$$
$$\dot{\xi}_1 = \mathbf{a}_1(\mathbf{x}, \xi_1) + \mathbf{b}_1(\mathbf{x}, \xi_1)y_2$$
$$y_1 = h_1(\xi_1)$$

$$\dot{\xi}_2 = \mathbf{a}_2(\mathbf{x}, \xi_1, \xi_2) + \mathbf{b}_2(\mathbf{x}, \xi_1, \xi_2)y_3$$
$$y_2 = h_2(\xi_2)$$
$$\vdots$$
$$\dot{\xi}_{k-1} = \mathbf{a}_{k-1}(\mathbf{x}, \xi_1, \cdots, \xi_{k-1}) + \mathbf{b}_{k-1}(\mathbf{x}, \xi_1, \cdots, \xi_{k-1})y_k$$
$$y_{k-1} = h_{k-1}(\xi_{k-1})$$

$$\dot{\xi}_k = \mathbf{a}_k(\mathbf{x}, \xi_1, \cdots, \xi_k) + \mathbf{b}_k(\mathbf{x}, \xi_1, \cdots, \xi_k)u$$
$$y_k = h_k(\xi_k)$$

where $\mathbf{x} \in \Re^n$, $u \in \Re$, $\mathbf{a}$ and $\mathbf{b} : \Re^n \mapsto \Re^n$, $\xi_i \in \Re^{n_i}$ where n_i denotes the order of the ith subsystem, $\mathbf{a}_i$ and $\mathbf{b}_i : \Re^{n+n_1+n_2+\cdots+n_i} \mapsto \Re^{n_i}$, $y_i \in \Re$, $h_i : \Re^{n_i} \mapsto \Re$.

In contrast to feedback linearization, backstepping is a Lyapunov-based design methodology. Backstepping handles the above three classes of nonlinear systems by initially considering only the dynamics formed by the $\mathbf{x}$ and

ξ_1 equations as if they described one subsystem having ξ_1 as input. A control law defined by a, so-called, *stabilizing function* α_1 for the "virtual input" ξ_1, is then determined. This stabilizing function should satisfy an intermediate Lyapunov function, often called the *control Lyapunov function* V_1, for stability of the $\mathbf{x}, \xi_1$ subsystem. This subsystem is then augmented with the ξ_2 dynamic equation and the procedure repeated for the new, augmented subsystem *i.e.* determine a stabilizing function α_2 for ξ_2 that satisfies a control Lyapunov function V_2 that is related to V_1. The process is repeated for all the ξ_i dynamics until at the last equation, the true system input $u(t)$ appears and an overall Lyapunov function V could be set up by chaining all the previous individual control Lyapunov functions V_i. From this, a stable control law is obtained for u that satisfies the stability conditions based on V. Effectively, each individual stabilizing function and control Lyapunov function are used to recursively synthesize an overall system Lyapunov function V. This method leads to stable control laws that are robust even in the presence of uncertain bounded nonlinearities. More details on backstepping could be found in [144].

2.2.3 Adaptive Control

Adaptive techniques are a particularly appealing solution to the problem of controlling systems that are subject to uncertainty. A good body of knowledge exists for adaptive control of linear systems, guaranteeing closed loop stability and robustness to unmodelled dynamics and disturbances [109]. On the other hand, nonlinear adaptive control is a relatively new field and a very active research area. It is important to draw a clear distinction between two types of uncertainty that could characterize a nonlinear system. With respect to the general system of Equation (2.1), these are given as follows:

- *Parametric uncertainty*: when $\mathbf{f} = \mathbf{f}(\mathbf{x}, u, \theta)$ and $h = h(\mathbf{x}, \theta)$ are *known* functions with *unknown* parameters θ
- *Functional uncertainty*: where the whole nonlinear functions $\mathbf{f}$ and $\mathbf{h}$ are unknown.

In this book (as in most of the publications on adaptive systems) we reserve the term *nonlinear adaptive control* to adaptive systems that deal with parametric uncertainty. These are usually adaptive versions of feedback linearization or backstepping methods. In particular, adaptive backstepping design has recently led to some very strong stability results for linearly parameterized uncertain systems [144]. Despite this progress however, nonlinear adaptive control techniques are unable to handle functional uncertainty and this is precisely the scenario where intelligent control has much to offer, particularly neural network based techniques. As mentioned previously, in this book we specifically use the term *functional adaptive* when referring to adaptive control techniques that handle functional uncertainty as opposed to parametric uncertainty. Since the level of uncertainty handled by functional adaptive systems is higher than that handled by nonlinear adaptive systems, it is not surprising that in the former case typically more stringent conditions have to be satisfied by the plant in order to guarantee closed loop stability. To make the distinction between the two approaches and their scope of application more clear, the next section will overview some basic results in nonlinear adaptive control. This is then followed by a section on functional adaptive control, specifically schemes that are based on neural networks.

Nonlinear Adaptive Control. Nonlinear adaptive control schemes typically handle parametric uncertain systems whose parameters appear linearly in the equations. As an example, a typical SISO system would be represented as

$$\dot{\mathbf{x}} = \mathbf{f}_o(\mathbf{x}) + \sum_{i=1}^{p} \theta_i \mathbf{f}_i(\mathbf{x}) + \left[\mathbf{g}_o(\mathbf{x}) + \sum_{i=1}^{p} \theta_i \mathbf{g}_i(\mathbf{x}) \right] u$$

$$y = h_o(\mathbf{x}) + \sum_{i=1}^{p} \theta_i h_i(\mathbf{x})$$

where $\mathbf{f}_i$, $\mathbf{g}_i$ are known smooth vector fields on $\Re^n$, h_i are known smooth scalar functions and θ_i are the unknown constant parameters.

During the early stages of development of stable adaptive designs for such systems, one of two types of constraints had to be imposed on the plant [142]:

1. *Uncertainty Constraints*; where the unknown parameter is restricted to appear in the same equation as the control (known as a *matching condition*) or at most one integrator away (*extended* matching condition), with no constraints on the nonlinearities [244].
2. *Nonlinearity Constraints*; where the location of the parameter is free but the nonlinearities are restricted by linear growth or Lipschitz conditions [224, 245].

By introducing an adaptive form of backstepping, Kanellakopoulos, Kokotović and Morse managed to remove both these limitations [142]. Their controller results in global stability without imposing any nonlinearity constraints, for the class of *parametric strict-feedback* systems. These are parameterized versions of strict-feedback systems that have the following form:

$$\dot{z}_1 = z_2 + \phi_1^T(z_1)\theta$$
$$\dot{z}_2 = z_3 + \phi_2^T(z_1, z_2)\theta$$
$$\vdots$$
$$\dot{z}_{n-1} = z_n + \phi_{n-1}^T(z_1, \cdots, z_{n-1})\theta$$
$$\dot{z}_n = \phi_n^T(\mathbf{z})\theta + \beta(\mathbf{z})u$$

where $\beta(\mathbf{z}) \neq 0 \ \forall \mathbf{z} \in \Re^n$ and θ is the vector of unknown parameters. They also developed a similar adaptive backstepping controller without nonlinearity constraints, for the more general *parametric pure-feedback* class. In this case the stability results are local but for this class, even if the parameters were known, a globally stable controller could not be attained in general.

Adaptive backstepping thus covers more general nonlinear functions. However it suffers from a problem related to over-parameterisation, because more than one estimate is calculated for the same parameter. This can be critical if the number of unknown parameters is large. The problem was resolved when Krstić introduced the idea of *tuning functions* [145], which leads to a minimal order controller. However when the number of parameters is large, this method is still plagued by very complex nonlinear expressions. Hence recent developments have departed from Lyapunov-based design and adopt techniques based on classical indirect adaptive control, but extended to cover the nonlinear case. These so-called *modular design* methods [146] consider the controller and identifier modules as almost separate design problems and are therefore much more flexible and versatile than their Lyapunov-based counterparts. Due to the nonlinear nature of the plant however, the certainty equivalence ideas that underpin classical indirect adaptive control are strengthened by a controller that is strongly robust to parameter uncertainty; to the extent that it guarantees boundedness even without adaptation.

Nonlinear adaptive control is a very active research area with scope for much more development, particularly on issues regarding robustness to unmodelled dynamics and disturbances. When applied to problems that fit inside its framework, such as induction motor control [104], nonlinear adaptive control has much to offer. However its reliance on rigid canonical forms and linear parameterisation renders it unsuitable to handle the levels of uncertainty normally associated with the use of intelligent control. Although a few results for systems whose parameters appear nonlinearly have been published, these are either restricted to specific classes of parameterizations [41, 88] or else address only the set-point regulation problem [168]. This leads us to consider the field of adaptive control by neural networks, which has emerged as a powerful method for handling the higher levels of uncertainty typified by unknown functionals instead of parameters.

Functional Adaptive Control by Neural Networks. The appeal of neural networks to functional adaptive control stems from their ability to approximate nonlinear mappings via a parameterized structure of interconnected processing elements. Various types of neural networks have been developed. The most popular are the Multi-layer Perceptron and Radial Basis Function networks, which are of the feedforward type. Recurrent (dynamic) Networks, which also include feedback interconnections, are also gaining wider usage. A large number of publications treating neural networks (NN) as a general tool for pattern recognition, optimization, signal classification and control are available for the interested reader [39, 101]. Due to space limitations, in this section we will only overview those aspects of NNs that are important for the control designs presented in this book.

Neural Networks. A general 2-layer feedforward NN has the configuration shown in Figure 2.1. The NN generates a mapping between an input $\mathbf{x} \in \Re^n$ and an output $\mathbf{y} \in \Re^l$. The mapping depends upon the characteristics of the parameterized nonlinear operators ϕ_{1i}, ϕ_{2j} which constitute the basic processing elements of the network. These are usually called *neurons, units* or simply *elements*. In this kind of network structure, signals are flowing only in the forward direction from the input layer $\mathbf{x}$, through the hidden layer $\mathbf{z}$, to the output layer $\mathbf{y}$. In general more than one hidden layer separating the input and output layers may be included. If the parameters of the ith and jth neurons in the hidden and output layers are respectively denoted by vectors $\mathbf{w}_{1i}$ and $\mathbf{w}_{2j}$, often called the network *weights*, then the jth output of the network is given as

$$y_j = \phi_{2j}(z_1, \cdots, z_m, \mathbf{w}_{2j})$$
$$z_i = \phi_{1i}(\mathbf{x}, \mathbf{w}_{1i}), \quad i = 1 \cdots m$$

where z_i denotes the output of the ith neuron in the hidden layer.

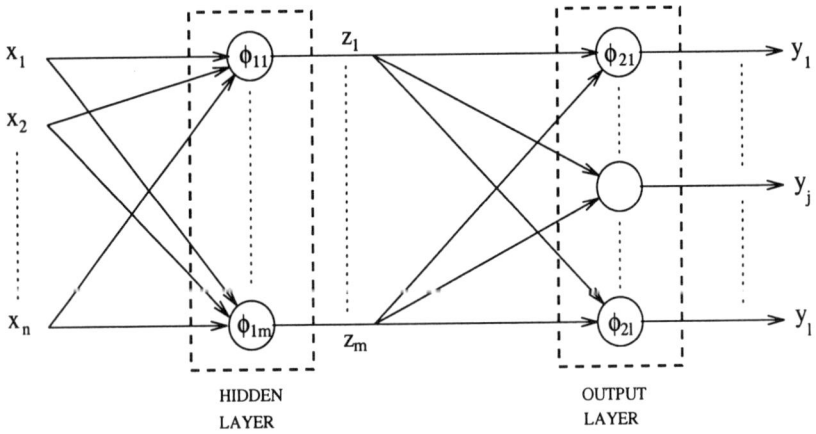

Fig. 2.1. Generic 2-layer feedforward neural network

Different functions for ϕ_1, ϕ_2 constitute different kinds of NNs. In the MLP network these are of the sigmoidal type *e.g.*:

$$\phi_{1i} = \frac{1}{1 + \exp(-\mathbf{w}_{1i}^T \mathbf{x}_a)}, \quad \phi_{2j} = \frac{1}{1 + \exp(-\mathbf{w}_{2j}^T \mathbf{z}_a)} \quad \text{or} \quad \phi_{2j} = \mathbf{w}_{2j}^T \mathbf{z}_a,$$

where $\mathbf{x}_a = [\mathbf{x}^T \ 1]^T$, $\mathbf{z}_a = [\mathbf{z}^T \ 1]^T$. The second option for ϕ_2 is linear rather than sigmoidal, normally used when the output range is required to exceed that of the sigmoid which is restricted between 0 and 1.

In RBF networks, ϕ_{1i} are called *basis functions*. Specifically, in Gaussian RBF networks they take the familiar Gaussian form:

$$\phi_{1_i} = \exp\left\{\frac{-\|\mathbf{x} - \mu_i\|^2}{2\sigma_i^2}\right\}, \tag{2.8}$$

where μ_i, σ_i, often called the *centre* and *width* respectively, constitute the elements of the parameter vector $\mathbf{w}_{1_i} := [\mu_i^T \ \sigma_i]^T$. The output layer functions ϕ_{2_j} take the linear form

$$\phi_{2_j} = \mathbf{w}_{2_j}^T \mathbf{z}_a.$$

The NN input-output mapping between $\mathbf{x}$ and $\mathbf{y}$ is determined by the values assigned to the weight vectors $\mathbf{w}_{1_i}$ and $\mathbf{w}_{2_j}; i = 1\ldots m, \ j = 1\cdots l$. These are typically adjusted such that the NN will model the mapping of some particular target system of interest. In control applications this is often some functional related to the plant. The process of weight adjustment is called network *training*. Training is often performed in a supervised manner, by first gathering input-output data from the target system and then adjusting the network weight vectors such that some norm of the difference between the network output and the target outputs is minimized. Various training algorithms have been developed, probably the most popular being the Back Propagation algorithm [218]. Following training, the NN is able to generate accurate estimates of the target outputs even for those inputs on which it was not trained. This feature is called *generalization* and is more accurately defined by the following Universal Approximation Property.

Definition 2.2.1. *A network is said to satisfy the Universal Approximation Property if on any compact subset $C \subset \Re^n$ of the input space, it is always possible to find an appropriate number of neurons and optimal weight vectors (not necessarily unique) such that any continuous function could be approximated to an arbitrary level of accuracy by the network.*

Note that this is an existence property, it does not convey any information on how to find the optimal number of neurons or weights. Nevertheless it is an important result because it gives confidence in the approximation capabilities of the network. It has been proved that both the 2-layer MLP and the GaRBF networks satisfy the Universal Approximation property [61, 196, 197].

GaRBF networks differ from MLP networks in that they perform local representations. This means that given a particular input, only a few of the basis functions whose parameter μ lies in the vicinity of the input, actually respond with a large output in the hidden layer. Hence, the network output is affected only by these localized basis functions. In the MLP network, on the other hand, the output is influenced by all the hidden layer functions and so even during training, *all* parameters need to be adjusted so as to reduce the approximation error to a minimum. Hence, in a supervised training scenario, the mapping of an MLP network would depend on the *order* of presentation of the training samples. On the contrary, the localized nature of the Gaussian basis functions ensures that only the weights of the basis functions located in the vicinity of the input need adjustment during training, making the network insensitive to the order of presentation of data. This feature makes GaRBF networks attractive for on-line applications where there is often no control on the order of presentation of data being used for training.

Another attractive feature of RBFs over MLPs is that very often the basis function parameters μ_i, σ_i could be pre-determined, leaving only the output layer weights $\mathbf{w}_{2j}$ to be adjusted according to the training data. Since these appear linearly in the dynamics, their estimation is a straightforward matter of using standard linear approximation techniques such as least squares. Pre-determination of the basis function parameters is done by choosing the centre parameters μ_i, such that the m neurons forming the hidden layer are located on regular points of a mesh covering the relevant input space C. The number of basis functions, and thus the distance between the mesh points, and the width parameters σ_i depend on the approximation accuracy required. A disadvantage of this technique is that for high dimensional inputs, paving of the input space in this way leads to a large number of basis functions. This is called the *curse of dimensionality*. On the contrary, for similar levels of accuracy, MLP networks typically require much less hidden neurons, making the MLP network less memory intensive [39].

Neural Control Systems. A large number of publications describing the use of neural networks for control of nonlinear systems have been published during this decade. The paper by Narendra and Parthasarathy published in 1990 [193], which introduced the concept of neural networks for identification and control of nonlinear dynamic systems, is often considered as the pioneering paper in this field. The range of solutions that have been proposed is very wide and it is not easy to categorize the various different approaches into specific classes. Nevertheless we will attempt to review these different approaches and identify the main differences between them. To make this task tractable, we limit ourselves to continuous-time neural control schemes that address the issue of closed loop stability, with the exception of a few publications that introduce original concepts that might be interesting.

The first stable neural control schemes to emerge, utilized RBF networks to control affine nonlinear plants of form (2.7) whose functions f and g are unknown. It is assumed that the basis function weights are known *a priori* and only the linear weights require updating. This leads to a situation of linear parameterization, allowing the use of powerful techniques from the theory of nonlinear adaptive control, *e.g.* [224]. However, issues related specifically to the features of the RBF neural network give rise to new problems, mainly:

- The effect on stability of the network's inherent approximation error, which is non zero even if the optimal weights were used.
- The prior choice of basis function parameters.
- The "compactness" of the region over which the network is able to approximate a function with a prescribed approximation error.
- The curse of dimensionality problem associated with RBF networks.

Most of these schemes utilize control laws based on feedback linearization. On one hand, this has the advantage that the system is not restricted to conform to one of the rather rigid canonical forms required in backstepping design, but on the other hand the nonlinearities are subject to satisfy some constraints. The adaptation laws are based on Lyapunov stability theory. Polycarpou and Ioannou [204] and Sanner and Slotine [219] consider systems with no internal dynamics, whilst Tzirkel-Hancock and Fallside [255] generalized this to systems having internal dynamics. It turns out that the network's inherent approximation error appears as a disturbance term in the systems error dynamics, which could lead to the well-known phenomenon of parameter drift [234]. This is compensated for by applying some robustifying techniques. For example, [204] utilize a projection algorithm in their indirect adaptive control method, whilst the direct adaptive scheme of [219] uses dead-zone adaptation. Reference [255] makes use of a low gain sliding mode control, augmenting the main feedback linearizing control signal. This is supplemented by the use of a boundary layer for the sliding mode component and dead-zone adaptation, to ensure further robustness to any high frequency unmodelled dynamics. Both [219] and [255] solve the problem associated with the compactness of the network approximation region by switching onto a high gain sliding mode control if the system's state exits this region.

Sanner and Slotine [219] address the problem of choosing the basis function parameters from a sampling theory point of view. In this technique, the Gaussian basis functions are assigned the same variance and are centred on a rectangular mesh covering the relevant region of input space. This is shown to be equivalent to a filtration of a sampled sequence of inputs by a low-pass filter having a Gaussian impulse response, whose characteristics depend on the basis function parameters. Hence if constraints on the degree of smoothness of the function being approximated are known, the basis function parameters could be determined from principles of signal reconstruction theory.

Fabri and Kadirkamanathan [70] address the curse of dimensionality problem by modifying the scheme of [255] with the inclusion of a dynamic structure neural network. Rather than using a fixed network whose basis functions scan the whole region of space where the state is expected to be contained, the dynamic structure network starts with no basis functions and assigns them sequentially in time according to the location of the trajectory of the system's state. By exploiting the localized representation of Gaussian basis functions and the fact that in most realistic situations the state trajectory only spans a small subset of the region of state space scanned by the fixed network, savings of up to 70% on the number of neurons were obtained and stability is still guaranteed by a direct adaptive control method. A similar dynamic structure methodology but based on different techniques, was later put forward in [162].

Spooner and Passino [239] modified both the indirect and direct adaptive approaches for the same class of affine plants considered above, to replace the GaRBF network with a class of Takagi-Sugeno fuzzy approximators. This allows the inclusion of *a priori* heuristic knowledge in the form of fuzzy rules.

Finally Polycarpou [205] addresses a special case of second order affine systems without internal dynamics, by deriving a control law from Lyapunov stability criteria rather than feedback linearization. Contrary to all the previous schemes, it does not require a bound on the optimal NN approximation error to be known. However the method only covers output regulation rather than the tracking problem and it is restricted to a very simple plant.

The use of MLP networks for stable adaptive control of the affine class of nonlinear systems was proposed in [54, 274]. Chen and Liu [54] consider relative degree 1 systems and derive adaptation and control laws that lead to a local convergence result, in the sense that the initial parameter errors must be small for convergence to be guaranteed. Yeşildirek and Lewis [274] consider relative degree n systems (*i.e.* no internal dynamics) and obtain more global convergence results, provided that a bound on the norm of the optimal network weights is known. The stability result obtained is based on the idea of Uniform Ultimate Boundedness (UUB) [139], and robustness to parameter drift is ensured by an e_1-modification scheme originally suggested in [185] for linear systems. The key contribution of this paper involves the use of a Taylor series expansion to determine bounds on the function approximation error in terms of the network parameter errors and the tracking error. These bounds are taken into consideration in the adaptation laws. A similar scheme was suggested to achieve tracking in n-link robotic manipulators [156].

Rovithakis and Christodoulou [216] utilized Recurrent Neural Networks (RNN) to effect state regulation of a multiple input affine system. UUB stability is guaranteed via a direct adaptive method. Kulawski and Brdyś [147] also applied RNNs to effect tracking in a general stable nonlinear system whose state is assumed unmeasurable. It assumes however that the RNN is able to model the plant perfectly well (*i.e.* no inherent approximation error) and stability is guaranteed provided that the optimal RNN parameters satisfy some conditions related to matrix invertibility and that the initial parameter error is small.

General nonlinear systems were also addressed by Zhang *et al.* [279], where networks were used to generate an "inverse controller" via Implicit Function Theory. This approach represents a very different philosophy from those mentioned previously because the controller is trained to approximate the control law rather than some functions associated with the plant. This idea was originally proposed in [89, 155] for discrete-time systems and will therefore be detailed in the following section.

2.3 Discrete-time Systems

In view of the fact that most modern controllers are implemented on digital computers, consideration of discrete-time systems is particularly important.

2.3 Discrete-time Systems

Even when the plant being controlled depends naturally on analogue signals, the sampling process for computer input/output constitutes a discrete-time function.

The state-space model of a general autonomous SISO nonlinear system in discrete-time is given by

$$\begin{aligned} \mathbf{x}(k+1) &= \mathbf{f}(\mathbf{x}(k), u(k)) \\ y(k) &= h(\mathbf{x}(k)) \end{aligned} \quad (2.9)$$

where $\mathbf{x}(k) \in \Re^n$, $u(k)$, $y(k) \in \Re$ respectively represent the state vector, the system input and the system output at time kT. T denotes the sampling period which is assumed constant and hence ignored in the above and subsequent equations. $\mathbf{f} : \Re^n \times \Re \mapsto \Re^n$ and $h : \Re^n \mapsto \Re$ are vector and scalar nonlinear functions respectively.

In contrast to the continuous-time case, most of the analysis on discrete-time systems is performed on the general form of equation (2.9) rather than on some subclass of systems that are affine in $u(k)$. The reason is that even if the plant being sampled (a) has a continuous-time nature (b) is affine in control and (c) admits a continuous-time feedback linearization controller; all these properties are lost as a result of sampling [96, 178]. Worse still, even in the case of purely discrete-time systems (*i.e.* discrete-time by nature rather than due to sampling) that have affine dynamics, reformulation of the state space equations to obtain a direct input-output relation destroys the affinity in $u(k)$. This renders the task of discrete-time controller design rather difficult and very often, approximate models that are affine in the control are used to make the task tractable, after important structural notions (such as the stability of zero dynamics) are established from the general form of equation (2.9) [191].

As in the continuous-time case, an important question to ask is *"In what way does the control input influence the output?"*. This is answered by defining the notion of relative degree for discrete-time systems. Consider reformulating equation (2.9) so as to obtain a direct IO relation. It is easy to see that:

$$\begin{aligned} y(k) &= h(\mathbf{x}(k)) := \phi_o(\mathbf{x}(k)) \\ y(k+1) &= h(\mathbf{x}(k+1)) = h(\mathbf{f}(\mathbf{x}(k), u(k))) := \phi_1(\mathbf{x}(k), u(k)) \\ &\vdots \\ y(k+i) &= h(\mathbf{x}(k+i)) \\ &= h(\mathbf{f}(\mathbf{f} \ldots \mathbf{f}(\mathbf{f}(\mathbf{x}(k), u(k)), u(k+1)), \ldots), u(k+i-1))) \\ &= h \circ \mathbf{f}^i(\ldots) := \phi_i(\mathbf{x}(k), u(k), \ldots u(k+i-1)) \end{aligned} \quad (2.10)$$

where $\mathbf{f}^i(\ldots)$ denotes an i-times iterated composition of $\mathbf{f}$, *i.e.*

$$\mathbf{f} \circ \mathbf{f} \circ \mathbf{f} \circ \ldots \circ \mathbf{f},$$

and $\mathbf{f} \circ \mathbf{f}$ denotes composition, *e.g.* $\mathbf{f}_1 \circ \mathbf{f}_2 \equiv \mathbf{f}_1(\mathbf{f}_2(\cdot))$.

Note that even for the special case of affine dynamics, where $\mathbf{f}(\cdot)$ in Equation (2.9) would take the form $\mathbf{f}(\mathbf{x}(k), u(k)) = \mathbf{a}(\mathbf{x}(k)) + \mathbf{b}(\mathbf{x}(k))u(k)$, the IO relations described above are no longer affine in u, e.g.:

$$y(k+1) = h[\mathbf{a}(\mathbf{x}(k)) + \mathbf{b}(\mathbf{x}(k))u(k)].$$

This is in sharp contrast with the continuous-time case where, because of the use of differentiation rather than composition, affinity in u is conserved (see Equation (2.3)).

To detect the time step at which the input $u(k)$ influences the output, and extend the notion of relative degree to the discrete-time case, we evaluate

$$\frac{\partial y(k+t)}{\partial u(k)} = \frac{\partial h \circ \mathbf{f}^t}{\partial u(k)}$$

and use the value of t at which this derivative is non-zero to determine the *relative degree*. The formal definition, due to Monaco and Normand-Cyrot [179], is given as follows:

Definition 2.3.1. *Suppose system (2.9) satisfies $\mathbf{f}(0,0) = \mathbf{0}$, $h(0) = 0$ (i.e., $(\mathbf{x}, u) = (\mathbf{0}, 0)$ is an equilibrium point) with $\mathbf{f}$, h being analytic functions. Then the system has a relative degree r if*

$$\frac{\partial h \circ \mathbf{f}^t(\cdots)}{\partial u(k)} = 0, \; \forall t = 1, \ldots, r-1$$

$$\frac{\partial h \circ \mathbf{f}^r(\cdots)}{\partial u(k)} \neq 0.$$

This means that the first output affected by the input $u(k)$ is $y(k+r)$. Consequently, subsequent inputs $u(k+1), \cdots, u(k+r-1)$ do *not* affect $y(k+r)$ and could be ignored in any equations for $y(k+r)$. Since $\mathbf{f}$, h are analytic, then $r \leq n$ or $r = \infty$ [48]. Cabrera and Narendra [46] offer a slightly different definition that could lead to the possibility of having a relative degree that is not well defined. However, for all cases where the relative degree of [46] is well defined, both definitions are equivalent.

If the relative degree $r < n$, then there exists a local and invertible mapping $\mathbf{T}(\cdot)$ transforming states $\mathbf{x}$ to a set of new states $\mathbf{z}$ (*i.e.* $\mathbf{z} = \mathbf{T}(\mathbf{x})$) such that the system could be represented in the following form [48]:

$$\begin{aligned}
z_{1_1}(k+1) &= z_{1_2}(k) \\
z_{1_2}(k+1) &= z_{1_3}(k) \\
&\vdots \\
z_{1_{r-1}}(k+1) &= z_{1_r}(k) \\
z_{1_r}(k+1) &= \gamma(\mathbf{z}_1(k), \mathbf{z}_2(k), u(k)) \\
\mathbf{z}_2(k+1) &= \mathbf{I}(\mathbf{z}_1(k), \mathbf{z}_2(k), u(k)) \\
y &= z_1(k).
\end{aligned} \quad (2.11)$$

2.3 Discrete-time Systems

Note that the transformed state vector $\mathbf{z}$ has been partitioned as $\mathbf{z} = [\mathbf{z}_1^T \ \mathbf{z}_2^T]^T$. Vector $\mathbf{z}_1(k) \in \Re^r = [h \ \ h \circ \mathbf{f} \ \ h \circ \mathbf{f}^2 \cdots h \circ \mathbf{f}^{r-1}]^T$ represents $y(k), \cdots, y(k+r-1)$, and z_{1_i} denotes the ith element of $\mathbf{z}_1$. Vector $\mathbf{z}_2 \in \Re^{n-r}$ is the vector of *internal states* that do not influence $\mathbf{z}_1$ and have dynamics $\mathbf{I}(\mathbf{z}_1(k), \mathbf{z}_2(k), u(k))$. Function $\gamma(\mathbf{z}_1(k), \mathbf{z}_2(k), u(k)) = h \circ \mathbf{f}^r(\mathbf{T}^{-1}(\mathbf{z}), u(k)) = y(k+r)$. Note that this representation is not referred to as being in *normal form* because, unlike the continuous-time case, $\mathbf{I}(\cdot)$ is not independent of u.

In a similar way to the continuous-time case, the *zero dynamics* are defined as the special case when the system outputs $y(k), \ldots, y(k+r-1)$ are constrained to be zero. This is satisfied if we start with initial conditions $\mathbf{z}_1(k) = \mathbf{0}$ and then find an input $u(k)$ that forces $y(k+r) = 0$, an input $u(k+1)$ that forces $y(k+r+1) = 0$ etc., to maintain $\mathbf{z}_1 = \mathbf{0}$ for all subsequent time-steps. The existence of these control inputs is ensured because from the definition of relative degree, $\frac{\partial y(k+r)}{\partial u(k)} \neq 0$. This allows the Implicit Function Theorem to be invoked and establish the existence of a unique $u(k) = v(\mathbf{x})$ that drives $y(k+r)$ to zero. The same argument applies for subsequent time steps involving $y(k+r-1), u(k+1)$ etc. Under this condition, the system dynamics (2.11) reduce to

$$\mathbf{z}_1(k+1) = \mathbf{0}$$
$$\mathbf{z}_2(k+1) = \mathbf{I}(\mathbf{0}, \mathbf{z}_2(k), v(\mathbf{T}^{-1}(\mathbf{z}))) \qquad (2.12)$$
$$y(k) = 0.$$

Equation (2.12) represents what are called the *zero output constrained dynamics* of the system, which exclusively characterize the internal states $\mathbf{z}_2$.

It has been shown in [46] that if system (2.9) has a well-defined relative degree (in the sense of Cabrera and Narendra) equal to r and its zero output constrained dynamics are uniformly asymptotically stable, then there exists a control law

$$u(k) = v(\mathbf{x}(k), y_d(k), \cdots, y_d(k+r)) \qquad (2.13)$$

which forces the output to asymptotically track the reference input y_d, i.e:

$$\lim_{k \to \infty} |y(k) - y_d(k)| = 0.$$

Note that this is an existence result, even for the case of known $\mathbf{f}$ and h, and in general the control law $v(\cdot)$ cannot be easily derived.

Although the state space representation (2.11) is very powerful for analysing the structural properties of the system, it is often more convenient to have a state-free representation for control design. This implies a pure IO representation between y and u, without any terms in $\mathbf{z}$. For a controllable and observable linear time invariant system, there exist equivalent representations in both state space and IO form. Unfortunately, this is not always the case for a nonlinear system.

Consider re-writing equations (2.10) in vector form, as follows:

$$\mathbf{y}_i(k) = \Phi_i(\mathbf{x}(k), \mathbf{u}_{i-1}(k))$$

where

- $\mathbf{y}_i(k) := [y(k)\ y(k+1)\ \cdots y(k+i)]^T \in \Re^{i+1}$ denotes the collection of outputs from time k till time $(k+i)$.
- $\mathbf{u}_{i-1}(k) := [u(k)\ u(k+1)\ \cdots u(k+i-1)]^T \in \Re^i$ denotes the inputs from time k till time $(k+i-1)$.
- $\Phi_i := [\phi_0(\mathbf{x}(k))\ \phi_1(\mathbf{x}(k), \mathbf{u}(k)) \cdots \phi_i(\mathbf{x}(k), u(k), \cdots, u(k+i-1))]^T$ whose components were defined in Equation (2.10).

Consider the particular case when $i = n - 1$, i.e.

$$\mathbf{y}_{n-1}(k) = \Phi_{n-1}(\mathbf{x}(k), \mathbf{u}_{n-2}(k)).$$

Then if the Jacobian $\frac{\partial \Phi_{n-1}}{\partial \mathbf{x}(k)}|_{(\mathbf{x}=0, \mathbf{u}_{n-2}=0)}$ is non-singular and the origin is an equilibrium point, the Implicit Function Theorem can be used to determine that *locally*, the state $\mathbf{x}(k)$ could be expressed in terms of the present and future outputs and inputs as follows:

$$\mathbf{x}(k) = \mathbf{F}(\mathbf{y}_{n-1}(k), \mathbf{u}_{n-2}(k))$$

where $\mathbf{F} : \Re^n \times \Re^{n-1} \mapsto \Re^n$ is a smooth function [191]. Hence, substituting for $\mathbf{x}(k)$ in the last of the series of Equations (2.10) with $i = n$, yields:

$$\begin{aligned} y(k+n) &= \phi_n[\mathbf{F}(\mathbf{y}_{n-1}(k), \mathbf{u}_{n-2}(k)), \mathbf{u}_{n-1}(k)] \\ &= H(\mathbf{y}_{n-1}(k), \mathbf{u}_{n-1}(k)) \end{aligned}$$

where $H : \Re^n \times \Re^n \mapsto \Re$ is a smooth scalar function of *past* outputs and inputs with respect to the time step $(k+n)$. Mapping $k \to k - n + 1$ yields the familiar NARMA model [154]:

$$y(k+1) = H(\mathbf{y}_{n-1}(k-n+1), \mathbf{u}_{n-1}(k-n+1)). \tag{2.14}$$

Note that Equation (2.14) relates $y(k+1)$ to $u(k)$ as the most recent input. However if the plant has a relative degree r, it is known that $y(k+1)$ depends on inputs appearing on and before the previous r time steps (*i.e.* time step $(k+1-r)$ and previous) and that subsequent inputs $u(k+2-r), u(k+3-r), \cdots, u(k)$ do not influence $y(k+1)$. This knowledge permits us to refine the IO relation (2.14) by including the effect of the relative degree as follows [155, 183]:

Lemma 2.3.1. *The NARMA representation of Equation (2.14) could be formulated in the following form if the system has a well-defined relative degree r in the region of interest where (2.14) is valid:*

$$y(k+r) = \overline{H}(\mathbf{y}_{n-1}(k-n+1), \mathbf{u}_{n-1}(k-n+1)) \tag{2.15}$$

where $\overline{H}$ is a smooth scalar function.

Proof. From Equation (2.14)

$$\begin{aligned} y(k+1) &= H(\mathbf{y}_{n-1}(k-n+1), \mathbf{u}_{n-1}(k-n+1)) \\ &= H(\mathbf{y}_{n-1}(k-n+1), \mathbf{u}_{n-r}(k-n+1)), \end{aligned} \tag{2.16}$$

since by the definition of relative degree, $\mathbf{u}(i)$, $i > k+1-r$ do not influence $y(k+1)$.

Similarly for $y(k+2)$

$$\begin{aligned} y(k+2) &= H(\mathbf{y}_{n-1}(k-n+2), \mathbf{u}_{n-r}(k-n+2)) \\ &= H(\mathbf{y}_{n-2}(k-n+2), y(k+1), \mathbf{u}_{n-r}(k-n+2)) \\ &= H(\mathbf{y}_{n-2}(k-n+2), H[\mathbf{y}_{n-1}(k-n+1), \\ & \quad \mathbf{u}_{n-r}(k-n+1)], \mathbf{u}_{n-r}(k-n+2)), \quad \text{by equation (2.16)} \\ &= H_1(\mathbf{y}_{n-1}(k-n+1), \mathbf{u}_{n-r+1}(k-n+1)) \end{aligned}$$

since $\mathbf{y}_{n-1}(k-n+1)$ subsumes $\mathbf{y}_{n-2}(k-n+2)$, and term $\mathbf{u}_{n-r+1}(k-n+1)$ covers both $\mathbf{u}_{n-r}(k-n+1)$ and $\mathbf{u}_{n-r}(k-n+2)$. By similar arguments we obtain that $y(k+3) = H_2(\mathbf{y}_{n-1}(k-n+1), \mathbf{u}_{n-r+2}(k-n+1))$ and the pattern is repeated such that at time step $(k+r)$ we get

$$y(k+r) = H_{r-1}(\mathbf{y}_{n-1}(k-n+1), \mathbf{u}_{n-1}(k-n+1)).$$

Denoting H_{r-1} by $\overline{H}$ proves the lemma. □

As shown in [191], Lemma 2.3.1 leads to the following important theorem on control of system (2.15):

Theorem 2.3.1. *A system that admits the IO representation of equation (2.15) can be forced to track a desired reference y_d by a unique control law:*

$$u(k) = \mu(\mathbf{y}_{n-1}(k-n+1), y_d(k+r), \mathbf{u}_{n-2}(k-n+1)) \quad (2.17)$$

where $\mu : \Re^n \times \Re \times \Re^{n-1} \mapsto \Re$ is a smooth function.

Proof. By definition, $\frac{\partial y(k+r)}{\partial u(k)} \neq 0$ due to the relative degree r. Equivalently, by using equation (2.15) and extracting $u(k)$,

$$\frac{\partial \overline{H}(\mathbf{y}_{n-1}(k-n+1), \mathbf{u}_{n-2}(k-n+1), u(k))}{\partial u(k)} \neq 0.$$

Since $y_d(k+r)$ is independent of $u(k)$, the above equation could be written as:

$$\frac{\partial \overline{H}(\mathbf{y}_{n-1}(k-n+1), \mathbf{u}_{n-2}(k-n+1), u(k)) - y_d(k+r)}{\partial u(k)} \neq 0.$$

Hence by the Implicit Function Theorem, there exists a unique control input $u(k) = \mu(\mathbf{y}_{n-1}(k-n+1), y_d(k+r), \mathbf{u}_{n-2}(k-n+1))$ that drives the term $(\overline{H}(\cdot) - y_d(k+r))$ to zero, which leads to $y(k+r) = y_d(k+r)$. □

2.3.1 Affine Approximations and Feedback Linearization

The results presented in the previous section represent interesting analogues to familiar concepts in linear systems theory, such as controllability, minimum phase and transfer function representation. Unfortunately however, although conditions for the existence of stable control laws have been established, it is not obvious at all how to derive them, even in the case of known dynamics. The main reason is that the appropriate control is a nonlinear function of previous inputs, outputs and the reference signal and there is no methodological way of determining this function in closed form [48]. Consequently, approximate solutions are often used. There are two main ways of setting up approximate schemes [183]:

- Finding an affine *approximation* to the IO representation of Equation (2.15) and designing an *accurate* controller based on this approximate model [191].
- Finding an *approximation* or an emulation of the control laws (2.13) or (2.17) that are based on the *accurate* plant representations (2.11) or (2.15) respectively [48, 155].

In practice Narendra and Mukhopadhyay found that the first approach is simpler and often more accurate than the second [191]. With this in mind, they developed two approximate models based on Taylor series expansions of the NARMA Equation (2.15). In one case, the expansion is performed around $\mathbf{y}_{n-1}(k-n+1), (\mathbf{u}_{n-1}(k-n+1) = \mathbf{0})$, which leads to the following representation, denoted as NARMA-L1:

$$y(k+r) = f_0(\mathbf{y}_{n-1}(k-n+1)) + \Sigma_{i=0}^{n-1} g_i(\mathbf{y}_{n-1}(k-n+1))u(k-i).$$

The second approximate model, denoted as NARMA-L2, depends on an expansion around $u(k)$ and leads to

$$\begin{aligned} y(k+r) = &\ \overline{f}(\mathbf{y}_{n-1}(k-n+1), \mathbf{u}_{n-2}(k-n+1)) + \\ &\ \overline{g}(\mathbf{y}_{n-1}(k-n+1), \mathbf{u}_{n-2}(k-n+1))u(k). \end{aligned}$$

Both these approximate models are affine in $u(k)$, and hence control laws could be determined in algebraic closed form on the basis of feedback linearization, as in continuous-time systems. The efficacy of the method depends on how accurate a representation of the plant is the NARMA-Lx model in the region of operation under consideration.

2.3.2 Adaptive Control

When compared to the continuous-time case, there are relatively few published results on nonlinear adaptive control of discrete-time systems. This is not only due to the problems of model representation discussed in the previous section, but also because adaptive control of discrete-time systems is not a straightforward modification of the continuous-time case. This is also true for the linear case, where specific analysis and design techniques had to be developed specifically to handle discrete-time systems. These include schemes based on boundedness of signals and convergence of sequences [69, 93], or Lyapunov-based schemes using error augmentation [111, 189].

In a similar manner, ideas for parametric uncertain nonlinear systems that proved to be so successful in continuous-time (*e.g.* adaptive backstepping) have not been readily extended to the discrete-time case and it is only recently that promising results have started to appear. Kanellakopoulos [138] considers a very simple scalar nonlinear system with one unknown parameter. He proposes a new least squares estimator with nonlinear data weighting to handle stability in the presence of nonlinear functions that are not necessarily linearly bounded. Yeh and Kokotović [275] introduced adaptive backstepping ideas for a discrete-time, output-feedback class of plants having unknown parameters. As opposed to the continuous-time backstepping case however, the stability and tracking results are only valid under the assumption of Lipschitz-type growth constraints on the nonlinearities, unless a particular rank condition is assumed. Madani *et al.* [165] introduced a different backstepping technique from [275] and obtained stable regulation results for a feedback-linearizable class of systems without imposing any nonlinearity constraints. Zhao and Kanellakopoulos [282] obtain similar results for a strict-feedback problem. Their method hinges on a nonlinear identification algorithm that computes projections of the unknown parameter vector on the basis of a subspace that depends on the plant nonlinear functions. The idea is extended to the output-feedback case in [281]. To sum up, nonlinear adaptive control of discrete-time systems has still not matured into a solid body of research, as reflected by the rather sparse publications mentioned above. Although a few promising stability results on regulation have recently started to appear, more work is necessary to achieve better and more general results that impose less rigid constraints on the nonlinearities.

Results published on functional uncertain discrete-time systems are also not very numerous. The early paper on neural control [193] by Narendra and Parthasarathy concentrated on discrete-time plants and introduced four different classes of models. It focused only on indirect adaptive methods and used MLP networks whose parameters were adjusted by gradient methods. Being a pioneering paper in the field, it only introduced basic concepts and definitions of adaptive control by neural networks. It did not address theoretical issues such as stability and how to establish the existence of important structural properties like internal dynamics and controllability.

An interesting offshoot of this paper were the results published in [89, 155], where the concept of what is sometimes known as *Implicit Function Emulation* was developed. The idea is to use a neural network N_c to learn the control law $\mu(\mathbf{y}_{n-1}(k-n+1), y_d(k+r), \mathbf{u}_{n-2}(k-n+1))$ previously mentioned in Equation (2.17), which is guaranteed to exist from the Implicit Function Theorem, subject to the plant admitting the IO representation (2.15). As pointed out before, even if the plant representation $y(k+r) = \overline{H}(\mathbf{y}_{n-1}(k - n + 1), \mathbf{u}_{n-1}(k - n + 1))$ is known, $\mu(\cdot)$ cannot be analytically determined in general. Hence, Implicit Function Emulation suggests to emulate it by training network N_c on the basis of data generated from a model of the plant. The use of a model to generate the training data is prescribed because if the training is performed on the actual plant, the closed-loop system might easily become unstable. It also has the advantage of simplifying the network weight adjustment algorithm. If a plant model is not available, it could be generated by another neural network N_p that is trained off-line to identify the plant from actual IO data. After N_c is trained, it is then safe to connect it to the actual plant and effect control action. Although this technique involves the use of neural networks and "learning", it cannot be considered as an adaptive control scheme in the usual sense. The reason is that it relies on off-line training of the neural controller N_c (and possibly even the plant model N_p when $\overline{H}$ is unknown) and once connected to the plant, N_c remains fixed, without being open to further adaptation. As such, this technique could be considered as a "neural controller design" scheme where a neural network is trained to synthesize a fixed controller, rather than an adaptive scheme. Its philosophy is related more to conventional, non-adaptive control design techniques rather than adaptive control. The difference is that it uses neural networks instead of precise mathematical plant models. Nevertheless it is an interesting approach that might be developed along "real" adaptive terms in the future. A step in this direction was recently taken for continuous-time systems [280] where it was suggested to use some *a priori* designed stabilizing controller in parallel with N_c, to take over the control task during the initial phase when N_c is training from actual IO data.

Another interesting and original paper by Yu and Annaswamy [276] addresses regulation of parametric uncertain systems whose parameters may appear nonlinearly in the plant equations. The network is used as a *parameter adjustment mechanism* instead of a controller or plant function identifier. Naturally the latter two are not really necessary in this problem because the plant nonlinearities and the structure of the control equations are assumed known. The neural network needs to be trained off-line, which requires that both the actual plant parameter values and corresponding IO response must be known during training. The idea is not too dissimilar from Implicit Function Emulation except that in this case, the neural network is emulating the adaptation law rather than the control law. Hence when the trained network is connected to the plant, the parameters used in the control law are adjusted, but the weights of the network itself remain fixed at the values learned during training.

Papers which consider the problem of adaptive control of functional uncertain systems in discrete-time include [53, 120, 160, 240]. Chen and Khalil [53] consider nonlinear SISO systems that are affine in the control, having form

$$y(k+1) = f(\cdot) + g(\cdot)u(k-d+1).$$

where $f(\cdot), g(\cdot)$ are unknown nonlinear functions of $(y(k-n+1), \cdots, y(k), u(k-d-m+1), \cdots, u(k-d))$. They present a stable, indirect adaptive control scheme that uses a feedback linearization type of control law and MLP networks. Robustness to the inherent network approximation error is handled by dead-zone adaptation, and stability is guaranteed only if the initial network parameter error is bounded by a predetermined value. Hence the convergence results of this scheme are not global with respect to the initial parameter values.

Jagannathan and Lewis [120] consider an affine multiple-input/multiple-output (MIMO) nonlinear system having a simpler form, because it does not include a nonlinear function multiplying $u(k)$, i.e.

$$\mathbf{y}(k+1) = \mathbf{f}(\mathbf{y}(k) \cdots \mathbf{y}(k-n+1)) + \mathbf{u}(k).$$

A direct adaptive control scheme using feedback linearization and a MLP network is presented. Robustness to the network approximation error is handled by adaptation laws based on the e_1-modification originally proposed in [185] for continuous-time systems. Convergence of tracking and parameter error is guaranteed in a Uniformly Ultimately Bounded (UUB) sense, i.e. they converge to a bounded region whose size depends on various design parameters. Unlike the previous case, the stability of this scheme is global with respect to the initial parameter error. However it requires prior knowledge of bounds on the norm of the optimal (unknown) network weights. This assumption is not always realistic in practice, because the parameters do not relate to any obvious physical features of the plant.

Sun et al. [240] consider an unknown, affine, MIMO continuous-time system that is sampled. They derive the equivalent discrete-time representation and develop an adaptive sliding sector-type variable structure control (VSC) system [86]. They use RBF networks and direct adaptation laws. The control law consists of a combination of feedback linearization and discontinuous variable structure terms. The VSC term covers both cases when the state exits outside the network approximation region and also when it is inside, but the error metric is outside a so-called *sliding sector*. The system is robust to the inherent network approximation error. The main problems of this scheme involve (a) the difficulty in determining correct values of a number of design parameters that are important for ensuring stability and (b) the assumption that the network Universal Approximation Property still holds if the network inputs involving $\mathbf{u}(k)$ are delayed by one time step. This latter assumption had to be included because otherwise the control law would not be causal and so $\mathbf{u}(k-1)$ is used in place of the theoretically correct $\mathbf{u}(k)$, as input to the network. The authors claim that the additional approximation error introduced by this assumption could be compensated for by the VSC law but it is not explained how bounds on the value of this error (that are required inside the control law) could be determined.

Liu *et al.* [160] develop a neural network scheme for a class of nonlinear discrete-time systems that is based on predictive control methods. In contrast to typical nonlinear predictive control schemes, they avoid the use of nonlinear programming methods for optimizing a performance index over the prediction horizon. Instead they apply a set of neural network-based predictors whose output is used directly inside a closed-form equation that determines the optimal control signal. Although this scheme guarantees the convergence of the network weights and the estimation error, unfortunately similar convergence results for the tracking error are not obtained.

2.4 Summary

A large number of practical systems operate under nonlinear conditions. Hence a generalization of the well-established techniques of linear control to the nonlinear case is an important research challenge. A considerable amount of nonlinear control problems have been solved during the last two decades, within both the non-adaptive and adaptive frameworks. In particular, intelligent control techniques have been successfully applied to design a number of stable adaptive solutions for particular classes of functional uncertain systems. All this is very encouraging, but a number of open questions still remain, especially for functional adaptive schemes. Unfortunately, several restrictive assumptions have to be imposed in order to derive schemes that guarantee stability.

This chapter has presented a number of important theoretical issues associated with nonlinear systems, as well as a review of the main publications on stable adaptive control for uncertain nonlinear plants. Parametric and functional uncertain systems have been considered in both continuous and discrete time. The achievements and limitations of the various schemes have been highlighted, particularly neural network-based functional adaptive strategies. This chapter has a dual purpose: it reflects the state of the art in adaptive control of nonlinear systems and secondly, it serves as a theoretical basis and motivation for the new control strategies presented in the rest of the chapters of the book.

3. Dynamic Structure Networks for Stable Adaptive Control

3.1 Introduction

The local representation properties and the possibility of predetermining the basis function parameters in Gaussian RBF networks, makes them ideal candidates for implementing functional adaptive control. However these attractive features are somewhat tarnished by the curse of dimensionality problem associated with GaRBF networks when used for high dimensional spaces.

In using RBF networks, one typically places the centre of the basis functions on regular points of say, a square mesh covering a relevant region of space where the state $\mathbf{x}$ is known to be contained, denoted by a compact set $\chi \subset \Re^n$. This region therefore constitutes the network approximation region which, in general, is known for a given system. The distance between the points, say μ, affects the number of basis functions required to span the region χ and hence determines the size of the neural network. On the other hand, μ depends on the approximation accuracy required of the neural network; a large μ would lead to a coarser approximation while a small μ would lead to a finer approximation. Neural network structure selection and, in particular, the centre allocation for radial basis functions have been discussed in the context of approximation theory in [169]. In fact, as developed in [219], there exists a systematic procedure for determining the variances and centres of the basis functions to ensure that the network approximation accuracy is uniformly bounded everywhere within a relevant and finite region of state-space.

When using a square mesh to cover the network approximation region, the number of RBFs increases exponentially with state dimensionality n. This often gives rise to implementation problems because it places excessive demands on memory. However for systems with order higher than one ($n > 1$), the state $\mathbf{x}(t)$ will typically span only a subset of the space in χ [219, 254]. For the Gaussian RBF network, with the basis functions having a localized receptive field, most of the basis functions lie outside this subset and are not used for function approximation and their parameters are barely changed during adaptation. This implies that placing basis functions on all mesh points within χ results in a lot of redundancy. An efficient scheme to overcome this problem was developed for time-series analysis and system identification [128, 136, 202] and for classification [135], where a Gaussian RBF network is 'grown' by sequentially placing units in those regions of state-space visited by the system during operation, giving rise to a *dynamic network structure*. This results in a much smaller network in which all of the basis functions are utilized effectively and with much reduced storage and computational requirements. A dynamic structure neural network has been developed for reinforcement learning controller design [10], which is based on the ideas of [202] used in time-series analysis. Similar suggestions for the direct adaptive controller have been made by Sanner and Slotine [219].

In this chapter, these suggestions and the growing network ideas from identification will be used in developing a dynamic structure neural network scheme for stable adaptive control of affine nonlinear systems. The scheme developed here follows closely and makes use of the theory developed in [219, 254], with the ideas from identification [127, 136, 202], and makes appropriate modifications to enhance the performance of the system and guarantee stability.

The use of a neural network with a finite number of basis functions will inherently result in the presence of an approximation error which gives rise to a *disturbance* signal [219, 254]. In turn, this might lead to *parameter drift* [184, 186, 224]. The robust adaptive control scheme used in [219] is restricted in that to ensure stability, at any one time a minimum number of nodes or units needs to be activated in the vicinity of the state trajectory. This minimum number increases with the desired tracking accuracy. This is due to that fact that to ensure robustness to parameter drift, [219] uses simple *dead-zone adaptation* [186, 204, 234]. This requires the disturbance signal to be uniformly bounded, with the size of the bound determining the size of the dead-zone [68, 200]. On the other hand, the dead-zone size also affects the tracking accuracy; the higher the tracking accuracy the smaller the dead-zone and hence the bound on the disturbance must be smaller to ensure robustness. The latter implies that more nodes would need activation at any one time. Hence in a dynamic network scheme, the need for network growth becomes ever more crucial with greater tracking accuracy. This in turn results in networks that grow unacceptably large, reducing the benefits of the dynamic network scheme.

In this chapter, we offer a solution to the above problem by introducing the ideas proposed in [254, 255], where the control law is augmented with a low gain *sliding control* [256, 257] term whilst $\mathbf{x} \in \chi$ to ensure robustness to the disturbance. The use of low gain sliding control results in a situation where the adaptation dead-zone, and hence the desired accuracy, is independent of the size of the disturbance term because the disturbance is compensated for by sliding control. Thus, it appears as though there is no restriction on the minimum number of nodes requiring activation at any one time. Hence the network growth rate becomes a *design parameter* whose choice does not affect the tracking accuracy. In practice this argument must be modified slightly because, as we shall see later, both the dead-zone width and the sliding control gain affect the bandwidth of the control signal [234, 254]. In practice, the allowed bandwidth is bounded above so that the excitation of high frequency unmodelled dynamics is avoided. This limitation on the bandwidth forces a compromise to be struck between the tracking accuracy and the growth rate of the neural networks, against the maximum control bandwidth.

In summary, the main contributions of this chapter are listed below:

1. The development and analysis of a stable adaptive control scheme that utilizes dynamic structure GaRBF networks for a class of functional uncertain, nonlinear continuous-time systems. This relieves the curse of dimensionality problem associated with RBF networks whilst guaranteeing stability and robustness in the presence of disturbances and approximation errors.
2. The use of the robustness ideas propsed in [254, 255] with the dynamic network structure scheme outlined in [219], which effectively leads to the decoupling of the tracking accuracy from the neural network growth rate, subject only to constraints on the maximum allowable system bandwidth. Given a desired level of tracking accuracy, this leads to more economic dynamic structure networks.
3. The introduction of an efficient RBF activation scheme that exploits the parallel architecture of neural networks to achieve faster computation when deciding whether new RBFs have to be introduced inside the network.

3.2 Problem Formulation

The control task is to ensure that the output of the affine class of nonlinear systems modelled by Equation (2.2), reproduced below for convenience, tracks a desired output.

$$\dot{\mathbf{x}} = \mathbf{a}(\mathbf{x}) + \mathbf{b}(\mathbf{x})u$$
$$y = h(\mathbf{x}).$$

The development of the ideas will be restricted to the case of single-input/single-output (SISO) systems, noting that the extension to multiple-input/multiple-output (MIMO) systems is straightforward.

The nonlinear functions $\mathbf{a}$, $\mathbf{b}$ and h are unknown and the control task is for the system output vector $\mathbf{y}(t) := [y\ y^{(1)} \ldots y^{(r-1)}]^T$ to track a desired output vector as specified by the reference input $\mathbf{y}_d(t) := [y_d\ y_d^{(1)} \ldots y_d^{(r-1)}]^T$, with the state vector $\mathbf{x}$ remaining bounded. It is assumed that the following conditions are satisfied:

Assumption 3.2.1 *The system has a globally defined relative degree r and it is known.*

Assumption 3.2.2 *A global diffeomorphism transforming the system dynamics to globally valid normal form exists.*

Assumption 3.2.3 *The system is minimum phase i.e., system's zero dynamics are globally exponentially stable.*

Assumption 3.2.4 *The system's internal dynamics* $\mathbf{I}(\mathbf{z}_1, \mathbf{z}_2)$ *are Lipschitz in* $\mathbf{z}_1$, $\mathbf{z}_2$.

Assumption 3.2.5 *The system state* $\mathbf{x}(t)$, *the output* $y(t)$ *and also its* $(r-1)$ *derivatives* y, $y^{(1)}, \cdots, y^{(r-1)}$ *are all available for measurement.*

Assumptions 3.2.1 and 3.2.2 imply that the system could be globally expressed in the input-output affine form of Equation (2.5) i.e.,

$$y^{(r)} = f(\mathbf{x}) + g(\mathbf{x})u(t), \tag{3.1}$$

where it is recalled that $f(\mathbf{x}) = L_\mathbf{a}^r h(\mathbf{x})$ and $g(\mathbf{x}) = L_\mathbf{b} L_\mathbf{a}^{r-1} h(\mathbf{x})$. Hence functions $f(\mathbf{x})$ and $g(\mathbf{x})$ are unknown. The following two assumptions are also required:

Assumption 3.2.6 *There exist a known function* $g_l(\mathbf{x})$ *and a known constant* g_l *that represent lower bounds on the system function* $g(\mathbf{x})$, *such that:*

$$g(\mathbf{x}) \geq g_l(\mathbf{x}) \geq g_l > 0 \ \forall \ \mathbf{x}.$$

Assumption 3.2.7 *The reference input and its r derivatives* y_d, $y_d^{(1)}, \cdots, y_d^{(r)}$ *are bounded and known.*

Functional adaptive control is required because of the uncertainty of functions $f(\mathbf{x})$ and $g(\mathbf{x})$. Neural networks can be used to approximate them by on-line adjustment of the network parameters. In this chapter we develop a stable scheme that utilizes dynamic structure GaRBF networks to attain this objective. Before proceeding however, a brief review of typical fixed-structure GaRBF network solutions is given in the next section.

3.3 Fixed-structure Network Solutions

As shown in Theorem 2.2.1, for the case of known system dynamics and given Assumptions 3.2.1 to 3.2.5 and 3.2.7, the feedback linearizing control law of Equation (2.6) would lead to state boundedness and stable asymptotic tracking. However this control law depends on the system Lie derivatives $f(\mathbf{x})$, $g(\mathbf{x})$, which in this case are unknown. When these nonlinear functions are known but are linearly parameterized with unknown parameters, adaptive control techniques that do not require the use of neural networks have been developed [142, 224, 245]. However, in the absence of known Lie derivatives, one must resort to functional adaptive control techniques whereby approximations $\hat{f}(\mathbf{x})$, $\hat{g}(\mathbf{x})$ to the actual Lie derivatives $f(\mathbf{x})$, $g(\mathbf{x})$ are determined via the neural networks, and used in the certainty equivalent control law:

$$u_{fl}(t) = \frac{-\hat{f}(\mathbf{x}) + v(t)}{\hat{g}(\mathbf{x})} \qquad (3.2)$$

where $v(t) = y_d^{(r)} - \alpha_r e^{(r-1)} - \cdots - \alpha_1 e$ has the same definition as in the feedback linearizing Equation (2.6), with coefficients α_i chosen so as to form a Hurwitz polynomial $\Gamma(s) = s^r + \alpha_r s^{r-1} + \cdots + \alpha_1$, where s denotes the familiar Laplace variable. Recall that e denotes the *output tracking error* $(y - y_d)$. To ensure global stability and convergence of the tracking error to zero [219, 254, 255], we augment the control law (3.2) by a sliding mode [256, 234] control term $u_{sl}(t)$, such that the system control input $u(t)$ becomes:

$$u(t) = u_{fl}(t) + u_{sl}(t). \qquad (3.3)$$

In [254, 255], a pair of Gaussian radial basis function neural networks are used to generate the approximations $\hat{f}(\mathbf{x})$, $\hat{g}(\mathbf{x})$ for $\mathbf{x} \in \chi$, where $\chi \subset \Re^n$ represent the network approximation region where the state vector is expected to be contained. The overall arrangement of this adaptive control scheme is shown in Figure 3.1.

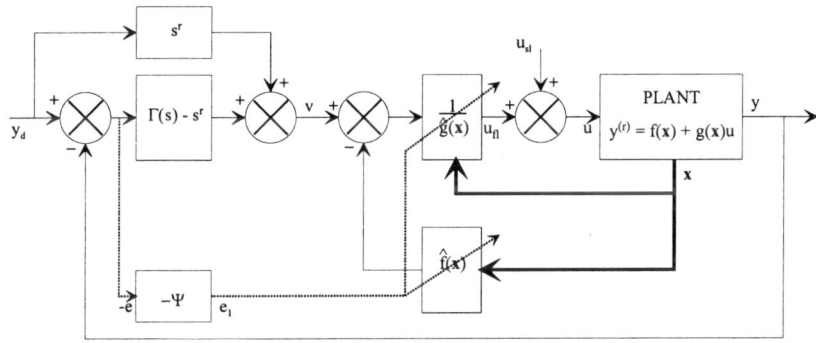

Fig. 3.1. Neural network based adaptive control scheme.

The network approximations are achieved via the functions:

$$\begin{aligned} \hat{f}(\mathbf{x}) &= \kappa(\mathbf{x})\hat{\mathbf{w}}_\mathbf{f}^T(t)\Phi_\mathbf{f}(\mathbf{x}) + f_o(\mathbf{x}) \\ \hat{g}(\mathbf{x}) &= \kappa(\mathbf{x})\hat{\mathbf{w}}_\mathbf{g}^T(t)\Phi_\mathbf{g}(\mathbf{x}) + g_o(\mathbf{x}) \end{aligned} \qquad (3.4)$$

where,

- $\hat{\mathbf{w}}_\mathbf{f}$, $\hat{\mathbf{w}}_\mathbf{g}$ are k-dimensional parameter vectors

- $\Phi_f(\mathbf{x})$, $\Phi_g(\mathbf{x})$ are k-dimensional basis function vectors, whose i^{th} element is a Gaussian function of the form:

$$\Phi_{f_i} = \exp\left\{\frac{-\|\mathbf{x}-\mathbf{m}_{f_i}\|^2}{2\sigma_f^2}\right\}$$
$$\Phi_{g_i} = \exp\left\{\frac{-\|\mathbf{x}-\mathbf{m}_{g_i}\|^2}{2\sigma_g^2}\right\} \quad (3.5)$$

such that:
  - $\mathbf{m}_{f_i}, \mathbf{m}_{g_i}$ are n-dimensional vectors representing the *centre* of the ith basis function
  - σ_f^2, σ_g^2 are the variances representing the *width* of the basis functions.
- $f_o(\mathbf{x})$, $g_o(\mathbf{x})$ are known prior estimates to $f(\mathbf{x})$, $g(\mathbf{x})$ respectively.
- $\kappa(\mathbf{x})$ is a state-dependent function. This entails the definition of χ^- as a subset of the network approximation region χ (being slightly smaller than χ) so that $\kappa(\mathbf{x})$ is defined as:

$$\kappa(\mathbf{x}) = \begin{cases} 1 & \text{if } \mathbf{x} \in \chi^- \\ 0 \leq \bullet \leq 1 & \text{if } \mathbf{x} \notin \chi^- \text{ and } \mathbf{x} \in \chi \\ 0 & \text{if } \mathbf{x} \notin \chi \end{cases} \quad (3.6)$$

Further, it is assumed that given any uniform bounds ϵ_f, ϵ_g and the prior estimates f_o and g_o, we could always find (non-unique) optimal parameter vectors $\mathbf{w}_f^*$, $\mathbf{w}_g^*$ and an optimal number k^* of basis functions, such that $\forall \mathbf{x} \in \chi^-$, the network approximation errors satisfy:

$$\begin{aligned} |\Delta_f| &= |f^*(\mathbf{x}) - f(\mathbf{x})| \leq \epsilon_f \\ |\Delta_g| &= |g^*(\mathbf{x}) - g(\mathbf{x})| \leq \epsilon_g \end{aligned} \quad (3.7)$$

where $f^*(\mathbf{x}) = \hat{f}(\mathbf{x}, \mathbf{w}_f^*, k^*)$ and $g^*(\mathbf{x}) = \hat{g}(\mathbf{x}, \mathbf{w}_g^*, k^*)$. This follows from the Universal Approximation Property exhibited by feedforward neural networks [61, 84, 103] and, in particular for our case, by radial basis functions [207].

The radial basis functions are placed on regular points of a square mesh in χ so that the mesh point separation, μ, depends directly on k^* and the size of χ. In this scheme, the Gaussian basis function parameters are assumed to be known *a priori*. This means that the network approximation region χ and the basis function centres and width parameters required to satisfy any desired ϵ_f, ϵ_g must be determined from prior information available about the system. Since χ should cover the space in which state $\mathbf{x}$ is contained, a fairly good estimate of it could be determined from the knowledge on the reference input y_d. As far as the basis function parameters are concerned, if bounds on the degree of smoothness of the functions being approximated are known, the technique of [219] could be employed to centre the basis functions on the points of a regular square mesh spanning χ, with the mesh point separation and basis function widths computed according to these smoothness bounds. The optimal parameter vectors, $\mathbf{w}_f^*$ and $\mathbf{w}_g^*$, that ensure the above accuracy, are unknown and must be estimated. This represents adaptation and is performed in an on-line manner, where the actual parameter estimates

$\hat{\mathbf{w}}_f, \hat{\mathbf{w}}_g$ are adjusted recursively in time according to the adaptation laws [254, 255]:

$$\begin{aligned} \dot{\hat{\mathbf{w}}}_f &= \eta_f \kappa(\mathbf{x}) e_{1\Delta}(t) \Phi_f(\mathbf{x}) \\ \dot{\hat{\mathbf{w}}}_g &= \eta_g \kappa(\mathbf{x}) e_{1\Delta}(t) \Phi_g(\mathbf{x}) u_{fl}(t) \end{aligned} \quad (3.8)$$

where:

- $e_{1\Delta}(t)$ represents a dead-zone function on $e_1(t)$ according to the equation

$$e_{1\Delta}(t) = \begin{cases} 0 & \text{if } -\phi < e_1 < \phi \\ e_1 - \phi\,\text{sign}(e_1) & \text{if } |e_1| > \phi \end{cases} \quad (3.9)$$

- $e_1(t)$ is a filtration of the tracking error $e(t)$ such that

$$e_1(t) = \beta_r e^{(r-1)} + \ldots + \beta_1 e,$$

or in the s-domain

$$e_1(s) = \Psi(s) e(s)$$

where $\Psi(s) := \beta_1 + \beta_2 s + \ldots + \beta_r s^{r-1}$. The coefficients β_i of the $(r-1)th$ order polynomial $\Psi(s)$ are chosen to make it Hurwitz and such that $\Psi(s)\Gamma^{-1}(s)$ is strictly positive real [23, 186].
- η_f, η_g are positive adaptation rates.
- ϕ is the boundary layer thickness associated with the sliding mode control augmentation $u_{sl}(t)$ [234, 254, 255].

In addition, a parameter resetting mechanism is included that ensures that:

$$\hat{g}(\mathbf{x}) \geq g_l(\mathbf{x}) - \epsilon^*(\mathbf{x}), \quad \forall\, \mathbf{x} \in \chi \quad (3.10)$$

where $\epsilon^*(\mathbf{x})$ is a small positive function such that $g_l(\mathbf{x}) - \epsilon^*(\mathbf{x}) > 0, \forall\, \mathbf{x} \in \chi$ and $g_o(\mathbf{x}) \geq g_l(\mathbf{x}), \forall\, \mathbf{x} \notin \chi^-$.

Use of adaptation laws (3.8) and the control law:

$$u(t) = u_{fl}(t) + u_{sl}(t) \quad (3.11)$$

where:

$$u_{sl}(t) = -\overline{k}(\mathbf{x}, t)\,\text{sat}\left(\frac{e_1}{\phi}\right) \quad (3.12)$$

with $\overline{k}(\mathbf{x}, t)$ being a state-dependent sliding gain and

$$\text{sat}(r) = \begin{cases} 1 & \text{if } r \geq 1 \\ r & \text{if } -1 < r < 1 \\ -1 & \text{if } r \leq -1 \end{cases} \quad (3.13)$$

results in closed-loop system stability, with the filtered tracking error $e_1(t)$ converging to a value such that $|e_1(t)| \leq \phi$. This is ensured by Lyapunov stability considerations [254, 255].

3.4 Dynamic Network Structure

Dynamic structure neural networks are based on the idea that one could start with a network having no basis functions at all. These are gradually allocated in response to some conditions being satisfied. This way, the network 'grows' sequentially according to some rules which ensure that the function being approximated is learned up to the required levels of accuracy, using only a minimal number of basis functions. Such an approach contrasts with the fixed structure neural networks described in the previous section, whereby the number of basis functions is determined *a priori* and remains fixed.

The use of such dynamic structure methods for Gaussian radial basis function networks in the context of system identification and function estimation is analysed in [127, 128, 136, 202]. For control applications, Sanner and Slotine [219] suggest a dynamic network scheme that is still linked to the concept of a mesh as utilized in a fixed structure neural network. In this technique each mesh point represents a *potential* location in state-space, where a basis function 'could' be placed if the need arises. Basis functions that are actually placed on such points represent *activated* basis functions, and the network grows as more of the nodes are activated, contributing to the dynamic structure neural network. This technique of radial basis function placement shall be referred to as *selective node activation*.

This scheme is justified by considering adaptation rules (3.8), expressed in discrete form, namely:

$$\Delta \hat{\mathbf{w}}_{\mathbf{f}} = \eta'_f \kappa(\mathbf{x}) e_{1\Delta}(t) \Phi_{\mathbf{f}}(\mathbf{x})$$
$$\Delta \hat{\mathbf{w}}_{\mathbf{g}} = \eta'_g \kappa(\mathbf{x}) e_{1\Delta}(t) \Phi_{\mathbf{g}}(\mathbf{x}) u_{fl} \tag{3.14}$$

where $\Delta \hat{\mathbf{w}}_{\mathbf{f}}$, $\Delta \hat{\mathbf{w}}_{\mathbf{g}}$ denote the change in the parameter vectors $\hat{\mathbf{w}}_{\mathbf{f}}$, $\hat{\mathbf{w}}_{\mathbf{g}}$.

Notice that if $\Phi_i(\mathbf{x})$ (the ith element of $\Phi_{\mathbf{f}}$ or $\Phi_{\mathbf{g}}$) is small, then $\Delta \hat{\mathbf{w}}_i$ will also be small. Thus, if the ith basis function is centered 'far away' from $\mathbf{x}(t)$, the change in the parameter for node i will be negligible and no great change in performance should be noticed if this change in parameter is ignored. Moreover, as shown in [219], if one starts with all parameters equal to zero (*i.e.*, $\hat{\mathbf{w}}_{\mathbf{f}} = \hat{\mathbf{w}}_{\mathbf{g}} = \mathbf{0}$ at time $t = 0$) and assuming that, as noted above, one need not change the parameter of basis functions that are centered 'far away' from the current state $\mathbf{x}(t)$, we shall have radial basis functions with non-zero parameters (referred to as *active nodes*) only in the localized regions of state space where $\mathbf{x}(t)$ has ventured, referred to as *active space*. In this scheme, the network output is due to the contribution of the nodes centered on mesh points (in χ) that are located in regions where $\Phi_i(\mathbf{x})$ has been significantly large at some time during the interval from the start of operation up to the present instant. The rest of the basis functions, referred to as *passive* nodes, need not have existed at all since their contribution to the network output is nil.

Thus, one could *activate* a basis function at any time t, only if its value is significantly large, unless otherwise it has already been activated previously.

3.4 Dynamic Network Structure

This *selective node activation* technique results in a dynamic structure basis function vector Φ_a and its corresponding dynamic parameter vector $\hat{\mathbf{w}}_a$, composed of basis functions and parameters of *active* nodes only, rather than all the nodes in χ, as in Equation (3.4). These vectors are *dynamic* in the sense that they can grow in size according to which nodes are activated. If the active space is small compared to χ, the number of nodes in such dynamic networks is significantly less than that of the non-dynamic case considered in the previous section, because the passive nodes are simply ignored.

Note that if the input is such that most regions of state-space in χ will be traversed, nearly all nodes in χ would be activated and the size of the dynamic network will be approximately equal to that of a non-dynamic scheme utilizing a 'full-size' network. Hence there is no guarantee that the size of the dynamic network will be significantly less than that used in a 'full-size' network scheme. However, in a typical scenario for systems of order higher than 1, it is often the case that active space spans only a small subset of χ, resulting in a much smaller dynamic network when compared to a 'full-size' network.

Consider the case in which a node i (in χ) is activated at time t, provided that it was not previously activated and that its basis function satisfies the condition:

$$\Phi_i(\mathbf{x}(t)) \geq \delta_{min} \tag{3.15}$$

where δ_{min} represents an activation threshold (δ_{min} is constant and satisfies $0 < \delta_{min} \leq 1$) below which the output of the basis function is considered negligible.

Hence, using the definition of Gaussian basis functions of Equation (3.5), it follows that for a basis function centered at $\mathbf{m}_{f_i}$ or $\mathbf{m}_{g_i}$ to be activated, then respectively:

$$\begin{aligned}\|\mathbf{x}(t) - \mathbf{m}_{f_i}\|^2 &\leq -2\sigma_f^2 \ln(\delta_{min}) \\ \|\mathbf{x}(t) - \mathbf{m}_{g_i}\|^2 &\leq -2\sigma_g^2 \ln(\delta_{min})\end{aligned} \tag{3.16}$$

In general, Equation (3.16) represents a space enclosed within an n-dimensional ball of radius $\sqrt{-2\sigma^2 \ln(\delta_{min})}$, that is centered around $\mathbf{x}(t)$, the present state. Hence, all nodes centered on mesh points within the ball must be activated, unless they have already been activated. Note that the larger the activation threshold, δ_{min}, the smaller the size of the ball, and so less nodes are activated at any one time, restricting the network growth rate further. Figure 3.2a shows the above ideas for a 2nd order system, where the ball is thus a circle.

As will be clarified subsequently, although slightly more nodes would be activated, it is more convenient to activate nodes that are located on mesh points within a hypercube that is centred on the nearest mesh point to $\mathbf{x}(t)$ with half-length equal to $l\mu$, where l is the closest integer to $\sqrt{-2\sigma^2 \ln(\delta_{min})}/\mu$, rather than a ball centred on $\mathbf{x}(t)$. This way, the hypercube edges coincide with the mesh points, as shown in Figure 3.2b. The

3. Dynamic Structure Networks for Stable Adaptive Control

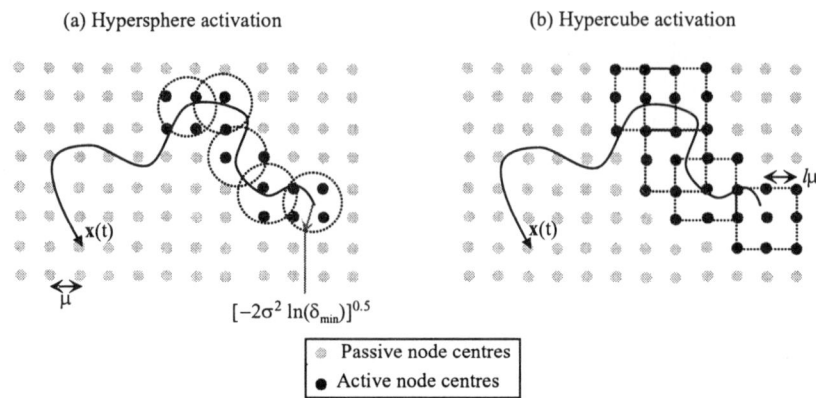

Fig. 3.2. Selective node activation technique: using a hypersphere or a hypercube.

main reason for this modification is the simplicity in evaluating the order of magnitude of that part of the disturbance term arising because of selective node activation. In addition, determination of the nodes to be activated at any one time is facilitated, an advantage whose benefits are felt more the higher the system dimensionality.

3.5 The Control Law and Error Dynamics

Using a dynamic network scheme, the output of the neural network at time t consists of the contribution of the active basis functions. Denoting the basis functions and parameters of such active nodes as dynamic vectors $\Phi_{\mathbf{fa}}$, $\Phi_{\mathbf{ga}}$, $\hat{\mathbf{w}}_{\mathbf{fa}}$, $\hat{\mathbf{w}}_{\mathbf{ga}}$; for the $f(\mathbf{x})$ and $g(\mathbf{x})$ networks respectively and the output of the dynamic networks as $\hat{f}_a(\mathbf{x})$, $\hat{g}_a(\mathbf{x})$ then:

$$\begin{aligned}\hat{f}_a(\mathbf{x}) &= \kappa(\mathbf{x})\hat{\mathbf{w}}_{\mathbf{fa}}^T(t)\Phi_{\mathbf{fa}}(\mathbf{x}) + f_o(\mathbf{x}) \\ \hat{g}_a(\mathbf{x}) &= \kappa(\mathbf{x})\hat{\mathbf{w}}_{\mathbf{ga}}^T(t)\Phi_{\mathbf{ga}}(\mathbf{x}) + g_o(\mathbf{x})\end{aligned} \qquad (3.17)$$

Hence, in this scheme, the control law becomes:

$$u(t) = u_{fl}(t) + u_{sl}(t) \qquad (3.18)$$

where:

$$\begin{aligned}u_{fl}(t) &= \frac{-\hat{f}_a(\mathbf{x}) + v(t)}{\hat{g}_a(\mathbf{x})} \\ u_{sl}(t) &= -\overline{k}_a \mathrm{sat}\left(\frac{e_1}{\phi}\right)\end{aligned} \qquad (3.19)$$

$\overline{k}_a$ being the sliding mode gain to be determined subsequently.

By definition of $f^*(\mathbf{x})$, $g^*(\mathbf{x})$ and Equations (3.4), (3.7) we get:

3.5 The Control Law and Error Dynamics

$$f(\mathbf{x}) = \kappa(\mathbf{x})\mathbf{w_f^*}^T \Phi_\mathbf{f} - \Delta_f + f_o$$
$$g(\mathbf{x}) = \kappa(\mathbf{x})\mathbf{w_g^*}^T \Phi_\mathbf{g} - \Delta_g + g_o \qquad (3.20)$$

where for convenience, the argument ($\mathbf{x}$) has been dropped for Δ_f, Δ_g, f_o, g_o. Consider indexing $\mathbf{w_f^*}$, $\mathbf{w_g^*}$, the optimal, 'full-network' parameter vectors, such that the first series of elements are the optimal parameters of the active nodes, denoted by sub-vectors $\mathbf{w_{fa}^*}$, $\mathbf{w_{ga}^*}$, and the next series of elements are formed from the optimal parameters of passive nodes, denoted by sub-vectors $\mathbf{w_{fp}^*}$, $\mathbf{w_{gp}^*}$, which represent nodes not yet activated. Similarly, $\hat{\mathbf{w}}_\mathbf{f}$ and $\hat{\mathbf{w}}_\mathbf{g}$, the full network parameter vectors are indexed accordingly into subvectors $\hat{\mathbf{w}}_\mathbf{fa}$, $\hat{\mathbf{w}}_\mathbf{ga}$, $\hat{\mathbf{w}}_\mathbf{fp}$, $\hat{\mathbf{w}}_\mathbf{gp}$. By definition, being passive, the last two subvectors and their time-derivative are zero vectors. Hence,

$$\hat{\mathbf{w}}_\mathbf{f} = \begin{bmatrix} \hat{\mathbf{w}}_\mathbf{fa} \\ \hat{\mathbf{w}}_\mathbf{fp} \end{bmatrix}, \quad \hat{\mathbf{w}}_\mathbf{g} = \begin{bmatrix} \hat{\mathbf{w}}_\mathbf{ga} \\ \hat{\mathbf{w}}_\mathbf{gp} \end{bmatrix}, \qquad (3.21)$$

where $\hat{\mathbf{w}}_\mathbf{fp}$ and $\hat{\mathbf{w}}_\mathbf{gp}$ are zero vectors. Vectors $\mathbf{w_f^*}$ and $\mathbf{w_g^*}$ can be decomposed likewise. Using the same indices to re-index $\Phi_\mathbf{f}$ and $\Phi_\mathbf{g}$, i.e.,

$$\Phi_\mathbf{f} = \begin{bmatrix} \Phi_\mathbf{fa} \\ \Phi_\mathbf{fp} \end{bmatrix}, \quad \Phi_\mathbf{g} = \begin{bmatrix} \Phi_\mathbf{ga} \\ \Phi_\mathbf{gp} \end{bmatrix} \qquad (3.22)$$

where:

- $\Phi_\mathbf{fa}(\mathbf{x})$, $\Phi_\mathbf{ga}(\mathbf{x})$ contain the basis functions of active nodes,
- $\Phi_\mathbf{fp}(\mathbf{x})$, $\Phi_\mathbf{gp}(\mathbf{x})$ contain the basis functions of passive nodes.

Thus, substitution in Equation (3.20) and subtraction from Equation (3.17) gives us:

$$\hat{f}_a(\mathbf{x}) - f(\mathbf{x}) = \kappa(\mathbf{x})\tilde{\mathbf{w}}_\mathbf{fa}^T \Phi_\mathbf{fa} - \kappa(\mathbf{x})\tilde{\mathbf{w}}_\mathbf{fp}^T \Phi_\mathbf{fp} + \Delta_f$$
$$\hat{g}_a(\mathbf{x}) - g(\mathbf{x}) = \kappa(\mathbf{x})\tilde{\mathbf{w}}_\mathbf{ga}^T \Phi_\mathbf{ga} - \kappa(\mathbf{x})\tilde{\mathbf{w}}_\mathbf{gp}^T \Phi_\mathbf{gp} + \Delta_g \qquad (3.23)$$

where, $\tilde{\mathbf{w}}_\mathbf{fa} := \hat{\mathbf{w}}_\mathbf{fa} - \mathbf{w}_\mathbf{fa}^*$, $\tilde{\mathbf{w}}_\mathbf{ga} := \hat{\mathbf{w}}_\mathbf{ga} - \mathbf{w}_\mathbf{ga}^*$, are the parameter errors for the active nodes and $\tilde{\mathbf{w}}_\mathbf{fp} := \hat{\mathbf{w}}_\mathbf{fp} - \mathbf{w}_\mathbf{fp}^* = -\mathbf{w}_\mathbf{fp}^*$, $\tilde{\mathbf{w}}_\mathbf{gp} := -\mathbf{w}_\mathbf{gp}^*$ are the parameter errors for the passive nodes.

Substitution of control law (3.18) in the system dynamics (3.1) and substituting for $v(t)$ and $e(t)$, yields:

$$e^{(r)} + \alpha_r e^{(r-1)} + \ldots + \alpha_1 e =$$
$$f(\mathbf{x}) - \hat{f}_a(\mathbf{x}) + (g(\mathbf{x}) - \hat{g}_a(\mathbf{x}))u_{fl} + g(\mathbf{x})u_{sl}. \qquad (3.24)$$

Also by definition of $\Gamma(s)$; $e^{(r)} + \alpha_r e^{(r-1)} + \ldots + \alpha_1 e \equiv \Gamma(s)e(s)$ in the Laplace domain, so that substitution of Equation (3.23) in (3.24) results in the following error dynamics:

$$\Gamma(s)e = -\kappa\tilde{\mathbf{w}}_\mathbf{fa}^T \Phi_\mathbf{fa} - \kappa\tilde{\mathbf{w}}_\mathbf{ga}^T \Phi_\mathbf{ga} u_{fl} + g(\mathbf{x})u_{sl}(t) + d_d(t) \qquad (3.25)$$

where,

$$d_d(t) = \kappa\mathbf{w_{fp}^*}^T \Phi_\mathbf{fp} + \kappa\mathbf{w_{gp}^*}^T \Phi_\mathbf{gp} u_{fl} - \Delta_f - \Delta_g u_{fl} \qquad (3.26)$$

represents a disturbance term that enters the error dynamics due to:

(i) The network inherent approximation errors $\Delta_f(\mathbf{x}), \Delta_g(\mathbf{x})$.
(ii) The selective node activation scheme, which ignores the contribution to the output of nodes that have not been activated, giving rise to errors in terms of the passive nodes.

Note that in [254], the disturbance was due only to (i), whilst in this case it is augmented by the errors due to selective node activation.

By definition $e_1(t) = \Psi e(t)$, so that the error dynamics Equation (3.25) expressed in terms of e_1 becomes:

$$e_1 = \Psi \Gamma^{-1} \left(-\kappa(\mathbf{x}) \tilde{\mathbf{w}}_{\mathbf{fa}}^T(t) \Phi_{\mathbf{fa}} - \kappa(\mathbf{x}) \tilde{\mathbf{w}}_{\mathbf{ga}}^T(t) \Phi_{\mathbf{ga}} u_{fl}(t) \right.$$
$$\left. + g(\mathbf{x}) u_{sl}(t) + d_d(t) \right) \quad (3.27)$$

As in [219, 254, 255], we could use sliding control in the system with e_1 defining a sliding surface if $\Psi \Gamma^{-1}(s)$ has a first-order lag transfer function, say:

$$\Psi \Gamma^{-1}(s) = \frac{1}{s + k_d}, \quad k_d > 0 \quad (3.28)$$

so that in this case:

$$\dot{e}_1 = -k_d e_1 - \kappa(\mathbf{x}) \tilde{\mathbf{w}}_{\mathbf{fa}}^T(t) \Phi_{\mathbf{fa}} - \kappa(\mathbf{x}) \tilde{\mathbf{w}}_{\mathbf{ga}}^T(t) \Phi_{\mathbf{ga}} u_{fl}(t)$$
$$+ g(\mathbf{x}) u_{sl}(t) + d_d(t). \quad (3.29)$$

3.6 The Adaptive System

We propose the use of adaptation laws (3.8) as in [254, 255], but based only on *active* nodes, i.e.,

$$\dot{\hat{\mathbf{w}}}_{\mathbf{fa}} = \eta_f \kappa(\mathbf{x}) e_{1\Delta}(t) \Phi_{\mathbf{fa}}(\mathbf{x})$$
$$\dot{\hat{\mathbf{w}}}_{\mathbf{ga}} = \eta_g \kappa(\mathbf{x}) e_{1\Delta}(t) \Phi_{\mathbf{ga}}(\mathbf{x}) u_{fl}(t) \quad (3.30)$$

together with a parameter-resetting mechanism:

$$\hat{\mathbf{w}}_{\mathbf{ga}}(t^+) = \hat{\mathbf{w}}_{\mathbf{ga}}(t) + (g_l - \hat{g}_{na}) \|\Phi_{\mathbf{ga}}(\mathbf{x})\|^{-2} \Phi_{\mathbf{ga}}(\mathbf{x}) \quad (3.31)$$

which is activated at any time t if:

$$\hat{g}_{na}(\mathbf{x}) \leq g_l(\mathbf{x}) - \epsilon^*(\mathbf{x}), \quad \forall \mathbf{x} \in \chi \quad (3.32)$$

where:

- $\hat{g}_{na}(\mathbf{x}) = \hat{\mathbf{w}}_{\mathbf{ga}}^T(t) \Phi_{\mathbf{ga}}(\mathbf{x}) + g_o$.
- $\epsilon^*(\mathbf{x})$ is a small, positive function such that $g_l(\mathbf{x}) - \epsilon^*(\mathbf{x}) > 0$, $\forall \mathbf{x} \in \chi$.
- $g_o(\mathbf{x}) \geq g_l(\mathbf{x})$, $\forall \mathbf{x} \notin \chi^-$.
- t^+ denotes the time just after the resetting mechanism is activated.

This resetting mechanism ensures that after a parameter reset, $\hat{g}_a(\mathbf{x}) \geq g_l(\mathbf{x})$, thus maintaining $\hat{g}_a(\mathbf{x})$ bounded below by $g_l(\mathbf{x}) - \epsilon^*(\mathbf{x})$.

3.7 Stability Analysis

System stability is analysed by use of the Lyapunov function

$$V = \frac{e_{1\Delta}^2}{2} + \frac{1}{2\eta_f}\tilde{\mathbf{w}}_\mathbf{f}^T\tilde{\mathbf{w}}_\mathbf{f} + \frac{1}{2\eta_g}\tilde{\mathbf{w}}_\mathbf{g}^T\tilde{\mathbf{w}}_\mathbf{g} \qquad (3.33)$$

where $\tilde{\mathbf{w}}_\mathbf{f} := \hat{\mathbf{w}}_\mathbf{f} - \mathbf{w}_\mathbf{f}^*$ and $\tilde{\mathbf{w}}_\mathbf{g} := \hat{\mathbf{w}}_\mathbf{g} - \mathbf{w}_\mathbf{g}^*$ are the parameter errors.

(a) For $|e_1| < \phi$: In this case $e_{1\Delta} = 0$ so that Equation (3.33) implies:

$$V = \frac{1}{2\eta_f}\tilde{\mathbf{w}}_\mathbf{f}^T\tilde{\mathbf{w}}_\mathbf{f} + \frac{1}{2\eta_g}\tilde{\mathbf{w}}_\mathbf{g}^T\tilde{\mathbf{w}}_\mathbf{g} > 0 \qquad (3.34)$$

Hence, V is a suitable, positive-definite Lyapunov function candidate for the given range of e_1.
Differentiation with respect to time gives:

$$\begin{aligned}\dot{V} &= \frac{1}{\eta_f}\tilde{\mathbf{w}}_\mathbf{f}^T\dot{\hat{\mathbf{w}}}_\mathbf{f} + \frac{1}{\eta_g}\tilde{\mathbf{w}}_\mathbf{g}^T\dot{\hat{\mathbf{w}}}_\mathbf{g} \\ &= \frac{1}{\eta_f}\begin{bmatrix}\tilde{\mathbf{w}}_\mathbf{fa}^T & \tilde{\mathbf{w}}_\mathbf{fp}^T\end{bmatrix}\begin{bmatrix}\dot{\hat{\mathbf{w}}}_\mathbf{fa}\\ \dot{\hat{\mathbf{w}}}_\mathbf{fp}\end{bmatrix} \\ &+ \frac{1}{\eta_g}\begin{bmatrix}\tilde{\mathbf{w}}_\mathbf{ga}^T & \tilde{\mathbf{w}}_\mathbf{gp}^T\end{bmatrix}\begin{bmatrix}\dot{\hat{\mathbf{w}}}_\mathbf{ga}\\ \dot{\hat{\mathbf{w}}}_\mathbf{gp}\end{bmatrix}\end{aligned} \qquad (3.35)$$

where re-indexing of terms into active and passive nodes as before has been performed. Being a dynamic structure network, $\dot{\hat{\mathbf{w}}}_\mathbf{fp} = \mathbf{0}$, $\dot{\hat{\mathbf{w}}}_\mathbf{gp} = \mathbf{0}$, by definition of passive nodes, and Equation (3.30) implies that for this range of e_1, $\dot{\hat{\mathbf{w}}}_\mathbf{fa}, \dot{\hat{\mathbf{w}}}_\mathbf{ga} = \mathbf{0}$, hence

$$\dot{V} = 0. \qquad (3.36)$$

(b) For $|e_1| \geq \phi$: In this case $e_{1\Delta} = (e_1 - \phi\mathrm{sign}(e_1))$ so that Equation (3.33) implies:

$$V = \frac{(e_1 - \phi\mathrm{sign}(e_1))^2}{2} + \frac{\tilde{\mathbf{w}}_\mathbf{f}^T\tilde{\mathbf{w}}_\mathbf{f}}{2\eta_f} + \frac{\tilde{\mathbf{w}}_\mathbf{g}^T\tilde{\mathbf{w}}_\mathbf{g}}{2\eta_g} > 0. \qquad (3.37)$$

Hence, V is a suitable, positive-definite Lyapunov function candidate. Differentiation of V and Equation (3.29) give:

$$\begin{aligned}\dot{V} = &-k_d e_{1\Delta}^2 - |e_{1\Delta}|(k_d\phi + \overline{k}_a g(\mathbf{x}) + \dot{\phi}) + e_{1\Delta}d_d(t) \\ &+ \tilde{\mathbf{w}}_\mathbf{fa}^T(\frac{1}{\eta_f}\dot{\hat{\mathbf{w}}}_\mathbf{fa} - e_{1\Delta}\kappa\Phi_\mathbf{fa}) \\ &+ \tilde{\mathbf{w}}_\mathbf{ga}^T\left(\frac{1}{\eta_g}\dot{\hat{\mathbf{w}}}_\mathbf{ga} - e_{1\Delta}\kappa\Phi_\mathbf{g}u_{fl}(t)\right).\end{aligned} \qquad (3.38)$$

Using adaptation laws (3.30) and letting ϕ be a constant, we get:

$$\dot{V} = -k_d e_{1\Delta}^2 - |e_{1\Delta}|(k_d\phi + \overline{k}_a g(\mathbf{x})) + e_{1\Delta}d_d(t). \qquad (3.39)$$

Choosing the sliding mode gain

$$\overline{k}_a = \max\left\{0, \frac{\overline{d}_d - k_d\phi}{g_l}\right\} \tag{3.40}$$

renders $\dot{V} \leq -k_d e_{1\Delta}^2$, where $\overline{d}_d(\mathbf{x})$ represents the bound on $d_d(t)$, the disturbance term of Equation (3.26), i.e., $|d_d(t)| \leq \overline{d}_d(\mathbf{x})$. Evaluation of this bound shall be expanded on subsequently.

Hence, adaptation laws (3.30) and sliding gain (3.40) ensure that for all $e_1(t)$, in the absence of parameter resets,

$$\dot{V} \leq -k_d e_{1\Delta}^2 \leq 0. \tag{3.41}$$

This, in turn, implies boundedness of $e_{1\Delta}(t)$, $\tilde{\mathbf{w}}_f$, $\tilde{\mathbf{w}}_g$ by Lyapunov stability theory.

In the event of parameter resets, the change in the Lyapunov function V of Equation (3.33) following a reset, denoted by ΔV, can be deduced to be:

$$\Delta V = \frac{1}{2\eta_g}(\hat{g}_{na} - g_l + 2(g_l - g_{na}^*))(g_l - \hat{g}_{na})\|\Phi_{\mathbf{ga}}\|^{-2} \tag{3.42}$$

where $g_{na}^* = \mathbf{w}_{\mathbf{ga}}^{*T}\Phi_{\mathbf{ga}}(\mathbf{x}) + g_o$ and $\hat{g}_{na}$ has been defined before. Defining g_n^* as the optimal full-network approximation to $g(\mathbf{x})$ when $\mathbf{x} \in \chi$;

$$g_n^* = g_o + \mathbf{w}_{\mathbf{g}}^{*T}\Phi_{\mathbf{g}}(\mathbf{x})$$
$$= g_{na}^* + \mathbf{w}_{\mathbf{gp}}^{*T}\Phi_{\mathbf{gp}}(\mathbf{x}), \tag{3.43}$$

where the definitions of $\mathbf{w}_{\mathbf{ga}}^*$, $\mathbf{w}_{\mathbf{gp}}^*$, g_{na}^* have been used. Then, it follows that

$$g_n^*(\mathbf{x}) - g_{na}^*(\mathbf{x}) \leq \max(\mathbf{w}_{\mathbf{gp}}^{*T}\Phi_{\mathbf{gp}}(\mathbf{x})) \tag{3.44}$$

where $\max(\mathbf{w}_{\mathbf{gp}}^{*T}\Phi_{\mathbf{gp}}(\mathbf{x}))$ is an upper bound on $\mathbf{w}_{\mathbf{gp}}^{*T}\Phi_{\mathbf{gp}}(\mathbf{x})$.

Assumption 3.7.1 *We assume that $g_l(\mathbf{x})$ is such that*

$$g_n^*(\mathbf{x}) - g_l(\mathbf{x}) \geq \max(\mathbf{w}_{\mathbf{gp}}^{*T}\Phi_{\mathbf{gp}}(\mathbf{x})). \tag{3.45}$$

Assumption 3.7.1 could be ascertained by choosing the lower bound $g_l(\mathbf{x})$ lower than $g_n^*(\mathbf{x})$ by at least $\max(\mathbf{w}_{\mathbf{gp}}^{*T}\Phi_{\mathbf{gp}}(\mathbf{x}))$, whose value is to be discussed later. Assumption 3.7.1 and Equation (3.44) thus lead to the inequality:

$$g_l(\mathbf{x}) - g_{na}^*(\mathbf{x}) \leq 0, \tag{3.46}$$

which ensures that

$$\Delta V \leq -\frac{1}{2\eta_g}(g_l - \hat{g}_{na})^2\|\Phi_{\mathbf{ga}}\|^{-2} \leq 0. \tag{3.47}$$

Hence the Lyapunov function is *reduced* as a result of parameter resetting, if (3.46) is satisfied.

Note that Assumption 3.7.1, introduced to ensure inequality (3.46), is different from that in [254] where a full-size network was used and it was sufficient to assume that $g_n^*(\mathbf{x}) - g_l(\mathbf{x}) \geq 0$ for maintaining $\Delta V \leq 0$. This difference in assumptions though, is necessary to ensure stability in the presence of parameter resets for the case of a dynamic structure network.

The above analysis leads to the following stability theorem.

3.7 Stability Analysis

Theorem 3.7.1. *Given Assumptions 3.2.1 to 3.2.7, 3.7.1 and control law (3.18), with u_{fl}, u_{sl} and the sliding mode gain given by Equations (3.19) and (3.40) respectively, adaptation law (3.30) and parameter resetting mechanism (3.31); then system (3.1) is stable with state vector x, the network parameter vectors and control input $u(t)$ remaining bounded and the tracking error and its $(r-1)$ derivatives asymptotically converging to within predetermined bounds that depend on $\Psi(s)$ and the boundary layer thickness ϕ.*

Proof. The relations (3.41) and (3.47) show that adaptation laws (3.30), sliding mode gain (3.40) and the proposed parameter resetting mechanism lead to boundedness of $e_{1\Delta}(t)$, $\tilde{\mathbf{w}}_\mathbf{f}$, $\tilde{\mathbf{w}}_\mathbf{g}$. In turn, following [254, 255], these considerations apply:

1. Boundedness of $e_{1\Delta}(t)$ implies boundedness of $e_1(t)$.
2. Boundedness of $\tilde{\mathbf{w}}_\mathbf{f}$, $\tilde{\mathbf{w}}_\mathbf{g}$ implies boundedness of $\tilde{\mathbf{w}}_\mathbf{fa}$, $\tilde{\mathbf{w}}_\mathbf{ga}$.
3. Boundedness of $\tilde{\mathbf{w}}_\mathbf{fa}$, $\tilde{\mathbf{w}}_\mathbf{ga}$ implies boundedness of $\hat{\mathbf{w}}_\mathbf{fa}$, $\hat{\mathbf{w}}_\mathbf{ga}$.
4. Tracking error $e(t)$ and its derivatives up to order $(r-1)$ are bounded since $e_1(t)$ is bounded and $\Psi(s)$ is Hurwitz. Hence by Assumption 3.2.7, it follows that $y, y^{(1)}, \cdots, y^{(r-1)}$ are bounded. This result, together with Assumptions 3.2.3 and 3.2.4, yields that $\mathbf{x}(t)$ is bounded [224].
5. Assuming $f(\mathbf{x})$, $g(\mathbf{x})$ to be continuous, boundedness of $\mathbf{x}$ implies that $f(\mathbf{x})$, $g(\mathbf{x})$ are bounded.
6. $\Phi_\mathbf{fa}(\mathbf{x})$, $\Phi_\mathbf{ga}(\mathbf{x})$ are bounded since $\mathbf{x}(t)$ is bounded and so, together with boundedness of $\hat{\mathbf{w}}_\mathbf{fa}$, $\hat{\mathbf{w}}_\mathbf{ga}$ we conclude that $\hat{f}_a$, $\hat{g}_a$ are bounded.
7. $v(t)$ is bounded since $e(t)$ and its derivatives up to order $(r-1)$ and $y_d(t)$ and its r derivatives, are bounded. Also, parameter resetting ensures that $\hat{g}_a(\mathbf{x})$ is bounded away from zero, so that u_{fl} is bounded.
8. Disturbance $d_d(t)$ is bounded (via $\bar{d}_d$) since all terms on the right of Equation (3.26) are bounded and $u_{sl}(t)$ is bounded since $\bar{k}_a$ is bounded. Hence, Equation (3.29) implies that $\dot{e}_1$ is bounded, since all terms on its right hand side are bounded.

Hence $\dot{e}_{1\Delta}$ is bounded as well, and so $e_{1\Delta}$ is uniformly continuous. Since $\dot{V} \leq -k_d e_{1\Delta}^2$, then $\int_0^t \dot{V} + k_d e_{1\Delta}^2 d\tau \leq 0$. Letting $V_1(t) = V(t) - \int_0^t \dot{V}(\tau) + k_d e_{1\Delta}^2 d\tau$, the above condition implies that $V_1(t) \geq 0$ and hence $V_1(t)$ is *bounded below*. Also, time differentiation of $V_1(t)$ implies that $\dot{V}_1 = -k_d e_{1\Delta}^2$, so that $\dot{V}_1(t) \leq 0$, making it *negative semidefinite*. Finally, since $e_{1\Delta}$ is uniformly continuous, $\dot{V}_1$ is also *uniformly continuous*. Hence, from these three conditions, by Barbalat's Lemma [186, 224, 234], we deduce that:

$$\dot{V}_1(t) \to 0 \text{ as } t \to \infty \text{ so that } e_{1\Delta}(t) \to 0 \text{ as } t \to \infty. \quad (3.48)$$

This implies that $e_1(t)$ converges to somewhere within $\pm\phi$ as $t \to \infty$ or;

$$|e_1(t)| \leq \phi \text{ as } t \to \infty. \quad (3.49)$$

Since $e(t)$ is a stable filtration of $e_1(t)$ via $\Psi^{-1}(s)$, boundedness of $e_1(t)$ implies boundedness of $e^{(i)}(t)$ for $i = 0, 1, \ldots, (r-1)$. Thus ϕ determines the value of the steady-state tracking error.

Note that if $\Psi(s) = (s+\lambda)^{r-1}$, where λ is a positive constant, then as shown in [234], in the steady-state:

$$|e^{(i)}| \leq 2^i \lambda^{i-r+1} \phi \qquad 0 \leq i \leq r-1. \tag{3.50}$$

□

Hence, if Equations (3.30),(3.31),(3.40) and the relevant assumptions are adhered to, use of selective node activation optimizes on network size and ensures stable closed-loop dynamics.

Let us now take a closer look at the function representing the disturbance bound, required for evaluating the sliding mode gain $\bar{k}_a$ given in Equation (3.40).

3.8 Evaluation of Control Parameters and Implementation

3.8.1 The Disturbance Bound

According to Equation (3.26), the disturbance is bounded by:

$$|d_d| \leq \kappa |\mathbf{w}_{\mathbf{fp}}^{*T} \Phi_{\mathbf{fp}}| + \kappa |\mathbf{w}_{\mathbf{gp}}^{*T} \Phi_{\mathbf{gp}}||u_{fl}| + |\Delta f(\mathbf{x})| + |\Delta g(\mathbf{x})||u_{fl}|.$$

In χ^-, the network approximation region, $\kappa(\mathbf{x}) = 1, |\Delta f(\mathbf{x})| \leq \epsilon_f$ and $|\Delta g(\mathbf{x})| \leq \epsilon_g$ by Equation (3.7). Hence,

$$|d_d| \leq |\mathbf{w}_{\mathbf{fp}}^{*T} \Phi_{\mathbf{fp}}| + |\mathbf{w}_{\mathbf{gp}}^{*T} \Phi_{\mathbf{gp}}||u_{fl}| + \epsilon_f + \epsilon_g |u_{fl}|. \tag{3.51}$$

Outside χ, $\kappa(\mathbf{x}) = 0$ so that by Equations (3.4), (3.17): $f^*(\mathbf{x}) = \hat{f}(\mathbf{x}) = \hat{f}_a(\mathbf{x}) = f_o$ and $g^*(\mathbf{x}) = \hat{g}(\mathbf{x}) = \hat{g}_a(\mathbf{x}) = g_o$. Hence, by Equation (3.7), we get:

$$\begin{aligned} \Delta_f &= |f_o - f(\mathbf{x})| \\ \Delta_g &= |g_o - g(\mathbf{x})|. \end{aligned} \tag{3.52}$$

As in [254], we assume the following conditions to hold:

Assumption 3.8.1 *The above differences are bounded by known functions $\bar{f}_o(\mathbf{x}), \bar{g}_o(\mathbf{x})$ such that:*

$$|f_o - f(\mathbf{x})| \leq \bar{f}_o(\mathbf{x}) \quad \text{and} \quad |g_o - g(\mathbf{x})| \leq \bar{g}_o(\mathbf{x}). \tag{3.53}$$

Hence outside χ,

$$|d_d| \leq \bar{f}_o + \bar{g}_o |u_{fl}|. \tag{3.54}$$

We therefore combine Equations (3.51) and (3.54), to obtain bounds for *both* inside and outside of χ as follows:

$$\begin{aligned} |d_d| \leq &\kappa(\mathbf{x})(|\mathbf{w}_{\mathbf{fp}}^{*T} \Phi_{\mathbf{fp}}| + |\mathbf{w}_{\mathbf{gp}}^{*T} \Phi_{\mathbf{gp}}||u_{fl}| + (\epsilon_f + \epsilon_g |u_{fl}|)) \\ &+ (1 - \kappa(\mathbf{x}))(\bar{f}_o + \bar{g}_o |u_{fl}|). \end{aligned} \tag{3.55}$$

3.8 Evaluation of Control Parameters and Implementation

Note that this equation is exact both inside χ^- and outside χ. In between χ^- and χ, it smoothly interpolates between the two bounds. Whilst ϵ_f, ϵ_g, $\bar{f}_o$, $\bar{g}_o$ are assumed to be known and $\kappa(\mathbf{x})$, $|u_{fl}(t)|$ could be determined since these are used in the controller; terms $|\mathbf{w}_{\mathbf{fp}}^{*T}\Phi_{\mathbf{fp}}(\mathbf{x})|$ and $|\mathbf{w}_{\mathbf{gp}}^{*T}\Phi_{\mathbf{gp}}(\mathbf{x})|$ are not yet known. In fact they depend on the number of passive nodes, which depends, amongst other things, on the *activation threshold* δ_{min}.

Note that

$$|\mathbf{w}_{\mathbf{fp}}^{*T}\Phi_{\mathbf{fp}}(\mathbf{x})| = |\sum_i w_{fp_i}^* \Phi_{fp_i}(\mathbf{x})|$$
$$\leq \bar{w}_{fb} \sum_i \Phi_{fp_i}(\mathbf{x}) \qquad (3.56)$$

where $\bar{w}_{fb} \geq \max(|w_{fp_i}^*|)$ represents a bound on the full network optimal parameters.

The term $\sum \Phi_{fp_i}$ is indexed by i, whose range is equal to the number of passive nodes, *i.e.*, the size of the passive nodes vector $\Phi_{\mathbf{fp}}$, denoted by R. This value varies with time, however it is always given by the difference between the total number of mesh points in χ, N, and the number of active nodes, A, *i.e.*, $R = N - A$.

Thus, the term,

$$\sum_{i=1}^{R} \Phi_{fp_i}(\mathbf{x})$$

is largest when R is a maximum, denoted by R_{max}. This occurs when A is at its smallest value, A_{min}, corresponding to the case at time $t = 0$ when only the nodes in one activation hypercube H_1 of half-length $l\mu$, centred around $\mathbf{x}_o$, the nearest mesh point to $\mathbf{x}(0)$, are activated. This is shown in Figure 3.3. Recall that l denotes the nearest integer to $\sqrt{-2\sigma^2 \ln(\delta_{min})}/\mu$. The number of nodes in this hypercube is given, in general by $A_{min} = (1 + 2l)^n$, where n is the dimension of the state-space. Thus, we may say that

$$\sum_{i=1}^{R} \Phi_{fp_i}(\mathbf{x}(t)) \leq \sum_{i=1}^{R_{max}} \Phi_{fp_i}(\mathbf{x}(0)). \qquad (3.57)$$

The right-hand term reflects a conservative bound for the left-hand term, for all times except at $t = 0$.

As in [219], the passive nodes can be viewed as forming the sides of nested hypercubes, H_{p_i}, surrounding the activation hypercube H_1, as shown by the dashed squares in Figure 3.3. This is why a hypercube activation zone centred around $\mathbf{x}_o$ was chosen, rather than a ball centred around $\mathbf{x}(t)$. The number of nodes on the perimeter of any hypercube of half-length $m\mu$ is thus given by $[(2m+1)^n - (2m-1)^n]$. On a given hypercube H_{p_i} of half-length $m\mu$, the radial basis functions contributing the largest value for an input at $\mathbf{x} = \mathbf{x}_o$, are those located at the mid-point of the sides and the corresponding function will be $\exp\left(\frac{-m^2\mu^2}{2\sigma^2}\right)$. Because all of the other nodes on the perimeter of H_{p_i} contribute less than this value, the summation of all basis functions for these

64   3. Dynamic Structure Networks for Stable Adaptive Control

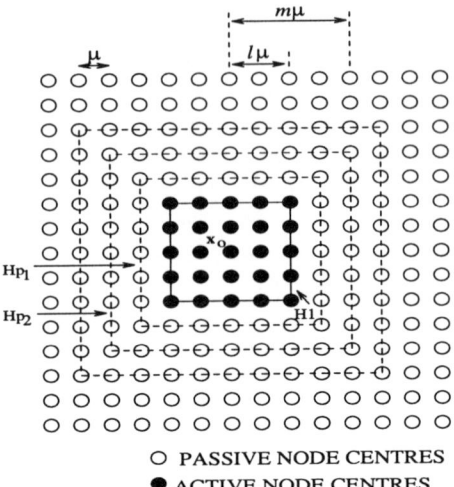

Fig. 3.3. Initial hypercube of activated units.

nodes is surely less than $[(2m+1)^n - (2m-1)^n]\exp\left(\frac{-m^2\mu^2}{2\sigma^2}\right)$. Again, this represents a conservative bound. Repeating for all H_{p_i} such that all passive nodes in χ are covered, we obtain:

$$\sum_{i=1}^{R} \Phi_{fp_i}(\mathbf{x}(t)) \leq \sum_{i=1}^{R_{max}} \Phi_{fp_i}(\mathbf{x}(0))$$
$$\leq \sum_{m=l+1}^{M}[(2m+1)^n - (2m-1)^n]\exp\left(\frac{-m^2\mu^2}{2\sigma^2}\right) \qquad (3.58)$$

where it is assumed that $\Phi_{fp_i}(\mathbf{x}(0)) \approx \Phi_{fp_i}(\mathbf{x}_o)$. This introduces some error but its value is small, especially for a mesh having small μ. In any case, the conservativeness of the bounds introduced previously more than compensates for this small error.

The upper limit of the last summation term reflects the radius (in terms of the number of mesh spacings) of the largest hypercube H_{p_i} required to span all χ. Clearly, this depends on the location of the initial centre $\mathbf{x}_o$. However, being conservative once again, M surely cannot exceed the number of nodes along the longest side of χ, so that M can be set to this value. In any case, the series on the right of the last inequality converges rapidly [219], so that the conservativeness of M does not really affect the value of this bound.

As shown in [219], the term $[(2m+1)^n - (2m-1)^n]$ can be expressed via the binomial theorem, as:

$$\sum_{j=0,\ldots,(n-1)|(n-j)\ is\ odd} 2^{j+1}\ {}^nC_j m^j$$

so that

3.8 Evaluation of Control Parameters and Implementation

$$\sum_{i=1}^{R} \Phi_{fp_i} \leq \sum_{j=0,\ldots,(n-1)} 2^{j+1} \, {}^nC_j \sum_{m=l+1}^{M} m^j exp\left(\frac{-m^2\mu^2}{2\sigma^2}\right) := \overline{\Phi}_{fb} \quad (3.59)$$

Similar considerations apply for $\sum \Phi_{gp_i}(\mathbf{x})$, the bound denoted by $\overline{\Phi}_{gb}$.

Whilst [219] provides a relation for determining the bound on the magnitude of the optimal parameters, $|w^*_{fp_i}|$, this is quite conservative, resulting in an unnecessarily larger value for $\overline{k}_a$. This goes against the aim of keeping the sliding control to its minimum whilst $\mathbf{x} \in \chi$. One way of determining the order of magnitude of the maximum value of $|w^*_{fp_i}|$ would be to subject the system to a *short* period of a persistently exciting input [23, 186, 224], so as to force the parameters to a value that is close to their optimal. From these values, one could obtain a hint on the order of magnitude of this bound, even though this would not be perfectly accurate due to the short period of application of the input. In practice, one could then use an overestimate based on this value, so as to ensure that $\max(|w^*_{fpi}|)$ is less than this estimate and use this value for $\overline{w}_{fb}$. Hence,

$$|\mathbf{w}^{*T}_{\mathbf{fp}} \Phi_{\mathbf{fp}}(\mathbf{x})| \leq \overline{w}_{fb} \overline{\Phi}_{fb} := \delta_f$$
$$|\mathbf{w}^{*T}_{\mathbf{gp}} \Phi_{\mathbf{gp}}(\mathbf{x})| \leq \overline{w}_{gb} \overline{\Phi}_{gb} := \delta_g. \quad (3.60)$$

Thus, Equation (3.55) implies:

$$|d_d| \leq \kappa(\delta_f + \delta_g|u_{fl}|) + \kappa(\epsilon_f + \epsilon_g|u_{fl}|)$$
$$+ (1-\kappa)(\overline{f}_o + \overline{g}_o|u_{fl}|) := \overline{d}_d. \quad (3.61)$$

The first term represent errors due to the fact that the contribution to the output by passive nodes is being ignored in the dynamic network scheme. The second term represents errors due to the fact that the network can only approximate the desired function to a particular accuracy that depends on μ, σ. The last term represents the error when the state falls outside the network approximation region, this having a typically large value making $\overline{d}_d$ large enough to force the state back into χ as quickly as possible. Note that whilst the state is in χ, $|\overline{d}_d|$ is limited by the size of the first two terms. This is typically much smaller than for the previously-mentioned case, so that use of crude, high-gain sliding control is avoided, keeping only a lower gain sliding control that is sufficient to ensure that parameter drift is avoided, thus guaranteeing robustness.

3.8.2 Choice of the Boundary Layer

Boundary layer sliding control is used to limit the bandwidth of the control signal so that high-frequency, unmodelled dynamics are not excited [235, 234]. Following the same line of development as in [254], with appropriate modifications for our dynamic network case, using Equation (3.29) and the fact that the system bandwidth is predominantly determined by the dynamics inside the boundary layer, where $|e_1| \leq \phi$, for this range of e_1 we get:

$$\dot{e}_1 = -e_1\left(k_d + g(\mathbf{x})\bar{k}_a \frac{1}{\phi}\right) + h \qquad (3.62)$$

where,

$$h = -\kappa\tilde{\mathbf{w}}_{\mathbf{fa}}^T \Phi_{\mathbf{fa}}(\mathbf{x}) - \kappa\tilde{\mathbf{w}}_{\mathbf{ga}}^T \Phi_{\mathbf{ga}}(\mathbf{x}) u_{fl} + d_d(t). \qquad (3.63)$$

These dynamics have a low-pass filter structure, with corner frequency

$$\left(k_d + \frac{\bar{k}_a g}{\phi}\right).$$

This could be taken as a measure of the control signal bandwidth, ν_c. The smaller the ϕ, the smaller is the tracking error, but the higher is the control bandwidth. This increases the chance of exciting higher frequency dynamics. For example, if the system is sampled with frequency ν_s, one can only ignore the effects of sampling if $\nu_s > 5\nu_c$ [219]. Hence, a trade-off between keeping a small tracking error and limiting the magnitude of the control bandwidth, must be settled for. If the maximum possible value of ν_c is ν_{max}, then

$$\left(k_d + \frac{\bar{k}_a g}{\phi}\right)_{max} \leq \nu_{max}. \qquad (3.64)$$

Substitution for $\bar{k}_a$ implies that:

$$\left(k_d + \frac{\bar{k}_a g}{\phi}\right)_{max} \leq \left(\frac{g\bar{d}_d}{g_l \phi}\right)_{max} \qquad (3.65)$$

so that the above is satisfied if $\left(\frac{g\bar{d}_d}{g_l \phi}\right)_{max} = \nu_{max}$. Substituting for $\bar{d}_d$ and assuming that $\kappa(\mathbf{x}) = 1$ when $|e_1| \leq \phi$, we obtain:

$$\frac{g_h}{g_l \phi_{min}}(\delta_f + \delta_g |u_{fl}|_{max} + \epsilon_f + \epsilon_g |u_{fl}|_{max}) = \nu_{max} \qquad (3.66)$$

where:

- $\bar{g}_h$ is an upper bound on $g_h(\mathbf{x})$, a function such that $g(\mathbf{x}) \leq g_h(\mathbf{x})$, $\forall \mathbf{x}$.
- g_l is a lower bound on $g_l(\mathbf{x})$.
- $|u_{fl}|_{max} = \frac{\hat{f}_{hm} + \bar{v}}{g_l}$ by definition of u_{fl}, where,
  - $\hat{f}_{hm}$ is an upper bound on $\hat{f}_h(\mathbf{x})$, where $\hat{f}_h(\mathbf{x}) \geq |\hat{f}_a(\mathbf{x})|$,
  - $\bar{v}$ is an upper bound on $v(t)$. As in [254, 255], $\bar{v} \approx \bar{y}_d^{(r)}$, the latter being an upper bound on $y_d^{(r)}$.

Hence,

$$\phi_{min} = \frac{g_h}{g_l \nu_{max}}(\delta_f + \delta_g |u_{fl}|_{max} + \epsilon_f + \epsilon_g |u_{fl}|_{max}) \qquad (3.67)$$

gives the smallest possible value of the boundary layer ϕ, ensuring a control bandwidth that does not exceed ν_{max}.

Note that this ϕ_{min} is larger than that derived in [254] where a full-size network has been used. This is due to the additional terms δ_f, $\delta_g |u_{fl}|_{max}$

3.8 Evaluation of Control Parameters and Implementation 67

introduced via the error in neglecting the contribution of the passive nodes in our networks. The difference obviously depends on the sizes of δ_f, δ_g which depend principally on the activation threshold δ_{min}. From Equation (3.67), it is straight forward to deduce that ϕ_{min} for a dynamic scheme is greater than that for a full network scheme by a factor of

$$\frac{|u_{fl}|_{max}(\epsilon_g + \delta_g) + \epsilon_f + \delta_f}{|u_{fl}|_{max}\epsilon_g + \epsilon_f}.$$

In most cases, the above term could be conveniently approximated to

$$\left(1 + \frac{\delta_g}{\epsilon_g}\right).$$

This implies a larger tracking error for the same ν_{max}, and reflects the price we have to pay in using smaller dynamic networks, for the high frequency unmodelled dynamics not to be excited. Note also, that if a boundary layer less than ϕ_{min} is actually used, the system is still guaranteed to be robust against parameter drift, but one gets a larger control bandwidth.

3.8.3 Comments

1. The larger the activation threshold, δ_{min}, the smaller is the size of the hypercube in which nodes are activated, resulting in a smaller network. However, this results in a larger $\overline{\Phi}_{fb}$, $\overline{\Phi}_{gb}$ (due to a smaller m in the exponential term of Equation (3.59)) and so larger δ_f, δ_g. In turn, these contribute to a larger $\overline{d}_d$ and hence bigger sliding gains, necessary to compensate for the larger disturbance arising from activation of a smaller number of nodes. Thus, activation threshold δ_{min}, can be viewed as a design parameter that determines the balance between the extent to which the network size is limited and the sliding gain that is required to ensure robustness to the disturbance term. The stronger the limitation on network growth (via large δ_{min}), the larger is the sliding gain; resulting in 'cruder' control having a larger closed-loop bandwidth and higher control activity. In all cases though, error convergence to the boundary layer and robustness to parameter drift are ensured.
2. ϕ appears both in the sliding control term as the width of the boundary layer, and also as the dead-zone width in the adaptation laws. Because of the latter, ϕ affects directly the steady-state tracking error. As in [254], if the effects of high frequency unmodelled dynamics are neglected, the boundary layer and hence the steady-state tracking error, can be made arbitrarily small without risking parameter drift, because of the use of sliding control within χ, which compensates for the disturbance term. Indeed as seen in Equation (3.67), if no limitation on the bandwidth were imposed (i.e., $\nu_{max} \to \infty$), ϕ_{min} could be set to zero. This contrasts with those schemes that use simple dead-zone adaptation (without sliding

control in χ) to avoid parameter drift, where the minimum dead-zone width is determined by the size of the disturbance.

In practice though, the bandwidth must always be constrained. Hence, even in our case, ϕ_{min} cannot be zero. Note however, that this limitation is imposed by the need to ensure a finite bandwidth and not to ensure robustness to the disturbance term. The argument applies equally well to the activation threshold δ_{min}: the larger it is, the bigger must be the sliding gain and so the bandwidth increases. If no limitation on the bandwidth existed, ϕ_{min} and δ_{min} could take arbitrary values within their ranges, *i.e.*, $\phi_{min} \geq 0$, $0 < \delta_{min} \leq 1$. Ideally, ϕ_{min} would be set to zero implying zero steady-state tracking error and δ_{min} close to one to give the greatest possible restriction on dynamic network growth, respectively. This reflects an improvement over the dynamic scheme suggested in [219], whereby to ensure robustness to the disturbance term, the activation threshold that limits network growth depends directly on the tracking accuracy required, because simple dead-zone adaptation is used.

Hence the advantages of using sliding mode control within χ, as proposed in [254, 255], are preserved in our dynamic network scheme; effectively decoupling the network growth rate and the tracking accuracy. Unfortunately, this decoupling is slightly weakened by the need to ensure that the closed-loop system bandwidth is less than ν_{max} so as to avoid the excitation of high frequency unmodelled dynamics. Thus, in actual design, one must seek a compromise between the steady-state tracking accuracy and the size of the dynamic structure neural network against the control bandwidth. This is reflected in Equation (3.67), showing the interaction between these three variables via the terms ϕ_{min}, δ_f, δ_g and ν_{max}.

3. In theory, if the entire state-space χ is being excited, then the dynamic structure network would still grow to a size that is nearly equal to that of a full network. In this case, the use of a full non-dynamic network is preferred because it requires a lower computational effort (due to the absence of activation schemes) and the gain in terms of network size will not be significant. In practice though, when the state dimensionality $n \geq 2$ and only a subset of state-space is being excited, the proposed dynamic structure network scheme is particularly effective because use of a full non-dynamic structure scheme would result in an excessively large network size due to the *curse of dimensionality*.

3.8.4 Implementation

For convenience, in this section, the arguments for implementation are developed in terms of the $\hat{f}$ network. Exactly similar considerations apply to the $\hat{g}$ network. The radial basis function network required to approximate $f(\mathbf{x})$ essentially consists of the active parameters vector, $\hat{\mathbf{w}}_{\mathbf{fa}}$, storing the output weights and a matrix $\mathbf{M}$, storing the coordinates of the centres of *active nodes*, $\mathbf{m}_{fa_i}$, *i.e.*,

3.8 Evaluation of Control Parameters and Implementation

$$\mathbf{M} = \begin{bmatrix} \mathbf{m}_{fa_1}^T \\ \mathbf{m}_{fa_2}^T \\ \vdots \end{bmatrix}, \qquad \hat{\mathbf{w}}_{\mathbf{fa}} = \begin{bmatrix} \hat{w}_{fa_1} \\ \hat{w}_{fa_2} \\ \vdots \end{bmatrix}.$$

The length of $\mathbf{M}$ and $\hat{\mathbf{w}}_{\mathbf{fa}}$ represents the number of *active* nodes, *i.e.*, the network size, and these vectors are *dynamic*, *i.e.*, they start from a size of zero and their length increases as nodes are activated. The activation occurs when the system state $\mathbf{x}(t)$ traverses 'new' mesh points in χ, whose basis function is currently in a hypercube, centered around the nearest mesh point to $\mathbf{x}(t)$, with half-length equal to $l\mu$ where l is the closest integer to $\sqrt{-2\sigma^2 ln(\delta_{min})}/\mu$.

Thus, when a state $\mathbf{x}(t)$ is observed, mesh nodes with coordinates $\mathbf{m}_j$ in the relevant hypercube must be considered for activation. It is a straight forward procedure to determine these coordinates. However before activating them, *ie*, storing them in $\mathbf{M}$ and augmenting $\hat{\mathbf{w}}_{\mathbf{fa}}$ accordingly, one must check if any of these nodes have already been activated before due to state $\mathbf{x}(t)$ passing 'near' them at some previous time. If this is the case, the coordinates of the node would already be stored in $\mathbf{M}$ and clearly, they must not be re-entered. Hence, each mesh node coordinate $\mathbf{m}_j^T$, must be matched with the rows in $\mathbf{M}$ to check for its presence, in which case it represents prior activation. Although this is a straight forward serial search routine, its execution is time consuming, especially when $\mathbf{M}$ grows appreciably large and if the system is to be used on-line.

However, a different technique can be used to check for the presence of $\mathbf{m}_j^T$ in $\mathbf{M}$. The idea centres on using the neural network itself to indicate if $\mathbf{m}_j$ is one of the centres of the current network basis functions, by temporarily buffering the current network output generating $\hat{f}_a(\mathbf{x})$ and applying consecutively, each coordinate $\mathbf{m}_j$ as input to the network. If one $\mathbf{m}_j$ is already present in $\mathbf{M}$ as the centre of say, the ith active node, then

$$\text{for} \quad \mathbf{m}_j = \mathbf{m}_i, \qquad \Phi_{fa_i} = \exp\left(\frac{-\|\mathbf{m}_j - \mathbf{m}_{fa_i}\|^2}{2\sigma_f^2}\right) = 1. \qquad (3.68)$$

If $\mathbf{m}_j$ is not present in $\mathbf{M}$, *none* of the basis functions will be equal to 1. Hence, if one of the basis functions responds with a 1 when the input is $\mathbf{m}_j$, the node centered at $\mathbf{m}_j$ has already been activated and the network need not grow. On the other hand, if *all* the basis functions are less than one, the node centred at $\mathbf{m}_j$ must be activated by augmenting $\mathbf{M}$ with $\mathbf{m}_j^T$ and $\hat{\mathbf{w}}_{\mathbf{fa}}$ with 0, resulting in network growth.

The advantage of this method is that given a parallel neural network architecture, detection of the presence of each $\mathbf{m}_j$ in $\mathbf{M}$ takes place in parallel, without having to perform a serial search of all the rows in $\mathbf{M}$ to test for a match with $\mathbf{m}_j^T$. This is achieved by augmenting the usual neural network structure with a Threshold Logic Unit (TLU) connected to the output of each radial basis function unit. The TLU outputs a 1 only if its input is greater

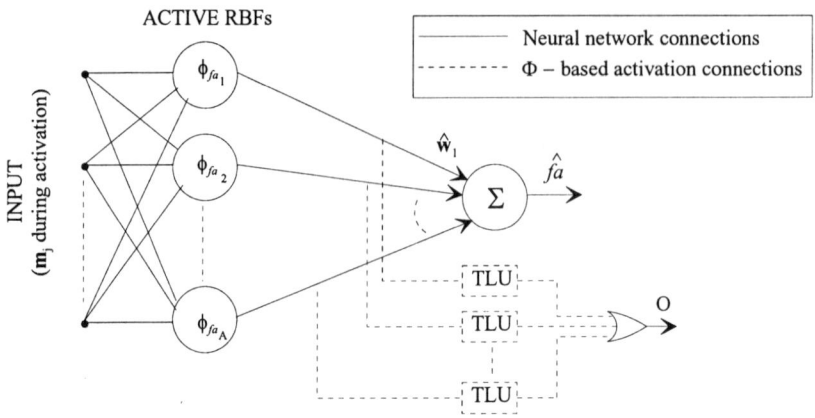

Fig. 3.4. Φ-based activation technique.

than or equal to one. Hence, if one of the basis function outputs is 1, only its own TLU will output a 1 and the rest will be at 0. Performing a logical OR of these TLU outputs will thus result in a logic 1 output only if $\mathbf{m}_j$ is already active and 0 if not. Hence, this output (O) is a binary signal indicating that the network needs to grow (by augmenting it with a node centered at $\mathbf{m}_j$) if O is zero, or need not otherwise. This activation-decision technique shall be referred to as *Φ-based activation* and is shown in Figure 3.4.

3.9 Simulation Examples

Two simulations, based on examples in [254] and [255] are carried out. The examples were chosen to evaluate the effectiveness of selective node activation by comparing the performances of the dynamic and non-dynamic structure networks. In the examples, with orders being 2 and 4 respectively, the input spans a relatively small subset of χ and hence a significant saving in terms of the size of network would be expected. Another point of interest is the tracking performance of the systems in the face of the parsimonious allocation of the neural network basis functions.

3.9.1 Example 1

The system used in the first simulation is taken from [255]. The dynamical system in this example is given by the equation,

$$\begin{aligned}\dot{x}_1 &= f(\mathbf{x}) + g(\mathbf{x})u \\ \dot{x}_2 &= x_1 - x_2 \\ y &= x_1\end{aligned} \quad (3.69)$$

3.9 Simulation Examples

where,

$$\begin{aligned} \mathbf{x} &= [x_1 \ x_2]^T \\ f(\mathbf{x}) &= \cos(7(x_1^2 + x_2^2))\exp(-(x_1^2 + x_2^2)) \\ g(\mathbf{x}) &= 2 + \cos(7x_1 x_2) \end{aligned} \quad (3.70)$$

This system is of order $n = 2$ and has degree $r = 1$. These dynamics are already expressed in global normal form and $\mathbf{I}(x_1, x_2) = x_1 - x_2$ has continuous and bounded partial derivatives in x_1, x_2, which implies that the system's internal dynamics satisfy the Lipschitz condition. The zero dynamics, $\dot{x}_2 = \mathbf{I}(0, x_2)$ are thus given by $\dot{x}_2 = -x_2$, which are globally exponentially stable. The reference input, y_d, is obtained by filtering a zero-average, 0.9 amplitude, 0.4Hz square wave by a filter $1/\left(1 + \frac{s}{10}\right)^3$, so that the derivative of the filter output is bounded. As shown in [255], for the given system and reference input, the desired state is bounded well within the interval $[-1, 1] \times [-1, 1]$ (along x_1, x_2 respectively). To cater for the fact that during the transient period the actual state may overshoot these bounds, the network approximation region is taken to be larger, namely $\chi = [-1.5, 1.5] \times [-1.5, 1.5]$. This ensures that the state is bounded within the approximation region, thus avoiding the use of crude, high-gain sliding control. By definition, χ^- must be a subset of χ and so it is set to $\chi^- = [-1, 1] \times [-1, 1]$.

It is assumed that the known prior estimates to the functions being approximated are $f_o = 0$, $g_o = 2$ and, as shown in [255], full network inherent approximation error bounds $\epsilon_f = \epsilon_g = 0.005$ would be obtained via basis functions with standard deviation σ_f, σ_g of 0.03 located on a mesh of spacing $\mu = 0.05$ within χ. Bounds $\overline{f}_o, \overline{g}_o$ are set to 2 and $g_l(\mathbf{x})$ is set equal to $g_l = 0.895$. The activation threshold is set to 0.2, resulting in an activation hypercube of half-length μ.

The control law is given by $v(t) = \dot{y}_d - 5(y - y_d)$ and $e_1(t) = e(t)$. η_f, η_g are set to 25 and ϵ^* to 0.1. Using Equation (3.59) and the above values, results in $\overline{\Phi}_{fb} = \overline{\Phi}_{gb} = 0.062$. The system was initially subjected to a 40s persistently exciting input to obtain the order of magnitude for bounds of the optimal parameters, from which over-estimates $\overline{w}_{fb} = \overline{w}_{gb} = 1.6$ were deduced. Using Equations (3.60), we obtain $\delta_f = \delta_g = 0.1$. Using $|v| = |y_d^{(1)}| \le 5$, $\hat{f}_{hm} = 2$, $\overline{g}_h = 4$, $\nu_s = 2000\pi$ rad/s, $\nu_{max} = 400$ rad/s, gives $\phi_{min} \approx 0.01$ which was used as the boundary layer. Note that the persistently exciting input is used to obtain δ_f, δ_g only; once these values are deduced, network activation is started afresh, with initially no nodes being activated.

The results of the simulation are shown in Figure 3.5. Note that the system is stable and that the error asymptotically converges to the expected value of 0.01, namely the size of the boundary layer. Moreover, after 150 seconds, the networks contained 556 nodes each, which is 15% of the 'full-size' networks having 3721 nodes used in [255]. This follows from the fact that the desired input forces the state to follow a subset of the space in χ, and thus the selective node activation technique places basis functions in

72   3. Dynamic Structure Networks for Stable Adaptive Control

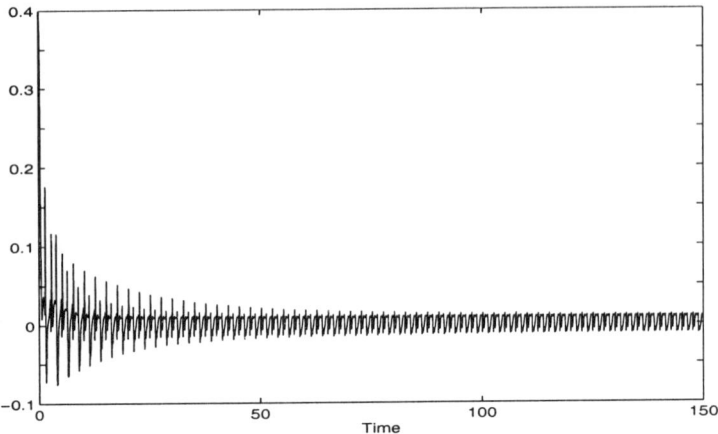

Fig. 3.5. Simulation 1: Tracking error.

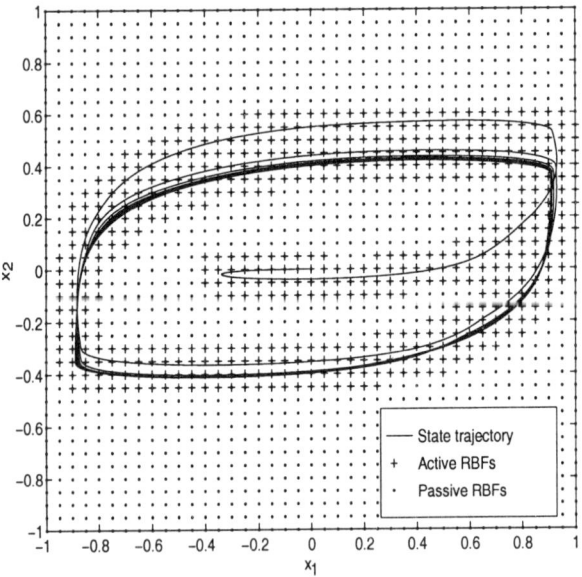

Fig. 3.6. Simulation 1: Activated nodes.

the neighbourhood of this subset only. This can be seen in Figure 3.6 which shows χ^-, the region where the state trajectory is contained, marking the positions of active nodes forming the dynamic networks with a cross and those of passive nodes, which were present in a 'full-size' network scheme, with a dot.

Moreover, as demonstrated in [254], even if a full-size network were used, the parameter vectors $\hat{\mathbf{w}}_f$, $\hat{\mathbf{w}}_g$ would only approach their optimal values $\mathbf{w}_f^*$, $\mathbf{w}_g^*$ if Φ_f, $\Phi_g u_{fl}$ are persistently excited. Otherwise, the network approximation accuracy is good only in state space regions visited during the operation of the system. This is the case for the dynamic structure network scheme as well. The role of the neural network is in learning the system dynamics while the adaptive control scheme guarantees stability and robustness. The advantage of using neural networks to learn the dynamics of the system are demonstrated in [255] in which it is shown, for the dynamic system used in this simulation, that the tracking errors are about a factor 20 higher for the scheme that does not use the neural network adaptive controller.

3.9.2 Example 2

The system for this second example, taken from [254], is a robotic manipulator having 2 degrees of freedom in a horizontal plane, as shown in Figure 3.7.

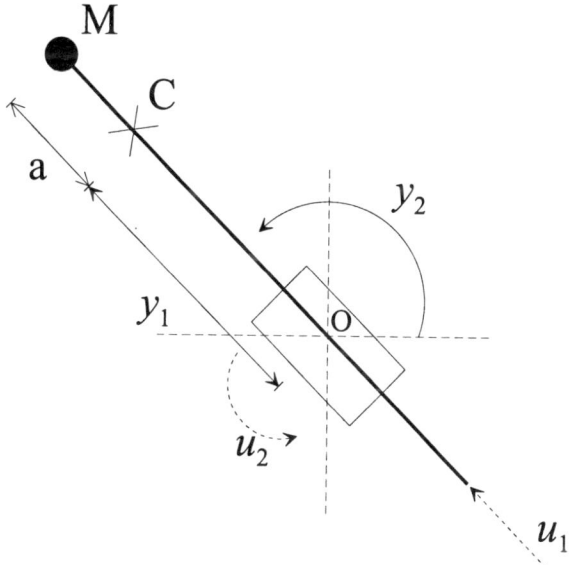

Fig. 3.7. Two degrees of freedom robotic manipulator.

The manipulator can rotate about axis O by application of torque u_2 which controls angle y_2 and move radially by application of force u_1, controlling the radial distance y_1. C is the centre of mass of the link, M_c is the link mass, M is the load mass and a is the distance from C to the load. J_1, J_2 are the moments of inertia of the link with respect to C and O respectively. The system dynamics are given by,

$$\begin{aligned}
\dot{x}_1 &= x_2 \\
\dot{x}_2 &= \frac{((M_c+M)x_1+Ma)x_4^2}{M_c+M} + \frac{u_1}{M_c+M} \\
\dot{x}_3 &= x_4 \\
\dot{x}_4 &= \frac{-2[(M_c+M)x_1+Ma]x_2x_4}{J_1+J_2+M_cx_1^2+M(x_1+a)^2} + \frac{u_2}{J_1+J_2+M_cx_1^2+M(x_1+a)^2} \\
y_1 &= x_1 \\
y_2 &= x_3
\end{aligned} \quad (3.71)$$

which could be re-written in terms of the radial and angular dynamics respectively, as

$$\begin{aligned}
\dot{x}_1 &= x_2 \\
\dot{x}_2 &= f_1(x_1,x_4) + g_1 u_1 \\
y_1 &= x_1 \\
\dot{x}_3 &= x_4 \\
\dot{x}_4 &= f_2(x_1,x_2,x_3) + g_2(x_1)u_2 \\
y_2 &= x_3
\end{aligned} \quad (3.72)$$

where,

$$\begin{aligned}
f_1(x_1,x_4) &= \frac{((M_c+M)x_1+Ma)x_4^2}{M_c+M} \\
g_1 &= \frac{1}{M_c+M} \\
f_2(x_1,x_2,x_3) &= \frac{-2[(M_c+M)x_1+Ma]x_2x_4}{J_1+J_2+M_cx_1^2+M(x_1+a)^2} \\
g_2(x_1) &= \frac{1}{J_1+J_2+M_cx_1^2+M(x_1+a)^2}
\end{aligned} \quad (3.73)$$

Functions f_1, f_2, g_1, g_2 represent the unknown system dynamics. Note that the two subsystems concerning radial movement (y_1) and angular movement (y_2), can be considered as two separate SISO affine systems, since the control gain matrix has a diagonal structure. Hence, two separate controllers are used, one for each subsystem.

The desired output is given by,

$$\begin{aligned}
y_{1d} &= 1 + 0.5\sin(2t) \\
y_{2d} &= \frac{\pi}{4}(1+\cos(2t))
\end{aligned} \quad (3.74)$$

so that the network approximation region, χ, is given by the interval $[0.3, 1.7] \times [-1.2, 1.2] \times [-0.1\pi, 0.6\pi] \times [-0.6\pi, 0.6\pi]$ along x_1, x_2, x_3, x_4 respectively. The prior estimates of the $\hat{f}_1$, $\hat{f}_2$ functions are 0 and those for the $\hat{g}_1$, $\hat{g}_2$ networks are 4.5×10^{-3} and 1×10^{-3}, respectively.

As in [254], full network inherent approximation error bounds ϵ_{f1}, ϵ_{f2}, ϵ_{g1}, ϵ_{g2} (for the networks approximating f_1, f_2, g_1, g_2 respectively) equal to 0.1, 0.05, 2×10^{-5}, 1×10^{-5} would be obtained by basis functions having standard deviations of 0.06, 0.09, 0.14 with respect to states x_1, x_2, x_4 for the $\hat{f}_1$, $\hat{f}_2$ functions, and 0.024 for the $\hat{g}_1$, $\hat{g}_2$ functions if mesh spacing of 0.1,

3.9 Simulation Examples

0.15, 0.23 in the directions of x_1, x_2, x_4 for the $\hat{f}_1$, $\hat{f}_2$ networks and 0.04 for the $\hat{g}_1$, $\hat{g}_2$ networks along x_1 are used. Since the $\hat{g}$ networks have the same input space, mesh spacing and standard deviation they could share the same nodes as that of $\hat{f}$. As mentioned before, not all of these basis functions are necessarily used in the dynamic scheme, and the mesh positions just indicate where a basis function could be placed if required.

The lower bounds on $\hat{g}_1$, $\hat{g}_2$ are set at $g_{l1} = 4.48 \times 10^{-3}$ and $g_{l2} = 8.4 \times 10^{-4}$, respectively. The activation threshold is set to 0.2. The control law is given by $\Psi(s) = (s+5)$ and k_d is set to 1 in the transfer function $\Psi \Gamma^{-1}$. η_{f1}, η_{f2}, η_{g1}, η_{g2} are set to 20, 20, 5×10^{-5}, 5×10^{-5} respectively and ϵ^* to 1×10^{-4}. Assuming that the functions are bounded by $\bar{g}_{h1} = 12.5 \times 10^{-3}$, $\bar{f}_{h1} = 5.18$, $\bar{g}_{h2} = 5.6 \times 10^{-3}$, $\bar{f}_{h2} = 6.2$, with the initial approximations used, the bounds follow as $\bar{f}_{o1} = 5.18$, $\bar{f}_{o2} = 6.2$, $\bar{g}_{o1} = 8 \times 10^{-3}$, $\bar{g}_{o2} = 5 \times 10^{-3}$ for $\hat{f}_1$, $\hat{f}_2$, $\hat{g}_1$, $\hat{g}_2$ respectively. From these values and use of Equation (3.59) adapted accordingly to accommodate for the fact that the mesh spacing and standard deviations are different along x_1, x_2, x_4, one obtains: $\bar{\Phi}_{f1b} = 0.065$, $\bar{\Phi}_{f2b} = 0.392$, $\bar{\Phi}_{g1b} = \bar{\Phi}_{g2b} = 0.0077$. Using overestimates for bounds of the optimal parameters as: $\bar{w}_{f1b} = 2$, $\bar{w}_{f2b} = 2$, $\bar{w}_{g1b} = 0.007$, $\bar{w}_{g2b} = 0.004$, one obtains $\delta_{f_1} = 0.13$, $\delta_{f_2} = 0.78$, $\delta_{g_1} = 5.4 \times 10^{-5}$, $\delta_{g_2} = 3.1 \times 10^{-5}$ for the $\hat{f}_1$, $\hat{f}_2$, $\hat{g}_1$, $\hat{g}_2$ networks, respectively. Using $\bar{y}_{1d}^{(r)} = 2$, $\bar{y}_{2d}^{(r)} = \pi$ (where for these dynamics $r = 2$), sampling frequency $\nu_s = 1000\pi$ rad/s, $\nu_{max} = 130$ rad/s; one obtains $\phi_{min1} = 0.0075$, $\phi_{min2} = 0.066$. Hence, boundary layers set to $\phi = 0.015$ and $\phi = 0.08$ were used for the radial and angular subsystems, respectively. Using Equation (3.50), these should result in steady-state error bounds of 0.003 metres and 0.016 radians respectively. The final design parameters are summarized in Table 3.1 for convenience.

Table 3.1. Design parameters for $\nu_s = 500$ Hz, $\nu_{max} = 130$ rad/s, $\delta_{min} = 0.2$.

	f_o	η_f	ϵ_f	σ_f	δ_f	ϕ_{min}	$\bar{f}_o$	μ_f
RADIAL	0	20	0.1	0.06-x_1 0.09-x_2 0.14-x_4	0.13	0.0075	5.18	0.10-x_1 0.15-x_2 0.23-x_4
ANGULAR	0	20	0.05	0.06-x_1 0.09-x_2 0.14-x_4	0.78	0.066	6.2	0.10 x_1 0.15-x_2 0.23-x_4

	g_o	η_g	ϵ_g	σ_g	δ_g	g_l	$\bar{g}_o$	μ_g
RADIAL $\times 10^{-3}$	4.5	0.05	0.02	24	0.054	4.48	8	40
ANGULAR $\times 10^{-3}$	1.0	0.05	0.01	24	0.031	0.84	5	40

The results of the simulation are shown in Figure 3.8. Parts (a) and (b) of Figure 3.8 show the radial and angular tracking errors respectively. Note that during the 100s simulation time, no instability was exhibited and the errors remained bounded. Furthermore, parts (c) and (d) of the same figure show that the radial and angular tracking errors converge to within their expected bounds in the steady-state. During the given simulation interval, the network parameters remained bounded, indicating robustness to parameter drift. For the given desired input, the size of the networks grew up to 89, 828 and 29 nodes for the $\hat{f}_1$, $\hat{f}_2$ and $\hat{g}$ networks, respectively. These represent a usage of 33%, 18% and 80.5% respectively, when compared to the corresponding non-dynamic structure (full-size) network scheme, demonstrating the effectiveness of the scheme. Note that the greatest saving is on the $\hat{f}_2$ network since its input is 3 dimensional and hence the active region represents a much smaller proportion of the complete approximation region, than for say the $\hat{g}$ networks whose nodes span a one dimensional space. Moreover, the tracking results confirm that there was no performance degradation arising from the use of the dynamic structure network when compared with the non-dynamic one.

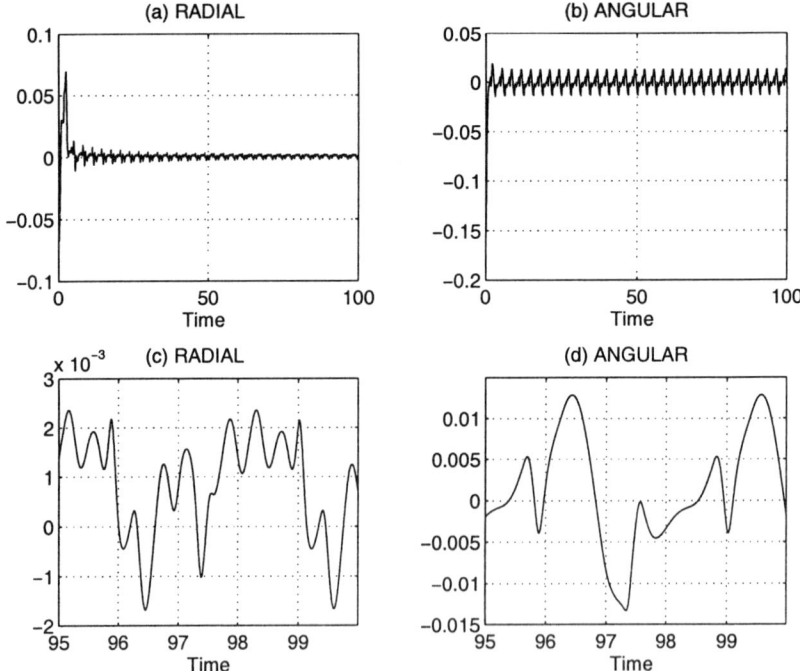

Fig. 3.8. Simulation 2: Tracking error.

3.10 Summary

A method, based on Lyapunov stability considerations, for using dynamic structure Gaussian radial basis function neural networks in adaptive control of affine nonlinear systems has been presented. The objective of the control scheme is to achieve good tracking performance in the absence of known system dynamics. The scheme is also expected to be robust against parameter drift and approximation errors. The task is carried out by using dynamic structure neural networks to approximate the underlying functions and generate control signals based on the feedback linearisation control law. The work presented is based on the fixed-structure neural control system developed in [254, 255], which was extended and modified to utilize dynamic structure neural networks, an idea suggested in [219] for adaptive control and developed for system identification in [202, 127].

The scope behind the use of dynamic structure networks is to keep the size of the neural networks limited, by ignoring potential basis functions that would be centred on mesh points located beyond a chosen distance from the active regions of state-space. The result is a network that 'grows' whilst the system is evolving, by activating basis functions that are centred on mesh points in appropriate locations. The adaptation laws that adjust the parameters of the neural networks are so designed to guarantee stability in the sense of Lyapunov.

The system is augmented by low-gain sliding control (as in [254]) and by high-gain sliding control (as in [219]) to ensure global stability of the tracking error and robustness to the presence of a disturbance term. The latter arises because of the network inherent approximation errors and those basis functions that are ignored in the dynamic structure scheme. This method allows for a certain degree of flexibility regarding the choice of the tracking accuracy and the node activation threshold; a flexibility that is limited only by the maximum allowed control bandwidth. A technique for the efficient implementation of the dynamic structure network within the adaptive control scheme has also been presented. This saves on the processing time required for node activation by avoiding lengthy serial search routines and exploiting the parallelism of the neural networks themselves.

The effectiveness of the dynamic structure scheme was put to test with two examples, used in [254, 255]. The performance results demonstrate that the scheme performs as predicted, by forming network sizes that are compact while achieving equally good tracking accuracy, evidenced by convergence to the predicted bounds, compared to the non-dynamic structure full-sized network scheme. The computational savings were very high where the order of the system was two or higher and where the input to the system was not persistently exciting over the entire network approximation region. The significant computational advantages offered by this scheme, together with the guaranteed levels of stability and robustness, render the use of GaRBF neural networks as a practical and feasible solution for effecting adaptive control of a wide variety of physical systems.

4. Composite Adaptive Control of Continuous-Time Systems

4.1 Introduction

In this chapter we introduce an improved adaptation scheme for control of functional uncertain nonlinear systems by neural networks. The proposed scheme, inspired from the composite adaptation law of Slotine and Li [233] that was originally developed for parametric uncertain systems, aims to improve the transient performance of RBF-based adaptive schemes for continuous time affine, nonlinear systems, such as those found in [204, 219, 255]. Although these papers do provide control and adaptation laws that ensure boundedness of all the system's signals and tracking error convergence, performance issues such as *rate* of convergence and general improvements in the transient response are not addressed at all. Indeed this topic has been largely neglected even in the classical literature on linear adaptive control, where not many results abound [144, 182]. The few principal works dealing with this problem in the linear, parametric uncertain case are briefly described next.

The main approaches for transient performance improvement of adaptive systems could be classified in two: (a) modifications of the parameter adaptation law and (b) modifications of the control law. A rather crude third alternative for enhancing parameter estimation involves the addition of an external signal that is persistently exciting in some way [110]. Although the latter brings about the beneficial effect of improving parameter convergence, the external signal may reflect itself as an undesirable disturbance superimposed on the system output, which is often not acceptable in practice.

In conventional adaptive systems, parameter estimation is handled either directly or indirectly. In the direct approach, the *tracking error* is used in the adaptation laws to adjust some unknown controller parameters. In the indirect approach, a plant identification model is set up and used to generate an *estimation error*, reflecting the difference between the model output and the corresponding plant output. This is then used to estimate the unknown plant parameters by adjusting the parameters of the identification model so as to reduce the estimation error. These estimates are then used in a certainty equivalent control law.

The realization that both tracking and estimation errors are a valid source of information for the adaptive system, prompted some researchers to use them together so as to enhance parameter estimation. This led to Slotine and

Li's *composite adaptive* scheme [233] and Duarte and Narendra's *combined adaptive* scheme [66], both of which utilize tracking and estimation errors in the adaptation laws, albeit in different ways.

Later work concentrated on modifying the control law, instead of the adaptation law. Sun [241] introduced a new control law for model reference adaptive systems that includes a term related to the estimation error, leading to an improved transient response. Datta and Ioannou [62] suggested two measures for quantifying transient behaviour: the mean square error and L_∞ error bounds, and modified further Sun's control law to improve performance on the basis of both these measures. Ortega's [194] analysis of Morse's (rather unconventional) adaptive controller [180] shows that it also leads to transient response improvement. More recently, Krstić *et al.* [144] showed that their tuning functions scheme, originally developed for parametric uncertain nonlinear systems, gives rise to much better transient performance when applied to linear systems. The main modification represented by tuning functions design lies in the control law, which is nonlinear both in terms of measured signals and parameter estimates.

Issues related to transient performance improvement in functional adaptive control schemes have not been given any attention. Hence this chapter, extending the preliminary work presented by Fabri and Kadirkamanathan in [71], will address the problem for an affine class of nonlinear systems. The scheme centres on the use of a new neural network parameter adaptation law, based on Slotine and Li's principle of composite adaptation. When compared with the direct adaptive control schemes developed for the same class of systems [254, 255], the new approach yields faster tracking error convergence and still guarantees closed-loop stability. The principal contributions of this chapter therefore concern the following aspects of the design:

1. The introduction of a new and robust composite adaptation law based on tracking and estimation errors, for functional adaptive control systems.
2. The setting up of a suitable identification model to generate the estimation error.
3. The enhancement of the parameter convergence rate, leading to a better transient response and guaranteeing system stability.

When compared with the original composite adaptive scheme of [233], the present approach includes the following important extensions:

1. It deals with functional uncertain, rather than parametric uncertain, systems and therefore needs re-formulation in terms of a neural network framework and affine nonlinear systems.
2. The parameter update laws are modified by a 'robustifying' technique to handle the effects of additive disturbance signals which, in our case, are inevitable because of the inherent neural network approximation error.

Hence this work leads to improvements in both functional adaptive control techniques, as well as the composite adaptation scheme of [233] that was developed for parametric uncertain systems under the assumption of no disturbance. Although the design presented in this chapter is based on fixed structure neural networks, it is straightforward to reformulate it for accommodating dynamic structure networks according to the methodology of the previous chapter. The presentation here concentrates only on fixed structure networks so as not to detract attention from the main objective of developing a scheme that exhibits improved parameter convergence rates.

4.2 Problem Formulation

The objective is to design a stable adaptive control scheme that provides an *improved* tracking performance, for the same affine class of nonlinear systems considered in Chapter 3. Hence the plant is modelled by Equation (2.2), reproduced below for convenience, where nonlinear functions **a**, **b** and h are unknown:

$$\dot{\mathbf{x}} = \mathbf{a}(\mathbf{x}) + \mathbf{b}(\mathbf{x})u$$
$$y = h(\mathbf{x}).$$

As in the previous chapter, the control task is for the output $y(t)$ to track a reference input $y_d(t)$, whilst the state vector **x** is to remain bounded. The conditions specified by Assumptions 3.2.1 to 3.2.7 are also assumed to hold, so that the system could be globally expressed in the input-output affine form of Equation (2.5), namely:

$$y^{(r)} = f(\mathbf{x}) + g(\mathbf{x})u(t). \tag{4.1}$$

As before, functional adaptive control is required because of the uncertainty of functions $f(\mathbf{x})$ and $g(\mathbf{x})$. Gaussian radial basis function (GaRBF) networks will be used to approximate them by on-line adjustment of the network parameters. The main emphasis of the work in this chapter covers this last-mentioned aspect, *i.e.* parameter adjustment, where composite adaptation laws are formulated so as to obtain an improved transient response, whilst ensuring overall stability.

4.3 The Neural Networks

The neural networks employed in this scheme are similar to the fixed-structure networks described in Section 3.3. Hence, utilizing the same notation as before, two GaRBF networks approximating $f(\mathbf{x})$ and $g(\mathbf{x})$ within a compact set $\chi \subset \Re^n$ where the state vector is expected to be contained, are required. These are respectively characterized by the equations:

$$\begin{aligned}\hat{f}(\mathbf{x}, \hat{\mathbf{w}}_\mathbf{f}) &= \kappa(\mathbf{x})\hat{\mathbf{w}}_\mathbf{f}^T \Phi_\mathbf{f}(\mathbf{x}) + f_o(\mathbf{x}) \\ \hat{g}(\mathbf{x}, \hat{\mathbf{w}}_\mathbf{g}) &= \kappa(\mathbf{x})\hat{\mathbf{w}}_\mathbf{g}^T \Phi_\mathbf{g}(\mathbf{x}) + g_o(\mathbf{x}).\end{aligned} \tag{4.2}$$

As usual, $\hat{\mathbf{w}}_\mathbf{f}$, $\hat{\mathbf{w}}_\mathbf{g}$ are the network output layer parameter vectors and $\Phi_\mathbf{f}(\mathbf{x})$, $\Phi_\mathbf{g}(\mathbf{x})$ are the Gaussian basis function vectors. Also, $f_o(\mathbf{x})$ and $g_o(\mathbf{x})$ represent any known prior estimates to $f(\mathbf{x})$, $g(\mathbf{x})$. Due to a number of design considerations that are explained in detail later on, $g_o(\mathbf{x})$ is chosen to satisfy the condition

$$g_o(\mathbf{x}) \geq g_l(\mathbf{x}). \tag{4.3}$$

Function $\kappa(\mathbf{x})$ is a state-dependent function defined as in Equation (3.6) and introduced for the same purpose as before.

The same notation and definitions are also used for the optimal network approximation errors and error bounds. This means that given any non-zero uniform network approximation error bounds ϵ_f, ϵ_g and the prior estimates f_o, g_o, there exist an optimal number of basis functions and some optimal output layer parameter vectors, denoted by $\mathbf{w_f}^*$ and $\mathbf{w_g}^*$, such that $\forall \mathbf{x} \in \chi^-$, the inherent network approximation errors defined as:

$$\begin{aligned} \Delta_f &:= f^*(\mathbf{x}) - f(\mathbf{x}) \\ \Delta_g &:= g^*(\mathbf{x}) - g(\mathbf{x}), \end{aligned} \tag{4.4}$$

where $f^*(\mathbf{x}) := \hat{f}(\mathbf{x}, \mathbf{w_f^*})$ and $g^*(\mathbf{x}) := \hat{g}(\mathbf{x}, \mathbf{w_g^*})$ denote the optimal approximations to $f(\mathbf{x})$ and $g(\mathbf{x})$ respectively, satisfy the given bounds, i.e. $|\Delta_f| \leq \epsilon_f$ and $|\Delta_g| \leq \epsilon_g$. Outside the network approximation region (i.e. $\mathbf{x} \notin \chi$), $\kappa(\mathbf{x}) = 0$ and so the approximation errors are given by $\Delta_f = (f_o(\mathbf{x}) - f(\mathbf{x}))$ and $\Delta_g = (g_o(\mathbf{x}) - g(\mathbf{x}))$. If it is assumed that these errors are bounded by known values $\overline{f}_o$ and $\overline{g}_o$ respectively, then the following optimal network approximation error bounds are assumed to hold for all $\mathbf{x}$:

$$\begin{aligned} |\Delta_f| &\leq \kappa(\mathbf{x})\epsilon_f + (1 - \kappa(\mathbf{x}))\overline{f}_o \\ |\Delta_g| &\leq \kappa(\mathbf{x})\epsilon_g + (1 - \kappa(\mathbf{x}))\overline{g}_o. \end{aligned} \tag{4.5}$$

Note that these bounds are exact for $\mathbf{x} \in \chi^-$ and $\mathbf{x} \notin \chi$. In between (i.e. $\mathbf{x}$ lying between χ^- and χ), a smooth interpolation between the two bounds is assumed. Knowledge of these bounds is crucial to make the system robust against the effect of the network approximation errors.

In this scheme, as in the previous chapter, the Gaussian basis function parameters are assumed to be known *a priori*. This means that the network approximation region χ and the basis function centres and width parameters required to satisfy any desired ϵ_f, ϵ_g must be determined from prior information available about the system. Hence the output layer parameter vectors $\hat{\mathbf{w}}_\mathbf{f}$, $\hat{\mathbf{w}}_\mathbf{g}$ are the only unknown parameters that require on-line estimation. In the composite adaptive method described in this chapter, adjustment of $\hat{\mathbf{w}}_f$, $\hat{\mathbf{w}}_g$ depends upon *both* the tracking error $e = (y - y_d)$ and an estimation error whose construction will be explained later on.

Using equations (4.2), where to simplify the notation we now write $\hat{f}(\mathbf{x}, \hat{\mathbf{w}}_\mathbf{f})$ as $\hat{f}(\mathbf{x})$ and the same for $\hat{g}$, together with the definitions of the optimal network approximation errors Δ_f and Δ_g in equations (4.4), we obtain that $\forall \mathbf{x}$:

$$\begin{aligned}\hat{f}(\mathbf{x}) - f(\mathbf{x}) &= \kappa \tilde{\mathbf{w}}_{\mathbf{f}}^T \Phi_{\mathbf{f}} + \Delta_f \\ \hat{g}(\mathbf{x}) - g(\mathbf{x}) &= \kappa \tilde{\mathbf{w}}_{\mathbf{g}}^T \Phi_{\mathbf{g}} + \Delta_g,\end{aligned} \qquad (4.6)$$

where $\tilde{\mathbf{w}}_{\mathbf{f}} := (\hat{\mathbf{w}}_{\mathbf{f}} - \mathbf{w}_{\mathbf{f}}^*)$ and $\tilde{\mathbf{w}}_{\mathbf{g}} := (\hat{\mathbf{w}}_{\mathbf{g}} - \mathbf{w}_{\mathbf{g}}^*)$ represent the parameter errors.

4.4 The Control Law

As shown in Theorem 2.2.1, for the case of known system dynamics, the feedback linearizing control law of equation (2.6) leads to stable asymptotic tracking. However this control law depends on the system functions $f(\mathbf{x})$, $g(\mathbf{x})$, which in this case are unknown. These will therefore be replaced by the neural network approximations $\hat{f}(\mathbf{x})$, $\hat{g}(\mathbf{x})$, giving rise to the certainty equivalent control law

$$u_{fl} = \frac{-\hat{f}(\mathbf{x}) + v(t)}{\hat{g}(\mathbf{x})}, \qquad (4.7)$$

where $v(t) = y_d^{(r)} - \alpha_r e^{(r-1)} - \cdots - \alpha_1 e$ has the same definition as in the feedback linearizing Equation (2.6), with coefficients α_i chosen so as to form a Hurwitz polynomial $\Gamma(s) = s^r + \alpha_r s^{r-1} + \cdots + \alpha_1$, where s denotes the familiar Laplace variable.

The approximation of the neural networks however, is limited to within χ only. Hence, as in [219], the above control is augmented with an additional sliding mode component u_{sl} to take over control whenever the state $\mathbf{x}$ moves outside χ. In addition, as proposed in [254], a low-gain version of this sliding mode component is also maintained whilst $\mathbf{x} \in \chi$ because it provides robustness to the network inherent approximation error. The exact equation for u_{sl} will be derived later on. The final control law thus has the form

$$u(t) = u_{fl}(t) + u_{sl}(t). \qquad (4.8)$$

Substituting Equations (4.7) and (4.8) in the system Equation (4.1) and using Equations (4.6), the following error dynamics equation is derived:

$$e = \Gamma^{-1}[-\kappa \tilde{\mathbf{w}}_{\mathbf{f}}^T \Phi_{\mathbf{f}} - \kappa \tilde{\mathbf{w}}_{\mathbf{g}}^T \Phi_{\mathbf{g}} u_{fl} + g u_{sl} + d], \qquad (4.9)$$

where $d := -\Delta_f - \Delta_g u_{fl}$. Note that in this equation we have utilized the convenient notation of mixing Laplace and time variables, as typically done in adaptive control theory. This equation clearly shows that the non-zero inherent network approximation errors Δ_f and Δ_g give rise to an additive disturbance term d affecting the error dynamics.

An essential condition for the design of *stable* direct adaptive control schemes involves the derivation of an error dynamics equation whose transfer function is strictly positive real (SPR) [23, 186]. Very often the transfer function of the error dynamics equation based on the tracking error, namely Γ^{-1} of Equation (4.9) in our case, is not SPR. Hence the tracking error e is filtered so as to obtain a signal e_1 such that the transfer function between e_1 and the right hand side of Equation (4.9) is SPR. Since the composite adaptive control scheme developed in this chapter includes also a direct adaptive component, the derivation of this filtered tracking error is important. From Assumptions 3.2.5 and 3.2.7, the output error $e(t)$ and its $(r-1)$ derivatives could be calculated. These could therefore be used to generate the filtered tracking error $e_1(t)$ as follows,

$$e_1(t) = \beta_r e^{(r-1)} + \cdots + \beta_1 e.$$

Note that in the s-domain, the above expression for e_1 could be expressed as

$$e_1(s) = \Psi(s) e(s), \qquad (4.10)$$

where $\Psi(s) = \beta_r s^{r-1} + \cdots + \beta_1$. The coefficients β_i are chosen such that $\Psi(s)$ is Hurwitz and that the transfer function $\Psi(s)\Gamma^{-1}(s)$ is SPR. In particular, if $\Psi(s)\Gamma^{-1}(s)$ is chosen to have the first-order lag form

$$\Psi \Gamma^{-1} = \frac{1}{s + k_d}, \qquad (4.11)$$

where $k_d > 0$, then $\Psi \Gamma^{-1}$ will be SPR as desired. Using this first-order lag equation, together with Equations (4.9) and (4.10), the dynamics of $e_1(t)$ could be expressed in the following form

$$\dot{e}_1 = -k_d e_1 - \kappa \tilde{\mathbf{w}}_f^T \Phi_f - \kappa \tilde{\mathbf{w}}_g^T \Phi_g u_{fl} + g u_{sl} + d. \qquad (4.12)$$

This equation will become useful in the stability analysis to be presented later on.

4.5 Composite Adaptation

In this section we develop a composite-type adaptation law for adjustment of the networks' parameter vectors. Since composite adaptation makes use not only of the tracking error but also of an estimation error, a suitably parameterized identification model for the plant must be set up. The identification model, which utilizes the network approximations, is developed in the following sub-section.

4.5.1 The Identification Model

When setting up an identification model for the plant, two conditions must be satisfied:

- *The linear parameterization condition*: This implies that ideally, a linear relation between the unknown parameters and some measurable signal must be sought, so as to use standard estimation techniques like gradient descent or least-squares.

- *The no-differentiation condition*: This means that one must not resort to differentiation of any measurable signals when formulating the above relation, because in practice differentiation is problematic. Very often, this situation can be resolved by suitable filtration.

For our case, Equations (4.1), (4.2) and (4.4) lead to a linearly parameterized relation in terms of the optimal output layer weights, as follows:

$$y^{(r)} = \kappa(\mathbf{x})\Phi_f^T(\mathbf{x})\mathbf{w}_f^* + f_o + (\kappa(\mathbf{x})\Phi_g^T(\mathbf{x})\mathbf{w}_g^* + g_o)u(t) - \Delta_f - \Delta_g u(t).$$

This equation, although linearly parameterized, does not satisfy the no-differentiation condition because of the $y^{(r)}$ term. However, by filtering with a network of transfer function $A_o^{-1}(s)$, where $A_o(s) = a_o + a_1 s + \ldots + a_{r-1}s^{r-1} + s^r$ is a Hurwitz and monic polynomial reflecting stable filtration, the following dynamics are obtained

$$y\left(\frac{s^r}{A_o}\right) = \frac{\kappa\Phi_f^T(\mathbf{x})\mathbf{w}_f^* + \kappa\Phi_g^T(\mathbf{x})\mathbf{w}_g^* u(t)}{A_o} + \frac{f_o + g_o u(t)}{A_o} - \frac{\Delta_f + \Delta_g u(t)}{A_o},$$

which can be re-written in linear form as

$$Y = \mathbf{p}^T \mathbf{w}^* + c + d_f \tag{4.13}$$

where

$$Y = y\left(\frac{s^r}{A_o}\right)$$

$$c = \frac{f_o + g_o u(t)}{A_o}$$

$$d_f = -\frac{\Delta_f + \Delta_g u(t)}{A_o}$$

$$\mathbf{p}^T = \left[\frac{\kappa\Phi_f^T(\mathbf{x})}{A_o} \quad \frac{\kappa\Phi_g^T(\mathbf{x})u}{A_o}\right]$$

$$\mathbf{w}^{*T} = \left[\mathbf{w}_f^{*T} \quad \mathbf{w}_g^{*T}\right].$$

Note that signal Y could be generated from the plant output y by suitable filtration without the use of differentiation, as shown in Figure 4.1. The term d_f depends on the disturbance due to the network inherent approximation errors, which is unknown except for its bounding values. Additionally, the optimal output layer parameter vector $\mathbf{w}^*$ is also unknown. Hence we propose an identification model where term d_f is not included and $\mathbf{w}^*$ is replaced by $\hat{\mathbf{w}} := \begin{bmatrix}\hat{\mathbf{w}}_f^T & \hat{\mathbf{w}}_g^T\end{bmatrix}^T$ (*i.e.* the network adjustable parameters), as follows:

$$\hat{Y} = \mathbf{p}^T \hat{\mathbf{w}} + c. \tag{4.14}$$

The estimation error, defined as $\epsilon := \hat{Y} - Y$, therefore satisfies

$$\epsilon = \mathbf{p}^T \tilde{\mathbf{w}} - d_f, \tag{4.15}$$

where $\tilde{\mathbf{w}} = \hat{\mathbf{w}} - \mathbf{w}^*$ reflects the error between the optimal network parameters and the actual parameter estimates. The estimation error is important because it is used in the indirect component of the composite adaptation law being proposed. The block diagram of Figure 4.1 illustrates how $\hat{Y}$ and ϵ are generated, and how the $A_o^{-1}(s)$ filters are implemented.

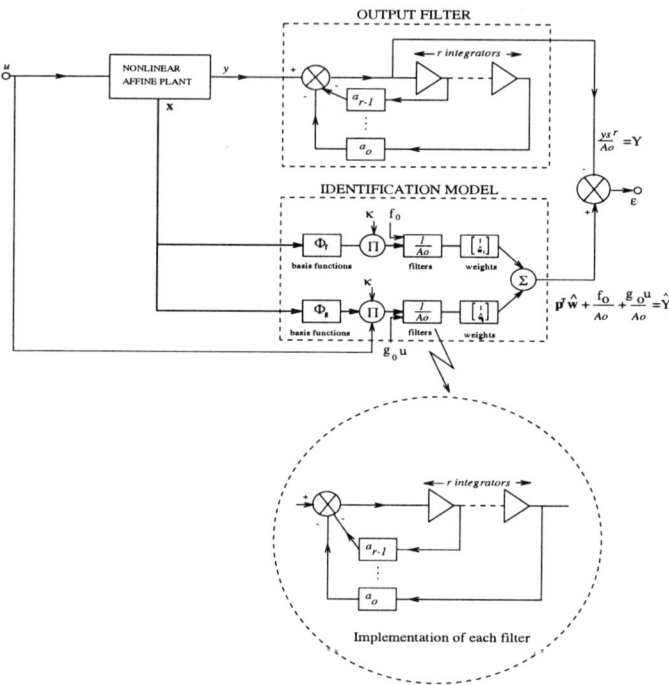

Fig. 4.1. Generation of the estimation error.

4.5.2 The Adaptation Law

Inspired from the composite adaptation idea of [233], but modified for the present case of adaptive control by neural networks that also includes a disturbance term d, we propose the following adaptation law:

$$\dot{\hat{\mathbf{w}}}_f = \Gamma_f [\kappa(\mathbf{x}) \Phi_f(\mathbf{x}) e_1(t) - k_e \mathbf{p}_f(\mathbf{x}) \epsilon_\Delta(t)]$$
$$\dot{\hat{\mathbf{w}}}_g = \Gamma_g [\kappa(\mathbf{x}) \Phi_g(\mathbf{x}) u_{fl}(t) e_1(t) - k_e \mathbf{p}_g(\mathbf{x}) \epsilon_\Delta(t)] \tag{4.16}$$

where

- $\epsilon_\Delta = \begin{cases} \epsilon & \text{if } |\epsilon| > \overline{d_f} \\ 0 & \text{otherwise} \end{cases}$ represents a dead-zone function on $\epsilon(t)$, with $\overline{d_f}$ defining a bound on d_f, such that $|d_f| \leq \overline{d_f}$.

- k_e is a positive constant that determines the extent to which the indirect component shall be used in the composite adaptation law. This is a design parameter.
- Γ_f, Γ_g are constant positive definite, symmetric gain matrices for the $\hat{\mathbf{w}}_f$, $\hat{\mathbf{w}}_g$ parameter update laws respectively.
- $\mathbf{p}_f = \kappa \Phi_f(\mathbf{x})/A_o$ and $\mathbf{p}_g = \kappa \Phi_g(\mathbf{x}) u(t)/A_o$. This means that the previously-defined vector $\mathbf{p}$ could be decomposed as $\mathbf{p}^T = [\mathbf{p}_f^T \quad \mathbf{p}_g^T]$.

Note the *composite* nature of adaptation law (4.16): it driven by *both* the filtered tracking error $e_1(t)$ (the direct component), as well as the estimation error $\epsilon(t)$ (the indirect component). The schemes of [219] and [254] utilize only the direct component and represent a special case of our more general adaptation law, corresponding to $k_e = 0$. Another generalization involves the use of adaptation gain *matrices* Γ_f, Γ_g rather than scalar gains that are the same for all components of the parameter vectors. Note that for the indirect component, dead-zone adaptation has been used to ensure signal boundedness despite the presence of the disturbance term d_f affecting the estimation error Equation (4.15).

To maintain the control term $u_{fl}(t)$ in Equation (4.7) bounded away from $\pm\infty$, the network approximation $\hat{g}(\mathbf{x})$ must be kept bounded away from zero. As in [255], this condition is maintained by including a parameter resetting mechanism for the $\hat{\mathbf{w}}_g$ vector. Whilst $\mathbf{x} \in \chi$, the parameter resetting mechanism involves setting

$$\hat{\mathbf{w}}_g(t^+) = \hat{\mathbf{w}}_g(t) + (g_l(\mathbf{x}) - \hat{g}_n(\mathbf{x}))\Gamma_g \Phi_g(\mathbf{x})/\Phi_g^T(\mathbf{x})\Gamma_g \Phi_g(\mathbf{x}) \quad (4.17)$$

at any time t whenever the following condition is detected

$$\hat{g}(\mathbf{x}, \hat{\mathbf{w}}_g(t)) \leq g_l(\mathbf{x}) - \delta^*(\mathbf{x}), \quad \forall \mathbf{x} \in \chi.$$

This resetting mechanism is a modified version of the one specified in [255] because it allows for the use of gain matrices, and not only scalar gains. In the above equations,

- $\delta^*(\mathbf{x})$ is a small positive function satisfying $g_l(\mathbf{x}) - \delta^*(\mathbf{x}) > 0$, $\forall \mathbf{x} \in \chi$.
- $\hat{g}_n(\mathbf{x}) := \hat{\mathbf{w}}_g^T(t)\Phi_g(\mathbf{x}) + g_o(\mathbf{x})$.
- t^+ denotes the time just after the resetting mechanism is activated.

Recall that by condition (4.3), $g_o(\mathbf{x}) \geq g_l(\mathbf{x})$ $\forall \mathbf{x}$. From Equations (4.2) and (4.17), this resetting mechanism yields that just *after* a parameter reset $\hat{g}(\mathbf{x})$ becomes,

$$\begin{aligned}
\hat{g}(\mathbf{x}, \hat{\mathbf{w}}_g(t^+)) &= \kappa(\mathbf{x})\Phi_g^T(\mathbf{x})\hat{\mathbf{w}}_g(t) + g_o(\mathbf{x}) + \kappa(\mathbf{x})(g_l(\mathbf{x}) - \hat{g}_n(\mathbf{x})) \\
&= \kappa(\mathbf{x})(g_l(\mathbf{x}) - g_o(\mathbf{x})) + g_o \\
&= (1 - \kappa(\mathbf{x}))g_o(\mathbf{x}) + \kappa(\mathbf{x})g_l(\mathbf{x}) \\
&\geq g_l(\mathbf{x}), \quad \forall \mathbf{x} \in \chi.
\end{aligned}$$

The last equation follows because (a) $\kappa = 1$ when $\mathbf{x} \in \chi^-$ and (b) $0 < \kappa < 1$ and $g_o \geq g_l > 0$ when $\mathbf{x}$ lies between χ^- and χ. Hence $\hat{g}(\mathbf{x})$ is bounded below by $g_l(\mathbf{x}) - \delta^*(\mathbf{x})$ $\forall \mathbf{x} \in \chi$, as a consequence of parameter resetting.

Outside χ, $\hat{g}(\mathbf{x}) = g_o(\mathbf{x})$ by definition, which is automatically bounded away from zero by the condition $g_o(\mathbf{x}) \geq g_l(\mathbf{x})$ $\forall \mathbf{x}$.

4.6 Stability Analysis

In this section we show how a suitable selection of the sliding mode component $u_{sl}(t)$ in control law (4.8), the value of the bound $\overline{d}_f$, and the use of adaptation law (4.16), results in a stable closed-loop system that guarantees asymptotic tracking of the output. Consider the Lyapunov function candidate

$$V(t) = \frac{1}{2}\left(e_1^2 + \tilde{\mathbf{w}}_f^T \Gamma_f^{-1} \tilde{\mathbf{w}}_f + \tilde{\mathbf{w}}_g^T \Gamma_g^{-1} \tilde{\mathbf{w}}_g\right). \tag{4.18}$$

Differentiating with respect to time and using Equation (4.12) yields

$$\dot{V} = -k_d e_1^2 - \kappa e_1 \Phi_f^T \tilde{\mathbf{w}}_f - \kappa e_1 \Phi_g^T \tilde{\mathbf{w}}_g u_{fl} + e_1 g(\mathbf{x}) u_{sl} + e_1 d + $$
$$\frac{1}{2}\left\{\tilde{\mathbf{w}}_f^T \Gamma_f^{-1} \dot{\tilde{\mathbf{w}}}_f + \dot{\tilde{\mathbf{w}}}_f^T \Gamma_f^{-1} \tilde{\mathbf{w}}_f + \tilde{\mathbf{w}}_g^T \Gamma_g^{-1} \dot{\tilde{\mathbf{w}}}_g + \dot{\tilde{\mathbf{w}}}_g^T \Gamma_g^{-1} \tilde{\mathbf{w}}_g\right\}.$$

Substituting adaptation law (4.16) results in

$$\dot{V} = -k_d e_1^2 + e_1(g u_{sl} + d) - k_e \mathbf{p}_f^T \tilde{\mathbf{w}}_f \epsilon_\Delta - k_e \mathbf{p}_g^T \tilde{\mathbf{w}}_g \epsilon_\Delta$$
$$= -k_d e_1^2 + e_1(g u_{sl} + d) - k_e \mathbf{p}^T \tilde{\mathbf{w}} \epsilon_\Delta, \tag{4.19}$$

where the definitions of $\mathbf{p}$ and $\tilde{\mathbf{w}}$ have been used.

If the sliding mode component of the control law is set to

$$u_{sl}(t) = -\frac{\overline{d}}{g_l(\mathbf{x})}\text{sign}(e_1), \tag{4.20}$$

where $\overline{d}$ is a bound on the disturbance term d satisfying $|d| \leq \overline{d}$, then $e_1(g u_{sl} + d) \leq 0$ and so from Equation (4.19) it follows that

$$\dot{V} \leq -k_d e_1^2 - k_e \mathbf{p}^T \tilde{\mathbf{w}} \epsilon_\Delta. \tag{4.21}$$

This helps us obtain some results on the stability of the system. Before doing so however, the following lemma is required.

Lemma 4.6.1. *For any real x, b, $\overline{b} \in \Re$, where $\overline{b} > |b|$ represents a bound on the magnitude of b,*

$$|x - b| > \overline{b} \Rightarrow x(x - b) \geq 0$$

Proof.

$$|x - b| > \bar{b} \Rightarrow |x - b| > |b|$$
Hence $\quad |x - b||b - x| > |b||b - x|$
$$\Rightarrow |x - b|^2 > |b(b - x)|$$
$$\Rightarrow (x - b)^2 \geq b(b - x)$$
$$\Rightarrow x(x - b) \geq 0.$$
$\square$

We are now in a position to prove the following stability theorem.

Theorem 4.6.1. *Given that Assumptions 3.2.1 to 3.2.7 hold, and that system (4.1) is subjected to control law (4.8), with u_{fl} and u_{sl} given by Equations (4.7) and (4.20) respectively, adaptation law (4.16) and parameter resetting mechanism (4.17); then the state vector $\mathbf{x}$ and the control input $u(t)$ remain bounded and the tracking error $e(t)$ converges asymptotically to zero.*

Proof. Using the definition of ϵ_Δ in Equation (4.21) yields,

$$\dot{V} \leq \begin{cases} -k_d e_1^2 - k_e \mathbf{p}^T \tilde{\mathbf{w}} \epsilon & \text{if } |\epsilon| > \overline{d_f} \\ -k_d e_1^2 & \text{otherwise} \end{cases}$$

$$= \begin{cases} -k_d e_1^2 - k_e \mathbf{p}^T \tilde{\mathbf{w}} (\mathbf{p}^T \tilde{\mathbf{w}} - d_f) & \text{if } |\mathbf{p}^T \tilde{\mathbf{w}} - d_f| > \overline{d_f} \\ -k_d e_1^2 & \text{otherwise} \end{cases}$$

by Equation (4.15). The above, and use of Lemma 4.6.1 with $x, b, \bar{b}$ substituted by $\mathbf{p}^T \tilde{\mathbf{w}}, d_f, \overline{d_f}$ respectively, imply that

$$\dot{V} \leq -k_d e_1^2 \leq 0.$$

In the event of parameter resets from Equation (4.17), straightforward but rather lengthy manipulations reveal that following a reset, the change in V of Equation (4.18) is

$$\Delta V = \frac{-(g_l(\mathbf{x}) - \hat{g}_n(\mathbf{x}))^2 + 2(g_l(\mathbf{x}) - g_n^*(\mathbf{x}))(g_l(\mathbf{x}) - \hat{g}_n(\mathbf{x}))}{2\Phi_\mathbf{g}^T(\mathbf{x})\Gamma_\mathbf{g}\Phi_\mathbf{g}^T(\mathbf{x})}, \quad (4.22)$$

where $g_n^* := \mathbf{w}^{*T}_\mathbf{g} \Phi_\mathbf{g}(\mathbf{x}) + g_o(\mathbf{x})$. When the condition for a reset is detected, $\hat{g}(\mathbf{x}, \hat{\mathbf{w}}_\mathbf{g}(t)) \leq g_l(\mathbf{x}) - \delta^*(\mathbf{x})$. By definition of $\hat{g}$, this implies that

$$\kappa(\mathbf{x})\Phi_\mathbf{g}^T(\mathbf{x})\hat{\mathbf{w}}_\mathbf{g}(t) + g_o(\mathbf{x}) < g_l(\mathbf{x}) - \delta^*(\mathbf{x}) < g_l(\mathbf{x})$$
$$\Rightarrow \kappa(\mathbf{x})\Phi_\mathbf{g}^T(\mathbf{x})\hat{\mathbf{w}}_\mathbf{g}(t) < g_l(\mathbf{x}) - g_o(\mathbf{x})$$
$$\leq 0 \text{ by condition (4.3).}$$

Using this, together with the fact that in χ (where parameter resetting can occur) $0 < \kappa(\mathbf{x}) \leq 1$, it follows that during a reset

$$\Phi_\mathbf{g}^T(\mathbf{x})\hat{\mathbf{w}}_\mathbf{g}(t) \leq \kappa(\mathbf{x})\Phi_\mathbf{g}^T(\mathbf{x})\hat{\mathbf{w}}_\mathbf{g}(t).$$

Therefore we may say that,

$$\Phi_{\mathbf{g}}^T(\mathbf{x})\hat{\mathbf{w}}_{\mathbf{g}}(t) + g_o(\mathbf{x}) \leq \kappa(\mathbf{x})\Phi_{\mathbf{g}}^T(\mathbf{x})\hat{\mathbf{w}}_{\mathbf{g}}(t) + g_o(\mathbf{x})$$
$$\leq g_l(\mathbf{x}) - \delta^*(\mathbf{x}) \text{ by the reset condition}$$
$$\Rightarrow \hat{g}_n(\mathbf{x}) \leq g_l(\mathbf{x}) - \delta^*(\mathbf{x})$$
$$\Rightarrow g_l(\mathbf{x}) - \hat{g}_n(\mathbf{x}) \geq 0.$$

Using this fact and restricting $g_l(\mathbf{x}) - g_n^*(\mathbf{x}) \leq 0$, Equation (4.22) yields that $\Delta V \leq 0$. Note that the last condition $g_l(\mathbf{x}) \leq g_n^*(\mathbf{x})$ places a further restriction on the value of $g_l(\mathbf{x})$ in addition to that of Assumption 3.2.6. In essence it is stating that $g_l(\mathbf{x})$ should lower bound not only the function $g(\mathbf{x})$, but also the optimal network approximation inside χ^-, which is given by $g_n^*(\mathbf{x})$.

Hence the Lyapunov function is reduced as a result of parameter resetting and its derivative is negative semidefinite, both of which ensure that $\tilde{\mathbf{w}}_{\mathbf{f}}$, $\tilde{\mathbf{w}}_{\mathbf{g}}$ and e_1 are bounded. In addition, the following consideration apply:

1. Boundedness of $\tilde{\mathbf{w}}_{\mathbf{f}}$, $\tilde{\mathbf{w}}_{\mathbf{g}}$ implies $\hat{\mathbf{w}}_{\mathbf{f}}$, $\hat{\mathbf{w}}_{\mathbf{g}}$ are bounded. Being Gaussians, $\Phi_{\mathbf{f}}(\mathbf{x})$, $\Phi_{\mathbf{g}}(\mathbf{x})$ are also bounded and so these two assertions imply that $\hat{f}$, $\hat{g}$ are bounded.
2. Boundedness of $e_1(t)$ implies that $e, e^{(1)}, \cdots, e^{(r-1)}$ are bounded since Ψ is Hurwitz. Together with Assumption 3.2.7, this implies that
   a) $v(t)$ is bounded,
   b) $y, y^{(1)}, \cdots, y^{(r-1)}$ are bounded.
   Assumptions 3.2.3, 3.2.4 and result 2(b) above imply that $\mathbf{x}$ is bounded.

   Use of parameter resetting and results 1 and 2(a) above imply that u_{fl} is bounded. Hence $d(t)$ is bounded, which implies that $\overline{d} < \infty$ and which in turn implies that u_{ol} is bounded. Hence control $u(t)$ is bounded, which also justifies the existence of bound $\overline{d_f} < \infty$.

   All these results mean that the terms on the right hand side of Equation (4.12) are bounded, and so $\dot{e}_1(t)$ is bounded.
3. Introduce
$$V_1(t) := V(0) - \int_0^t k_d e_1^2(\tau) d\tau.$$
   a) From this definition, note that $\dot{V}_1 = \dot{V}(0) - k_d e_1^2 \leq -k_d(e_1^2(0) + e_1^2) \leq 0$, since $\dot{V} \leq -k_d e_1^2$. In addition, $\dot{V} \leq -k_d e_1^2$ also implies that $\int_0^t -k_d e_1^2(\tau) d\tau \geq V(t) - V(0)$. Hence
   $$V_1(t) \geq V(t) \geq 0.$$
   Thus V_1 is bounded below and its derivative is negative semidefinite. These imply that V_1 converges to a finite limit.
   b) Since both e_1, $\dot{e}_1$ are bounded, then $de_1^2/dt = 2e_1\dot{e}_1$ is bounded and so e_1^2 is uniformly continuous, which implies that $\dot{V}_1$ is also *uniformly continuous*.

Hence, from Barbalat's Lemma [234], results 3(a) and 3(b) above imply that V_1 asymptotically converges to zero, and so

$$\lim_{t \to \infty} e_1(t) = 0.$$

Since $e(t)$ is a filtration of $e_1(t)$ by a stable and strictly proper transfer function $\Psi^{-1}(s)$, then the tracking error $e(t)$ converges asymptotically to zero as well, *i.e.*:

$$\lim_{t \to \infty} e(t) = 0.$$ □

4.7 Determination of the Disturbance Bounds

The sliding mode component of the control law requires knowledge of an upper bound $\bar{d}$ on the size of the disturbance term $d := -\Delta_f - \Delta_g u_{fl}$, appearing in the error dynamics Equation (4.12). From Equation (4.5) and the definition of d it follows that $\forall \mathbf{x}$:

$$|d| \leq \kappa(\mathbf{x})(\epsilon_f + \epsilon_g |u_{fl}|) + (1 - \kappa(\mathbf{x}))(\overline{f_o} + \overline{g_o}|u_{fl}|).$$

The right hand side of this equation is therefore a good evaluation for the bound on $|d|$ and so $\bar{d}$ is defined as,

$$\bar{d} := \kappa(\mathbf{x})(\epsilon_f + \epsilon_g |u_{fl}|) + (1 - \kappa(\mathbf{x}))(\overline{f_o} + \overline{g_o}|u_{fl}|).$$

Note that $\bar{d}$ affects the gain of the sliding mode component u_{sl}, which is typically small whilst $\mathbf{x} \in \chi^-$ ($\kappa = 1$), just enough to overcome the effect of the inherent network approximation errors ϵ_f, ϵ_g, and increases appreciably as $\mathbf{x}$ ventures outside the network approximation region ($\kappa(\mathbf{x}) = 0$), taking over control so as to pull $\mathbf{x}$ inside χ.

The dead-zone region of adaptation law (4.16) requires knowledge of another upper bound $\bar{d}_f$, characterizing the size of the disturbance term d_f appearing in the estimation error Equation (4.15). From the definitions of d_f, d and $u(t)$ it follows that

$$d_f = \frac{d - \Delta_g u_{sl}}{A_o}. \tag{4.23}$$

From Equation (4.5) and the definition of $u_{sl}(t)$ it follows that

$$|\Delta_g u_{sl}| \leq \frac{\bar{d}}{g_l(\mathbf{x})}[\kappa(\mathbf{x})\epsilon_g + (1 - \kappa(\mathbf{x}))\overline{g_o}]. \tag{4.24}$$

But

$$\begin{aligned}
|d - \Delta_g u_{sl}| &\leq |d| + |\Delta_g u_{sl}| \\
&\leq \bar{d} + |\Delta_g u_{sl}| \\
&\leq \bar{d} + \frac{\bar{d}}{g_l(\mathbf{x})}[\kappa(\mathbf{x})\epsilon_g + (1 - \kappa(\mathbf{x}))\overline{g_o}], \quad \text{by Equation (4.24)}.
\end{aligned}$$

This term therefore represents an upper bound on $|d - \Delta_g u_{sl}|$, which shall be denoted by $\bar{d}_1$, i.e.:

$$\bar{d}_1 := \bar{d} + \frac{\bar{d}}{g_l(\mathbf{x})}[\kappa(\mathbf{x})\epsilon_g + (1 - \kappa(\mathbf{x}))\bar{g}_o].$$

But Equation (4.23) shows that d_f is a stable filtration of $(d - \Delta_g u_{sl})$ via $A_o^{-1}(s)$. Hence if $A_o(s)$ is chosen to have the form $A_o(s) = (s + \lambda)^r$, where $\lambda > 0$, then as shown in [234],

$$|d_f^{(i)}| \leq 2^i \lambda^{i-r} \bar{d}_1 \text{ for } 0 \leq i \leq r.$$

Hence for our case of interest, $i = 0$,

$$|d_f| \leq \lambda^{-r} \bar{d}_1 := \bar{d}_f.$$

This defines the required bound $\bar{d}_f$ in terms of known parameters.

4.8 Simulation Examples

4.8.1 Example 1

The first example is based on the following nonlinear affine system [254]:

$$\dot{x} = \sin(x) + 0.5\cos(3x) + u$$
$$y = x$$

where $g(\mathbf{x}) = 1$ is assumed known, and $f(\mathbf{x}) = \sin(x) + 0.5\cos(3x)$ represents the unknown dynamics. The system is of order $n = 1$ and degree $r = 1$. The desired output y_d is a unity amplitude, $0.1Hz$ square wave filtered by a network of transfer function $1/(s+1)$. The network approximation region is chosen as $\chi = [-1.7, 1.7]$ and $\chi^- = [-1.1, 1.1]$. It is assumed that no prior estimate to $f(\mathbf{x})$ is known, so that $f_o = 0$. As shown in [254], for this case, an optimal network approximation error bound $\epsilon_f = 0.02$ is possible when using radial basis functions having $\sigma = 0.06$ placed on a mesh of spacing 0.05 inside χ. This represents a network of 69 basis functions. Assuming that it is known that $f(x)$ is bounded by 1.5, then since $f_o = 0$ it follows that $\bar{f}_o = 1.5$. The choice of $\Gamma(s) = (s+1)$ results in $v(t) = \dot{y}(t) - (y - y_d)$. Hence, $\Psi(s)$ could be set to unity because this way $\Psi\Gamma^{-1}$ is a first order lag transfer function corresponding to $k_d = 1$ in Equation (4.11). The identification model filter characteristic equation is chosen as $A_o = (s+5)$, so that $\lambda = 5$.

Four different trials were carried out in this experiment. Figure 4.2 shows the results of the trial 1, where the gain matrix $\Gamma_f = 5\mathbf{I}$ ($\mathbf{I}$ denotes the identity matrix) and $k_e = 0$. This trial therefore corresponds to the direct adaptive case of [254]. Note that the output y (solid) follows the reference input y_d (dashed) with the tracking error converging asymptotically to zero. Figure 4.3 (a) shows the actual function $f(\mathbf{x})$ (solid) plotted together with the network approximation $\hat{f}(\mathbf{x})$ (dashed) using the parameters obtained after 50 time units of simulation. Figure 4.3 (b) shows that the corresponding function approximation error (*i.e.* the difference between $f(\mathbf{x})$ and $\hat{f}(\mathbf{x})$), lies within the range ± 0.25 for the values of $\mathbf{x}$ inside χ^-.

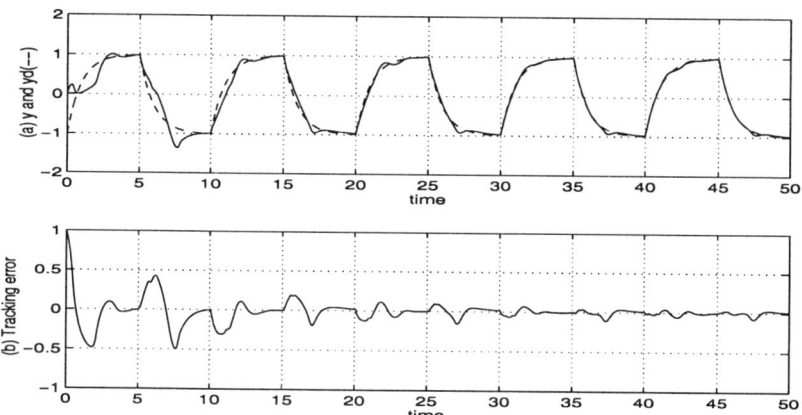

Fig. 4.2. Trial 1 of Example 1; (a) System output (b) Tracking error with direct adaptation.

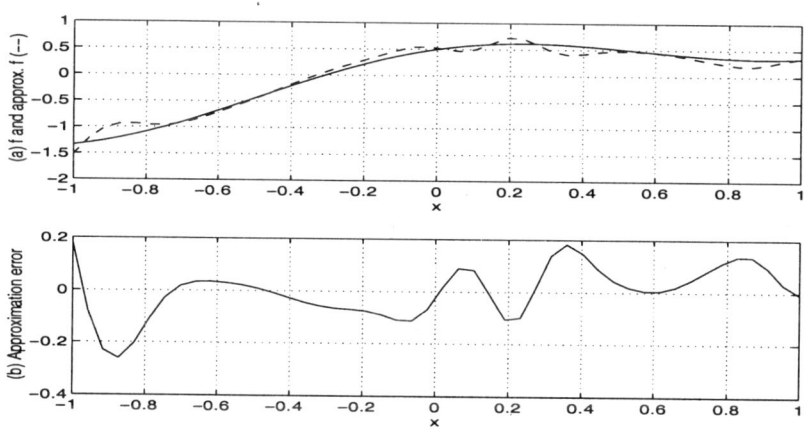

Fig. 4.3. Trial 1 of Example 1; (a) Function approximation (b) Approximation error with direct adaptation.

94   4. Composite Adaptive Control of Continuous-Time Systems

Figures 4.4 and 4.5 show the results of trial 2, where the gain matrix $\Gamma_f = 5\mathbf{I}$ as before, but $k_e = 10$. This corresponds to the use of a composite adaptive law. Note that the system is also stable but the tracking error converges to zero much faster than for the direct adaptive case of trial 1. In fact, after 50 time units, the tracking error of the composite adaptive scheme converges to within ± 0.01 as opposed to ± 0.08 for the direct adaptive scheme. This represents a reduction of 87.5% in tracking error. This improved transient performance is attributed to the faster parameter convergence of the composite adaptive scheme. In fact, as seen in Figure 4.5, the network approximation after 50 time units is superior to that of the first trial, with the approximation error being well within ± 0.1 for most values of $\mathbf{x}$ inside χ^-. This represents a reduction of 60%.

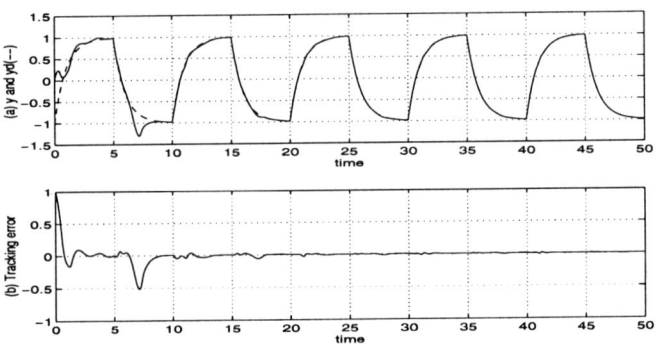

Fig. 4.4. Trial 2 of Example 1; (a) System output (b) Tracking error with composite adaptation.

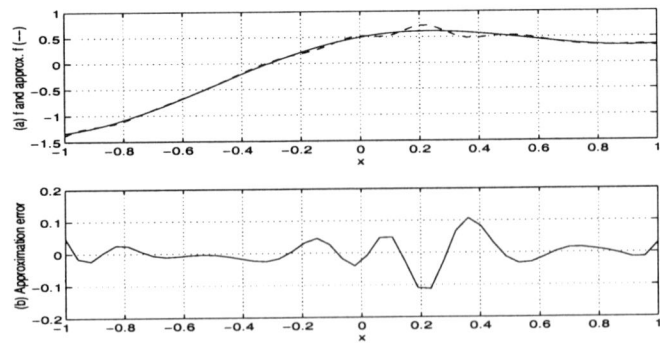

Fig. 4.5. Trial 2 of Example 1; (a) Function approximation (b) Approximation error with composite adaptation.

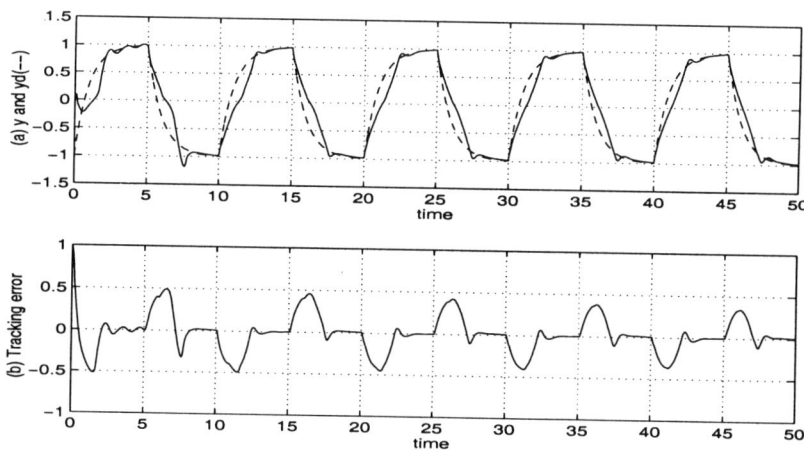

Fig. 4.6. Trial 3 of Example 1; (a) System output (b) Tracking error with *high gain* direct adaptation.

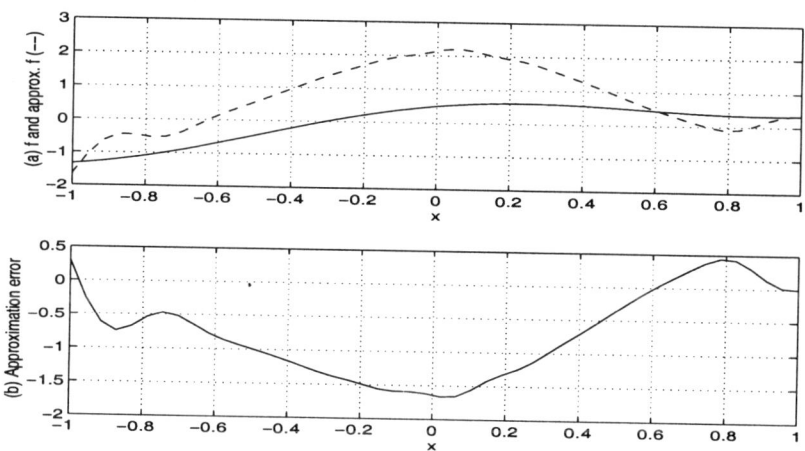

Fig. 4.7. Trial 3 of Example 1; (a) Function approximation (b) Approximation error with *high gain* direct adaptation.

Figure 4.6 shows the results of trial 3. This trial is an attempt to match the performance of the direct adaptive scheme of trial 1 with that of the composite adaptive scheme of trial 2, by simply increasing the gain matrix Γ_f of the direct adaptive scheme from $5\mathbf{I}$ to $15\mathbf{I}$. Contrary to what might be expected, Figure 4.6 shows that the system exhibits worse tracking performance than in trial 1 because too high an adaptation gain leads to worse function approximation, as seen in Figure 4.7. This phenomenon is well-known in adaptive control and is attributed to the fact that large adaptation gains tend to

96   4. Composite Adaptive Control of Continuous-Time Systems

encourage an oscillatory behaviour in the estimated parameters, leading to slower convergence [234]. This is confirmed in Figure 4.8 which shows the convergence of one arbitrarily chosen parameter as being very oscillatory and slow.

The benefits of composite adaptation are definitely seen in the fourth trial, where the indirect adaptive component is introduced for the same gain matrix as in trial 3, by setting $k_e = 10$. As shown in Figures 4.9 and 4.10, composite adaptation restores a very good transient response and function approximation. This shows that with composite adaptation, the use of higher adaptation gains is possible because parameter oscillations tend to be smoothed out. Hence composite adaptation not only improves the speed of convergence, but also dampens out undesirable oscillations during parameter adjustment [234]. This is confirmed in Figure 4.11, showing the variation of the same parameter plotted in Figure 4.8, but which is clearly non-oscillatory and faster to converge.

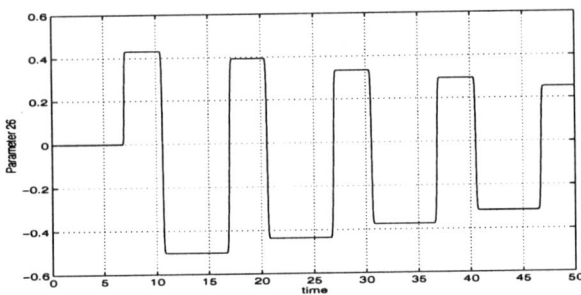

Fig. 4.8. Trial 3 of Example 1; time variation of an arbitrarily chosen network parameter with *high gain* direct adaptation.

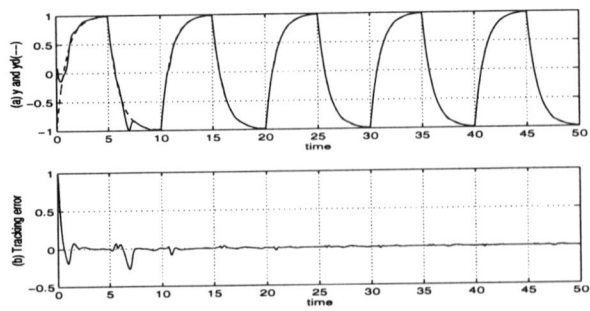

Fig. 4.9. Trial 4 of Example 1; (a) System output (b) Tracking error with *high gain* composite adaptation.

4.8 Simulation Examples   97

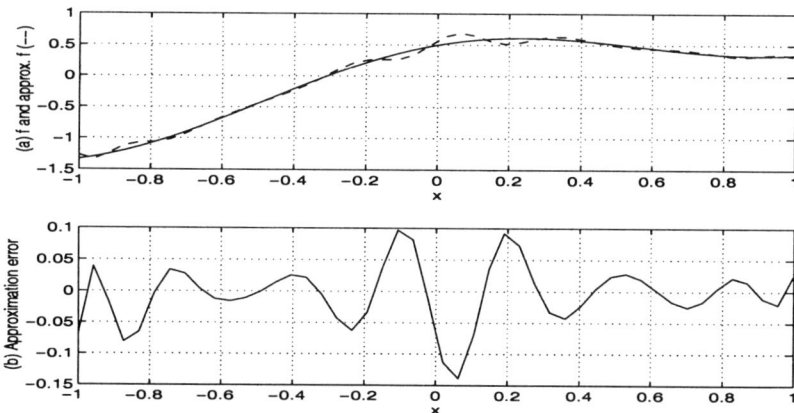

Fig. 4.10. Trial 4 of Example 1; (a) Function approximation (b) Approximation error with *high gain* composite adaptation.

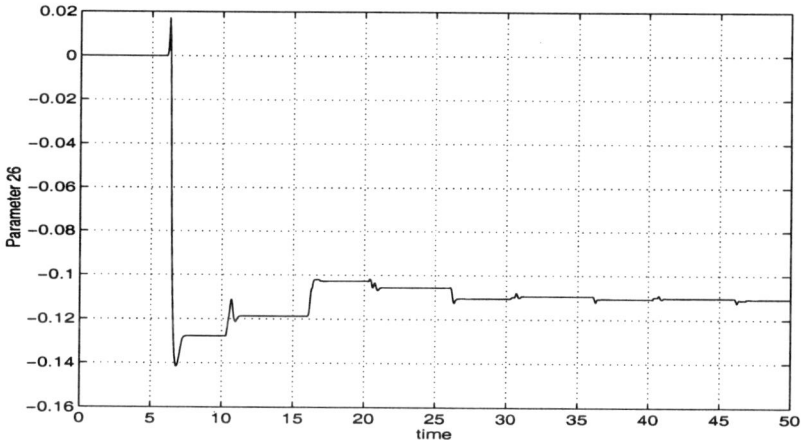

Fig. 4.11. Trial 4 of Example 1; time variation of an arbitrarily chosen network parameter with *high gain* composite adaptation.

4.8.2 Example 2

The plant used in the second example is taken from [255]. The plant model is given by the dynamics:

$$\dot{x}_1 = f(\mathbf{x}) + g(\mathbf{x})u$$
$$\dot{x}_2 = x_1 - x_2$$
$$y = x_1$$

where

$$\mathbf{x} = [x_1 \ x_2]^T$$
$$f(\mathbf{x}) = \cos(7(x_1^2 + x_2^2)) \exp(-(x_1^2 + x_2^2))$$
$$g(\mathbf{x}) = 2 + \cos(7x_1 x_2).$$

This system is of order $n = 2$ and has degree $r = 1$. The dynamics are already expressed in global normal form and the internal dynamics $\mathbf{I}(\mathbf{x}) = x_1 - x_2$ satisfy the appropriate Lipschitz conditions, because they have continuous and bounded partial derivatives in x_1, x_2. The zero dynamics are given by $\dot{x}_2 = -x_2$ which is globally exponentially stable. The desired output, y_d, is obtained by filtering a zero-average, 0.9 amplitude, 0.4 Hz square wave by a filter $1/(1 + 0.1s)^3$, so that the derivative of the filter output is bounded. For the given plant and reference input, the desired state is bounded well within the interval $[-1, 1] \times [-1, 1]$, along x_1, x_2 respectively. To cater for the fact that during the transient period the actual state may overshoot these bounds, the network approximation region is taken to be larger, namely $\chi = [-1.5, 1.5] \times [-1.5, 1.5]$. This ensures that the state is bounded within the approximation region, thus avoiding the use of crude, high-gain sliding control. χ^-, by definition, must be a subset of χ so that it is set to $\chi^- = [-1, 1] \times [-1, 1]$. It is assumed that the known prior estimates to the functions being approximated are $f_o = 0$, $g_o = 2$ and, as shown in [255], full network inherent approximation error bounds $\epsilon_f = \epsilon_g = 0.005$ are possible via basis functions having a standard deviation σ_f, σ_g of 0.03 located on a mesh of spacing 0.05 within χ. Note that in this example, the basis functions of the $\hat{f}$ and $\hat{g}$ networks are shared because they have the same parameters. These settings lead to neural networks of 3721 basis functions. Bounds $\bar{f}_o$, $\bar{g}_o$ are set to 2 and $g_l(\mathbf{x})$ to 0.895. The control law signal $v(t)$ is chosen as $v(t) = \dot{y}_d - 5(y - y_d)$ and the tracking error filter polynomial $\Psi(s) = 1$, leading to $e_1(t) \equiv e(t)$. The adaptation gain matrices $\Gamma_\mathbf{f}$, $\Gamma_\mathbf{g}$ were both set to $25\mathbf{I}$, and the identification model filter characteristic equation $A_o = (s + 5)$.

Two simulations were performed: one with $k_e = 0$, corresponding to direct adaptation, and another with $k_e = 15$, corresponding to composite adaptation. The results are shown in Figures 4.12(a) and 4.12(b) respectively. Once again, note that in both cases the system is stable and that the tracking error converges asymptotically to zero. However the composite law gives better performance, with the error converging much more rapidly to zero. In fact, between 90 and 100 time units, the steady-state tracking error for the composite law is well within ±0.004. This represents a reduction of 75% when compared with the direct law, as shown in Figure 4.13.

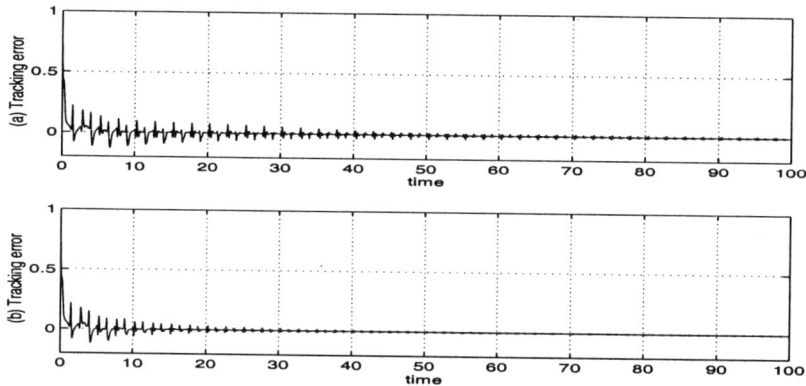

Fig. 4.12. Results of Example 2; (a) Direct adaptation (b) Composite adaptation

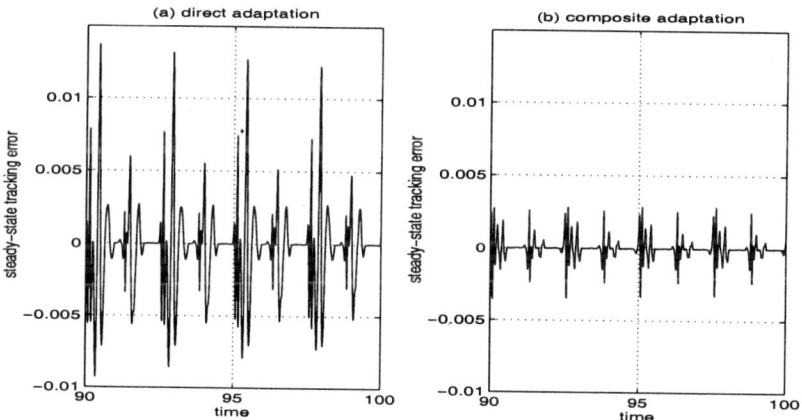

Fig. 4.13. Steady-state tracking errors of Example 2; (a) Direct adaptation (b) Composite adaptation

4.9 Summary

This chapter has addressed the important issue of transient performance improvement for functional adaptive control of a class of continuous-time, uncertain nonlinear systems. The problem has been tackled by deriving a novel neural network adaptation law that utilizes both the tracking and estimation errors for parameter adjustment. This represents a generalization of the composite adaptive concept originally developed for parametric uncertain systems [233]. The proposed scheme shows how the benefits of composite adaptation could also be extended to the case of neural network adaptive control of functional uncertain systems.

The scheme guarantees closed-loop stability and tracking error convergence by Lyapunov techniques. The presence of the inherent neural network approximation error, which in general might lead to instability, is handled by including a dead-zone function in the adaptation laws and a low-gain sliding mode component in the control.

When compared with the direct adaptive scheme developed in [254, 255], the composite scheme presented here shows an improved transient response with faster tracking error and parameter convergence. In addition, the composite adaptation law permits the inclusion of large adaptation gains without triggering undesirable parameter oscillations. Hence the speed of response of the composite adaptive scheme could be improved to much better levels than the corresponding direct adaptive scheme, whose transient response suffers as a result of the parameter oscillations.

The importance of this work to intelligent control stems from the fact that complex nonlinear systems typically require a neural network consisting of a large amount of parameters. This is particularly true for localized approximation methods such as GaRBF neural networks, and it often leads to a slow transient response. The functional adaptive composite scheme presented here, therefore represents an effective way of improving the situation, whilst guaranteeing overall system stability.

5. Functional Adaptive Control of Discrete-Time Systems

5.1 Introduction

In general, Lyapunov-based adaptive designs for discrete-time systems are not a straightforward translation of their continuous-time counterpart [138]. Given a linearly parameterized model, continuous-time systems yield a Lyapunov function derivative that is linear in the derivative of the parameter estimate. By contrast, for discrete-time systems, the difference in the Lyapunov function also includes a term involving the *square* of the difference of the parameter estimate.

In this chapter we develop a novel, stable adaptive control scheme for a class of functional uncertain, SISO discrete-time nonlinear systems [74]. The proposed scheme utilizes a GaRBF network within a direct adaptive control methodology, and it ensures signal convergence and stable closed-loop tracking of a reference input by deriving adaptation laws based on discrete-time Lyapunov stability theory. An augmented error method that was originally introduced within the context of adaptive control for linear, parametric uncertain systems [111, 186, 189], is suitably modified and applied to derive appropriate adaptation and control laws for this class of functional uncertain nonlinear systems. As will be shown in more detail later on, the augmented error technique provides a way of eliminating the nonlinear squared term appearing in the Lyapunov difference equation. This leads to a simpler control scheme and more global stability results than alternative techniques that have been proposed so far for functional uncertain, discrete-time nonlinear systems [53, 120]. A dead-zone nonlinearity is also included inside the adaptation law to ensure robustness to the inherent network approximation error.

The class of systems under consideration is defined by the following equation:

$$y(k+1) = f[\mathbf{x}(k)] + u(k), \qquad (5.1)$$

where $\mathbf{x}(k) = [y(k)\ y(k-1)\ \cdots\ y(k-n+1)]^T$ is the state vector, $u(k)$ is the control input and $y(k)$ is the plant output. The nonlinear, smooth function $f : \Re^n \mapsto \Re$ is assumed unknown.

This functional form represents a sub-class of the NARMA-L2 model described in Section 2.3.1 where function $\bar{g}$, multiplying $u(k)$, is equal to 1.

Although the class of systems under consideration is less general than the one treated in [53], which is of the same type as NARMA-L2, the stability results obtained in this chapter are more global. The reason is that no restriction on the size of the initial network parameter errors has to be imposed to ensure stability. Hence our scheme possesses the desirable feature of not requiring any off-line network training prior to the application of on-line adaptive control. The class of plants considered in [120] is similar to the one treated in this chapter, in the sense that the nonlinear coefficient function $\bar{g} = 1$, and the stability results do not depend upon any constraints on the initial parameter errors. However the control scheme of [120] requires more prior knowledge than the method proposed here; namely that the network's optimal output layer parameters are bounded by known values.

The main contributions of this chapter could therefore be summed up as follows:

1. The derivation of an adaptive control scheme for a class of functional uncertain, nonlinear discrete-time systems that: (a) exhibits more global stability results than those obtained in [53], albeit for a less general class of systems, (b) requires less prior knowledge on the value of the unknown optimal network parameters than [120].
2. The extension of the augmented error adaptive control technique, originally developed for disturbance-free linear systems, to nonlinear systems subjected to a disturbance-like signal arising from the neural network modelling error.
3. An original error convergence analysis technique introduced to take into consideration the effect of dead-zone adaptation laws that were included to mitigate the de-stabilizing effect of the above-mentioned disturbance signal.
4. Extension of the proposed scheme to effect stable, adaptive *discrete-time sliding mode* control of the same class of nonlinear systems.

5.2 Problem Formulation

The objective is to design a controller that gives good tracking performance when applied to the plant of Equation (5.1), whose nonlinear dynamics $f[\mathbf{x}(k)]$ are unknown. The desired output is specified in terms of a reference input signal $y_d(k)$ that denotes the value required at the plant output during time instant k. The following conditions regarding the system and reference input are assumed to hold:

Assumption 5.2.1 *The order of the system, n, is known.*

Assumption 5.2.2 *The reference input is known at least one time step ahead (i.e., $y_d(k+1)$ is known at time k) and is bounded.*

Functional adaptive control techniques are used to deal with the uncertainty arising from the lack of knowledge of function $f[\mathbf{x}(k)]$. Being nonlinear, a GaRBF neural network is used to approximate it, and the network parameters are adjusted in an on-line manner.

5.3 The Neural Network

By the Universal Approximation Property, the GaRBF network is able to approximate $f[\mathbf{x}(k)]$ up to an arbitrary level of accuracy within a compact subset of state-space, $\chi \subset \Re^n$. The network output is given as:

$$\hat{f}[\mathbf{x}(k), \hat{\mathbf{w}}(k)] = \hat{\mathbf{w}}^T(k) \Phi[\mathbf{x}(k)], \tag{5.2}$$

where $\hat{\mathbf{w}}(k)$ denotes the network output layer parameter vector and $\Phi[\mathbf{x}(k)]$ denotes the vector of Gaussian basis functions. Hence, given any arbitrary non-zero approximation error bound ϵ_f, there exist some *optimal* basis function parameters and output layer parameters $\hat{\mathbf{w}} = \mathbf{w}^*$, such that the network satisfies the condition that $\forall \mathbf{x} \in \chi$,

$$|\Delta_f(\mathbf{x})| \leq \epsilon_f, \tag{5.3}$$

where $\Delta_f(\mathbf{x}) := \hat{f}(\mathbf{x}, \mathbf{w}^*) - f(\mathbf{x})$ denotes the *optimal* network approximation error. The following conditions are assumed to hold for the neural network:

Assumption 5.3.1 *The state is always confined within the network approximation region defined by subset χ, whose boundaries are known. Note that χ is a design parameter and could be made arbitrarily large so as to satisfy this assumption.*

Assumption 5.3.2 *The basis function centres and width parameters ensuring that condition (5.3) is satisfied for a given value of ϵ_f, are known a priori.*

As explained in the previous chapter, the technique of [219] could be used to satisfy the last assumption. Hence the optimal value $\mathbf{w}^*$ of the output layer parameter vector is unknown, and so $\hat{\mathbf{w}}(k)$ is recursively adjusted via suitable on-line adaptation laws that ensure system stability and tracking error convergence.

Defining the *general* network approximation error as

$$\tilde{f}[\mathbf{x}(k)] := f[\mathbf{x}(k)] - \hat{f}[\mathbf{x}(k), \hat{\mathbf{w}}(k)]$$

it follows from the definition of $\Delta_f(\mathbf{x})$ and Equation (5.2) that:

$$\tilde{f}[\mathbf{x}(k)] = -\tilde{\mathbf{w}}^T(k) \Phi[\mathbf{x}(k)] - \Delta_f[\mathbf{x}(k)] \tag{5.4}$$

where $\tilde{\mathbf{w}}(k) := \hat{\mathbf{w}}(k) - \mathbf{w}^*$ represents the parameter error.

5.4 The Control Law

The tracking error $e(k)$, defined as $e(k) := y(k) - y_d(k)$, asymptotically converges to zero from any initial condition if it satisfies the first order difference equation

$$e(k+1) + c_1 e(k) = 0,$$

provided that $|c_1| < 1$. If $f(\mathbf{x})$ were known, the plant Equation (5.1) clearly shows that the above relation is satisfied when the following feedback linearization-type control law is applied

$$u(k) = y_d(k+1) - f[\mathbf{x}(k)] - c_1 e(k).$$

In the absence of knowledge on function f, the network approximation $\hat{f}$ will be used on a certainty equivalence basis to replace f inside this control law, yielding the following law for the adaptive case:

$$u(k) = y_d(k+1) - \hat{f}[\mathbf{x}(k), \hat{\mathbf{w}}(k)] - c_1 e(k). \tag{5.5}$$

Substituting Equation (5.5) in the plant Equation (5.1) and using the definition of $\tilde{f}$, we obtain the following *error dynamics equation*

$$e(k) + c_1 e(k-1) = \tilde{f}[\mathbf{x}(k-1)].$$

Note, from Equation (5.4), that the non-zero inherent network approximation error Δ_f directly affects the above error dynamics equation via the term $\tilde{f}[\mathbf{x}(k-1)]$, where it effectively appears as an additive disturbance term. The error dynamics equation may be equivalently represented in the following transfer function form:

$$e(k) = \Gamma^{-1}(z^{-1})[\tilde{f}[\mathbf{x}(k-1)]], \tag{5.6}$$

where $\Gamma(z^{-1}) = 1 + c_1 z^{-1}$, and z^{-1} denotes the discrete-time delay operator.

Note that control law (5.5) results in a first order relationship between the tracking error $e(k)$ and the network approximation error $\tilde{f}[\mathbf{x}(k-1)]$. This considerably simplifies the derivation of stable adaptation laws by Lyapunov techniques.

5.5 The Adaptive System

To enable derivation of adaptation laws that do not just ensure stability, but are also causal, an *augmented error* signal $e_1(k)$ is generated from the tracking error $e(k)$ together with an appropriately designed *auxiliary signal* $v(k)$. The following augmented error is proposed for this system:

$$e_1(k) = \beta(e(k) - \Gamma^{-1}(z^{-1})[v(k)]), \tag{5.7}$$

where β is a positive, non-zero constant. Substituting for $e(k)$ from Equation (5.6) gives

$$e_1(k) = \beta \Gamma^{-1}(z^{-1})[\tilde{f}[\mathbf{x}(k-1)] - v(k)],$$

which leads to the time-domain relation

$$e_1(k-1) = \frac{\beta(\tilde{f}[\mathbf{x}(k-1)] - v(k)) - e_1(k)}{c_1}. \tag{5.8}$$

The auxiliary signal $v(k)$ will be determined subsequently, as part of the design procedure. The augmented error signal $e_1(k)$ is used to drive the following adaptation law, which includes also a dead-zone nonlinearity to ensure robustness to the network inherent approximation error Δ_f:

$$\Delta \hat{\mathbf{w}}(k) = \begin{cases} \frac{\beta}{\gamma c_1^2} \Phi[\mathbf{x}(k-1)] e_1(k) & \text{if } |e_1(k)| > \epsilon_f/G \\ 0 & \text{if } |e_1(k)| \leq \epsilon_f/G \end{cases} \tag{5.9}$$

where $\Delta \hat{\mathbf{w}}(k) := \hat{\mathbf{w}}(k) - \hat{\mathbf{w}}(k-1)$, γ and G are strictly positive constants and c_1 is a non-zero constant with magnitude less than 1.

5.6 Stability Analysis

Consider the discrete-time Lyapunov function candidate

$$V(k) = e_1^2(k) + \gamma \tilde{\mathbf{w}}^T(k)\tilde{\mathbf{w}}(k).$$

The first difference

$$\Delta V(k) := V(k) - V(k-1)$$
$$= e_1^2(k) - e_1^2(k-1) + \gamma(\tilde{\mathbf{w}}^T(k) + \tilde{\mathbf{w}}^T(k-1))(\tilde{\mathbf{w}}(k) - \tilde{\mathbf{w}}(k-1)).$$

Using Equation (5.8) for $e_1(k-1)$ and the definitions of $\tilde{\mathbf{w}}$ and $\Delta \hat{\mathbf{w}}$, it follows that

$$\Delta V(k) = -V_1 + \frac{2\beta(\tilde{f}[\mathbf{x}(k-1)] - v(k))e_1(k)}{c_1^2} +$$
$$\gamma(\Delta \hat{\mathbf{w}}^T(k) + 2\tilde{\mathbf{w}}^T(k-1))\Delta \hat{\mathbf{w}}(k)$$

where

$$V_1 = \left[\frac{e_1^2(k)(1-c_1^2)}{c_1^2} + \frac{\beta^2(\tilde{f}[\mathbf{x}(k-1)] - v(k))^2}{c_1^2} \right] \geq 0,$$

since $0 < c_1^2 < 1$.

Substituting for $\tilde{f}[\mathbf{x}(k-1)]$ via Equation (5.4) yields:

$$\Delta V(k) = -V_1 + 2\tilde{\mathbf{w}}^T(k-1)\left[\gamma \Delta \hat{\mathbf{w}}(k) - \frac{\beta}{c_1^2}\Phi[\mathbf{x}(k-1)]e_1(k)\right] -$$
$$\frac{2\beta}{c_1^2}[\Delta_f[\mathbf{x}(k-1)] + v(k)]e_1(k) + \gamma \Delta \hat{\mathbf{w}}^T(k)\Delta \hat{\mathbf{w}}(k). \tag{5.10}$$

Substituting the adaptation law Equation (5.9), we obtain that $\Delta V(k)$ is:

$$\begin{cases} -V_1 - \frac{2\beta}{c_1^2}[\Delta_f[\mathbf{x}(k-1)] + v(k)]\, e_1(k) + \\ \left(\frac{\beta}{\sqrt{\gamma}c_1^2}\right)^2 \Phi^T[\mathbf{x}(k-1)]\Phi[\mathbf{x}(k-1)]e_1^2(k) & \text{if } |e_1(k)| > \epsilon_f/G \\[2mm] -V_1 - \frac{2\beta}{c_1^2}\{\tilde{\mathbf{w}}^T(k-1)\Phi[\mathbf{x}(k-1)] + \\ v(k) + \Delta_f[\mathbf{x}(k-1)]\}\, e_1(k) & \text{if } |e_1(k)| \le \epsilon_f/G \end{cases} \quad (5.11)$$

Our aim is to force $\Delta V(k)$ to be negative definite. One problem to satisfy this condition was highlighted in the introduction to this chapter. It concerns the appearance of the positive squared term $\Delta \hat{\mathbf{w}}^T(k)\Delta \hat{\mathbf{w}}(k)$ in the Lyapunov function difference equation, as clearly seen in Equation (5.10). This results in the positive semidefinite term $\left(\frac{\beta}{\sqrt{\gamma}c_1^2}\right)^2 \Phi^T[\mathbf{x}(k-1)]\Phi[\mathbf{x}(k-1)]e_1^2(k)$ to appear in Expression (5.11) for the condition $|e_1(k)| > \epsilon_f/G$, which complicates the process of obtaining a negative definite $\Delta V(k)$.

The augmented error technique helps to resolve the situation however, because it leads to the appearance of the auxiliary signal $v(k)$ in the equations. Choosing the auxiliary signal as:

$$v(k) = v_1(k) + v_2(k) \quad (5.12)$$

with

$$v_1(k) := \frac{\beta}{2\gamma c_1^2}\Phi^T[\mathbf{x}(k-1)]\Phi[\mathbf{x}(k-1)]e_1(k) \quad (5.13)$$

$$v_2(k) := Ge_1(k), \quad (5.14)$$

has two beneficial effects with respect to Expression (5.11):

- $v_1(k)$ cancels the positive semidefinite term $\left(\frac{\beta}{\sqrt{\gamma}c_1^2}\right)^2 \Phi^T[\mathbf{x}(k-1)]\Phi[\mathbf{x}(k-1)]e_1^2(k)$ appearing in the first equation.
- $v_2(k)$ mitigates the effect of term $\Delta_f[\mathbf{x}(k-1)]$ that appears in both Expressions (5.11).

Note that since the inherent network approximation error Δ_f is unknown, it is not possible to choose $v_2(k)$ such as to completely eliminate this term. However, as will be shown in the next lemma, the suggested choice of v_2 still renders $\Delta V(k)$ negative definite when $|e_1(k)| > \epsilon_f/G$.

Lemma 5.6.1. *Subject to all the relevant previously-stated assumptions; control law (5.5), adaptation law (5.9) and auxiliary signal (5.12) applied to the system of Equation (5.1), result in $\Delta V(k) < 0$ for $|e_1(k)| > \epsilon_f/G$.*

Proof. Substituting for $v(k)$ in the first expression of (5.11), corresponding to $|e_1(k)| > \epsilon_f/G$, and noting that under this condition $-V_1$ is strictly negative, it follows that

$$\Delta V(k) < \frac{-2\beta}{c_1^2}[\Delta_f[\mathbf{x}(k-1)] + Ge_1(k)]\, e_1(k). \quad (5.15)$$

But $|\Delta_f(\mathbf{x})| \leq \epsilon_f$ by definition. Hence under the condition $|e_1(k)| > \epsilon_f/G$:

$$|e_1(k)| > \frac{|\Delta_f[\mathbf{x}(k-1)]|}{G}$$
$$\Rightarrow e_1^2(k) > \frac{|\Delta_f[\mathbf{x}(k-1)]e_1(k)|}{G} \geq \frac{-\Delta_f[\mathbf{x}(k-1)]e_1(k)}{G}$$
$$\Rightarrow [\Delta_f[\mathbf{x}(k-1)] + Ge_1(k)]\,e_1(k) > 0.$$

Using this in Equation (5.15) proves the lemma since $\beta, c_1^2 > 0$. □

When $|e_1(k)| \leq \epsilon_f/G$, nothing particular could be deduced about $\Delta V(k)$ from the second of Expressions (5.11) because it can take on any value, positive or negative. Recall however that under this condition, $\Delta\hat{\mathbf{w}}(k)$ is restricted to zero by the dead-zone nonlinearity in adaptation law (5.9). This fact and Lemma 5.6.1 allow us to deduce some interesting signal boundedness properties that will now be illustrated qualitatively prior to proving them formally.

The Lyapunov function $V(k)$ could be equivalently written as follows

$$V(k) = e_1^2(k) + \gamma\|\tilde{\mathbf{w}}(k)\|^2$$

where $\|\cdot\|$ denotes the Euclidean norm of a vector. Hence, the level curves of $V(k)$ could be graphically represented as semi-ellipses on the $(e_1, \|\tilde{\mathbf{w}}\|)$ plane, as shown in Figure 5.1. Note that the outer level curves correspond to larger values of $V(k)$. The boundaries of the region $|e_1(k)| \leq \epsilon_f/G$ are shown as dashed vertical lines.

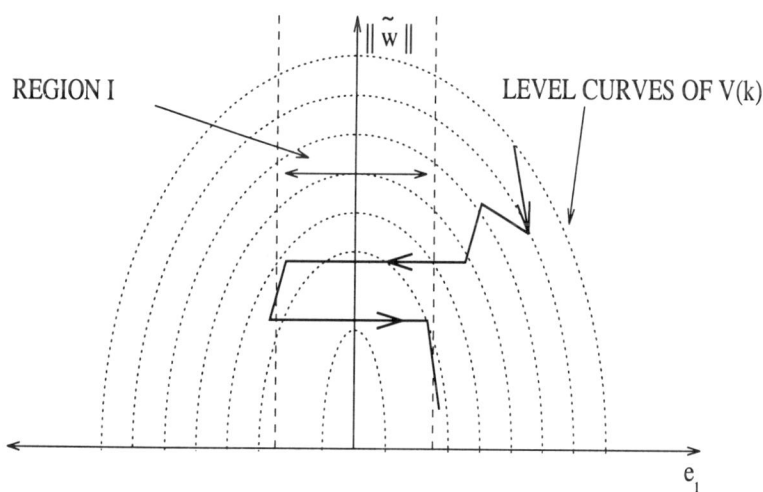

Fig. 5.1. The $(e_1, \|\tilde{\mathbf{w}}\|)$ plane: the semi-ellipses represent the level curves of $V(k)$ with the outer ones corresponding to larger values; the vertical dashed lines denote the boundaries of region I; and the solid line reflects the constraints of a typical trajectory.

Let us define the set

$$I := \left\{ (\|\tilde{\mathbf{w}}\|, e_1) \mid |e_1| \leq \epsilon_f/G \right\}$$

to denote points within this region, which shall also be referred to as region I. From the dead-zone adaptation law, $\Delta \hat{\mathbf{w}}(k) = 0$ inside region I and so $\|\tilde{\mathbf{w}}(k)\| = \|\tilde{\mathbf{w}}(k-1)\|$, since $\mathbf{w}^*$ is constant. Hence when entering inside region I from outside or moving within it, the trajectory of points $(\|\tilde{\mathbf{w}}\|, e_1)$ must be parallel to the e_1-axis, *i.e.*, along a horizontal direction in Figure 5.1. Outside region I, Lemma 5.6.1 shows that $V(k) < V(k-1)$, which means that the trajectory of $(\|\tilde{\mathbf{w}}\|, e_1)$ must have come from a *higher* valued level curve at the previous time instant.

Hence if at time k the trajectory is outside region I, at time $(k+1)$ it can only:

1. go to a smaller valued level curve outside region I or
2. go inside region I but *only* along a horizontal direction (which also implies a smaller-valued level curve).

On the other hand, if at time k the trajectory is inside region I, at time $(k+1)$ it can only:

1. go to a smaller valued level curve outside region I or
2. remain inside region I but moves along a horizontal direction.

All these possibilities are shown in the trajectory plotted in Figure 5.1. They indicate that if the system starts with finite values for $(\|\tilde{\mathbf{w}}\|, e_1)$, then $e_1(k)$ and $\|\tilde{\mathbf{w}}(k)\|$ remain bounded for all subsequent time steps k. This notion is formally stated and proved in the following theorem.

Theorem 5.6.1. *Subject to all the relevant previously-stated assumptions, when plant (5.1) is subjected to control law (5.5), adaptation law (5.9) and auxiliary signal (5.12) and $\|\tilde{\mathbf{w}}(0)\|^2$, $e_1^2(0) < \infty$; then $e_1^2(k)$ and $\|\tilde{\mathbf{w}}(k)\|^2$ are bounded $\forall k$. In addition, $\lim_{k \to \infty} |e_1(k)| \leq \epsilon_f/G$.*

Proof. The proof is divided into two parts: boundedness of $\|\tilde{\mathbf{w}}\|$ and e_1 will be proved in Part 1 and the asymptotic convergence of $|e_1|$ in Part 2.

Part 1: From the definition of $V(k)$, $V(0) < \infty$ if $\|\tilde{\mathbf{w}}(0)\|^2$, $e_1(0) < \infty$. With reference to set I, suppose that at some point in time V becomes unbounded when $(\|\tilde{\mathbf{w}}\|, e_1) \in I$. This could only take place if $\|\tilde{\mathbf{w}}\|$ is unbounded, since e_1 is bounded in I by definition. However $\|\tilde{\mathbf{w}}\|^2$ could only have become unbounded previously, when $(\|\tilde{\mathbf{w}}\|, e_1) \in I^c$ (where I^c denotes the complement set of I) because from Equation (5.9), $\|\tilde{\mathbf{w}}\|$ cannot change in I and $\|\tilde{\mathbf{w}}(0)\| < \infty$. Hence it follows that V must have become unbounded in I^c, which is impossible since under this condition $\Delta V(k) < 0$ by Lemma 5.6.1 and $V(0) < \infty$. Hence $V(k)$ can never be unbounded; neither within set I when $|e_1(k)| \leq \epsilon_f/G$, nor within I^c, when $|e_1(k)| > \epsilon_f/G$. Hence $e_1(k)^2$,

$\|\tilde{\mathbf{w}}(k)\|^2$ are bounded $\forall k$.

Part 2: To show that $|e_1(k)| \leq \epsilon_f/G$ as $k \to \infty$, consider the following three cases:

1. If the dynamics evolve such that $(\|\tilde{\mathbf{w}}\|, e_1) \in I$ $\forall k$, then $|e_1| \leq \epsilon_f/G$ $\forall k$, by definition.
2. If $(\|\tilde{\mathbf{w}}\|, e_1) \in I^c$ $\forall k$, then by Lemma 5.6.1 $\Delta V_k < 0$ $\forall k$. Hence $V(k)$ reduces to its smallest possible value under the condition that $|e_1(k)| > \epsilon_f/G$. This occurs when $\|\tilde{\mathbf{w}}\|^2 \to 0$ and $e_1^2 \to (\epsilon_f/G)^2$. Hence $\lim_{k \to \infty} |e_1(k)| = \epsilon_f/G$.
3. When $(\|\tilde{\mathbf{w}}\|, e_1)$ traverses from I^c into I and returns to I^c, the *net* value of V is reduced *i.e.*, the value of V when $(\|\tilde{\mathbf{w}}\|, e_1)$ emerges from I to I^c is less than V when $(\|\tilde{\mathbf{w}}\|, e_1)$ had previously entered from I^c into I.

The property of case 3 follows from the following arguments. Consider an $I^c \to I \to I^c$ traversal such that $(\|\tilde{\mathbf{w}}(k)\|, e_1(k)) \in I^c$, $(\|\tilde{\mathbf{w}}(k+i)\|, e_1(k+i)) \in I$ for $i = 1\ldots m$, and $(\|\tilde{\mathbf{w}}(k+m+1)\|, e_1(k+m+1)) \in I^c$. Then

$$V(k) = \gamma\|\tilde{\mathbf{w}}(k)\|^2 + e_1^2(k)$$
$$> \gamma\|\tilde{\mathbf{w}}(k)\|^2 + (\epsilon_f/G)^2 \tag{5.16}$$

since $|e_1(k)| > \epsilon_f/G$ in I^c. In addition, whilst the trajectory is inside region I (*i.e.*, $i = 1\ldots m$),

$$\begin{aligned}
V(k+i) &= \gamma\|\tilde{\mathbf{w}}(k+i)\|^2 + e_1^2(k+i) \\
&= \gamma\|\tilde{\mathbf{w}}(k)\|^2 + e_1^2(k+i) \quad \text{by the dead-zone adaptation} \\
&\leq \gamma\|\tilde{\mathbf{w}}(k)\|^2 + (\epsilon_f/G)^2, \; \forall\, i = 1\ldots m, \\
&\quad \text{by the definition of region } I \\
&< V(k) \; \forall\, i = 1\ldots m \quad \text{by (5.16).} \tag{5.17}
\end{aligned}$$

Also, $V(k+m+1) < V(k+m)$ since $\Delta V(k+m+1) < 0$ by Lemma 5.6.1. Hence, together with (5.17) it is concluded that:

$$V(k+m+1) < V(k+m) < V(k). \tag{5.18}$$

This proves the property of case 3 regarding the net value of V.

Now, if after time $(k+m+1)$, $(\|\tilde{\mathbf{w}}(k)\|, e_1(k))$ remains in I^c forever, the situation reduces to case (2) and so $\lim_{k \to \infty} |e_1(k)| = \epsilon_f/G$. If on the other hand, it goes back into I and remains there forever, the situation becomes equivalent to case (1) with $|e_1(k)| \leq \epsilon_f/G$. If none of these is true, the only possibility is that case (3) repeats itself indefinitely, so that the net value of V reduces every time $(\|\tilde{\mathbf{w}}(k)\|, e_1(k))$ emerges into I^c. Hence, after emerging into I^c a sufficient number of times, V will ultimately reach its minimum under the constraint $|e_1(k)| > \epsilon_f/G$; which means that $e_i^2(k) \to (\epsilon_f/G)^2, \|\tilde{\mathbf{w}}\| \to 0$. Now, if $(\|\tilde{\mathbf{w}}\|, e_1)$ goes back into I at some future time step *after* reaching the minimum in I^c, it would be unable to emerge into I^c any more, because otherwise condition (5.18) would be violated since V had already reached its minimum value the last time that $(\|\tilde{\mathbf{w}}\|, e_1) \in I^c$. Therefore in this case, $(\|\tilde{\mathbf{w}}(k)\|, e_1(k))$ subsequently remains in I, with $|e_1| \leq \epsilon_f/G$.

Hence in all cases 1, 2 and 3; $|e_1|$ is ultimately bounded by ϵ_f/G, which proves the theorem. □

Remarks 5.6.1

1. Inside region I, the Lyapunov function $V(k)$ is both upper and lower bounded. The lower bound is given by the following argumentation:
   Using the same time notation as before, where k denotes the time step just before $(\|\tilde{\mathbf{w}}\|, e_1)$ entered into region I and $(k+i)$ denotes all subsequent time steps during which $(\|\tilde{\mathbf{w}}\|, e_1)$ stays in I, then by the definition of I and adaptation law (5.9):

   $$e_1^2(k+i) \leq (\epsilon_f/G)^2 \text{ and } \|\tilde{\mathbf{w}}(k+i)\| = \|\tilde{\mathbf{w}}(k)\|.$$

   Using these conditions in the definition of $V(k)$, the following two relations hold true:

   $$V(k+i) + (\epsilon_f/G)^2 = \gamma\|\tilde{\mathbf{w}}(k)\|^2 + e_1^2(k+i) + (\epsilon_f/G)^2$$
   $$\geq \gamma\|\tilde{\mathbf{w}}(k)\| + (\epsilon_f/G)^2, \tag{5.19}$$
   $$\text{and } V(k+1) \leq \gamma\|\tilde{\mathbf{w}}(k)\|^2 + (\epsilon_f/G)^2. \tag{5.20}$$

   Hence (5.19) and (5.20) $\Rightarrow$

   $$V(k+i) \geq V(k+1) - (\epsilon_f/G)^2,$$

   which gives the lower bound of V within region I. The upper bound is given by Equation (5.17), showing that it is less than the value of V *just* before entering region I, which is surely bounded. Hence, combining these results yields:

   $$V(k+1) - (\epsilon_f/G)^2 \leq V(k+i) < V(k) < \infty.$$

   This remark confirms that even though ΔV is not guaranteed to be negative definite when $|e_1| < \epsilon_f/G$, the upper bound ensures that V will not drift to infinity. At the same time however, the lower bound indicates that V can only reduce up to a certain point. This has implications on the smallest value that $\|\tilde{\mathbf{w}}\|$ can attain whilst $(\|\tilde{\mathbf{w}}\|, e_1) \in I$, which leads to the next remark.

2. On a transition from inside region I to the outside, $\|\tilde{\mathbf{w}}\|$ reduces, i.e., the parameters are learned a little "better" in some sense. From the proof of Theorem 5.6.1 and using the same time notation, since $V(k+m+1) < V(k+m)$ it follows that

   $$\gamma(\|\tilde{\mathbf{w}}(k+m+1)\|^2 - \|\tilde{\mathbf{w}}(k+m)\|^2) < e_1^2(k+m) - e_1^2(k+m+1).$$

   But since $e_1^2(k+m+1) > (\epsilon_f/G)^2 \geq e_1^2(k+m)$ by definition, then $e_1^2(k+m) - e_1^2(k+m+1) < 0$ and so,

   $$\|\tilde{\mathbf{w}}(k+m+1)\|^2 < \|\tilde{\mathbf{w}}(k+m)\|^2.$$

This relation indicates that for the parameter error to converge to zero, the trajectories of $(\|\tilde{\mathbf{w}}\|, e_1)$ must exit and enter region I for a sufficient number of times. The fulfilment of this condition largely depends upon the richness and excitation of the input signals. This illustrates the well-known concept of *persistent excitation* in adaptive control and estimation [23, 186].

5.7 Tracking Error Convergence

In this section it is shown that the convergence of augmented error $e_1(k)$ also results in the tracking error $e(k)$ to converge to a bound that could be predetermined by proper choice of design parameters c_1, G and the network approximation accuracy ϵ_f.

The norm of the network RBF vector, $\Phi^T[\mathbf{x}(k-1)]\Phi[\mathbf{x}(k-1)]$, is upper bounded by some value, denoted as $\bar{\Phi}$. This follows because the basis functions are Gaussians whose value does not exceed 1. Bound $\bar{\Phi}$ can be easily determined from the known basis function parameters. From this property, Equations (5.12), (5.13), (5.14), and the convergence of $e_1(k)$, it follows that $v(k)$ is also bounded and converges asymptotically as follows:

$$\lim_{k \to \infty} |v(k)| \leq \left[\frac{\beta\bar{\Phi}}{2\gamma c_1^2} + G\right] \frac{\epsilon_f}{G}$$

$$= \left[\frac{\beta\bar{\Phi}}{2\gamma c_1^2 G} + 1\right] \epsilon_f. \tag{5.21}$$

Hence, from Equation (5.7), it is concluded that $e(k)$ is bounded as well, since both $e_1(k)$ and $v(k)$ are bounded and $\Gamma^{-1}(z^{-1})$ is a stable filter. In addition, $e(k)$ asymptotically converges to a circular region centred at the origin, whose radius is evaluated next.

Let us first re-write Equation (5.7) as

$$e(k) = e_1(k)/\beta + \Gamma^{-1}[v(k)]. \tag{5.22}$$

Denoting the rightmost term $\Gamma^{-1}[v(k)]$ by $\alpha(k)$, it follows that

$$\alpha(k) + c_1\alpha(k-1) = v(k). \tag{5.23}$$

Using successive substitution, with $\alpha(k)$ as initial condition, it follows that for any initial $k > 0$ and $t \geq 1$,

$$\alpha(k+t) = v(k+t) + (-c_1)v(k+t-1) + \ldots$$
$$+ (-c_1)^{t-1}v(k+1) + (-c_1)^t \alpha(k).$$

Applying the triangle inequality and noting that $|(-c_1)^r| = |c_1|^r$, irrespective of whether r is odd or even, we get:

112   5. Functional Adaptive Control of Discrete-Time Systems

$$|\alpha(k+t)| \le |v(k+t)| + |c_1||v(k+t-1)| + \ldots$$
$$+ |c_1|^{t-1}|v(k+1)| + |c_1|^t|\alpha(k)|$$
$$\le B\left[1 + |c_1| + \ldots + |c_1|^{t-1}\right] + |c_1|^t|\alpha(k)|, \quad (5.24)$$

where $B = \sup_{i=1,\cdots,t} |v(k+i)|$ represents a finite upper bound on $|v(k+1)|, \ldots, |v(k+t)|$. As $t \to \infty$, the following conditions are deduced:

- Since $|c_1| < 1$, the first term on the right hand side of Equation (5.24) is a convergent geometric series that tends to $1/(1-|c_1|)$.
- The second term on the right hand side of equation (5.24) vanishes since $|c_1|^t \to 0$ and $|\alpha(k)| < \infty$ because α is a stable filtration of v, that is bounded.

Hence,

$$\lim_{t \to \infty} |\alpha(k+t)| \le \frac{B}{1-|c_1|}. \quad (5.25)$$

But Equation (5.21) shows that if k is sufficiently large, $B \to \left[\frac{\beta\bar{\Phi}}{2\gamma c_1^2 G} + 1\right]\epsilon_f$. Hence B can be replaced by this term to give,

$$\lim_{k \to \infty} |\alpha(k)| \le \left[\frac{\beta\bar{\Phi}}{2\gamma c_1^2 G} + 1\right] \frac{\epsilon_f}{(1-|c_1|)}. \quad (5.26)$$

From Theorem 5.6.1 and Equations (5.22) and (5.26), we therefore obtain

$$\lim_{k \to \infty} |e(k)| \le \frac{1}{\beta}\left(\frac{\epsilon_f}{G}\right) + \left[\frac{\beta\bar{\Phi}}{2\gamma c_1^2 G} + 1\right]\frac{\epsilon_f}{(1-|c_1|)}$$
$$= \left[1 + \frac{1-|c_1|}{\beta G} + \frac{\beta\bar{\Phi}}{2\gamma c_1^2 G}\right]\frac{\epsilon_f}{(1-|c_1|)} \quad (5.27)$$
$$\approx \frac{\epsilon_f}{1-|c_1|} \quad \text{if } G \text{ is chosen such that } G >> \frac{1-|c_1|}{\beta} + \frac{\beta\bar{\Phi}}{2\gamma c_1^2}.$$

The final equation shows that with this choice of G, the tracking error $e(k)$ converges to within $\pm\epsilon_f/(1-|c_1|)$ of the ideal value of 0, in the steady-state. Note that this convergence region is proportional to the network approximation accuracy ϵ_f and it comes about as a result of the dead-zone in the adaptation laws, included to ensure stability in the presence of the inevitable network approximation error. If there were no network approximation error (i.e., $\epsilon_f = 0$), the dead-zone could be removed and $e(k) = 0$ is approached asymptotically.

5.8 Simulation Examples

In this section we present two simulation examples of the stable, discrete-time adaptive controller proposed in this chapter to demonstrate that the claimed convergence properties and error bounds are actually satisfied.

5.8.1 Example 1

The plant under consideration is a first order, nonlinear system

$$y(k) = \frac{5y(k-1)(1-y(k-1))}{1+\exp(-0.25y(k-1))} + u(k-1). \tag{5.28}$$

This model has the same form as the example of [276], where only parametric uncertainty was considered, related to the coefficients 5 and 0.25 appearing in the model equation. In this example we consider the more complex case of functional uncertainty where the whole function

$$f[\mathbf{x}(k-1)] = \frac{5y(k-1)(1-y(k-1))}{1+\exp(-0.25y(k-1))}$$

is unknown. The system order, $n = 1$, is assumed known. The reference input $y_d(k)$ is a square wave of amplitude 1. The network approximation region is chosen to cover the range $[-50, 50]$, with Gaussian basis functions of width parameter 0.4 centred on a regular mesh of spacing 0.2. This represents a network of 501 basis functions. Approximate off-line experiments indicate that an optimal network approximation error of less than 0.003 is attainable in the range $[-45, 40]$ with these RBF parameters. Hence bound ϵ_f is set to 0.003. The network output layer weights are initially set to zero. Settings for the rest of the design parameters are shown in Table 5.1. With these settings, Equation (5.27) shows that $|e(k)|$ is expected to converge within the range $[0, 0.0037]$.

Table 5.1. Design parameters for Example 1

c_1	β	γ	G
-0.01	0.001	0.001	50000

The simulation results are shown in Figures 5.2 to 5.4. Figure 5.2(a) shows that after an initial transient, the output satisfactorily tracks the reference input. The sequential learning features of the neural network are clearly seen by noting how during the first few cycles of the simulation, the tracking performance progressively improves until the error becomes barely noticeable as soon as the initial transitory period is over. Figure 5.2(b) shows the signal generated by the control law in order to achieve this tracking. Figure 5.3(a) shows the tracking error for the initial part of the simulation, indicating that it is truly bounded and Figure 5.3(b) confirms that it reduces to well within the predicted steady-state convergence region of ±0.0037. Finally, Figure 5.4 shows the neural network approximation $\hat{f}(\mathbf{x}, \hat{\mathbf{w}})$ of the unknown nonlinear function $f(\mathbf{x})$ when utilizing the network parameters obtained at the end of the simulation. Note that over the region of state space that was excited by the reference input, the network closely approximates the actual function,

114   5. Functional Adaptive Control of Discrete-Time Systems

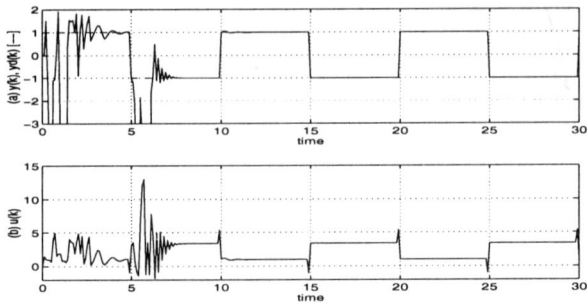

Fig. 5.2. Results of Example 1; (a) Output (solid) and reference (dashed) (b) Control input

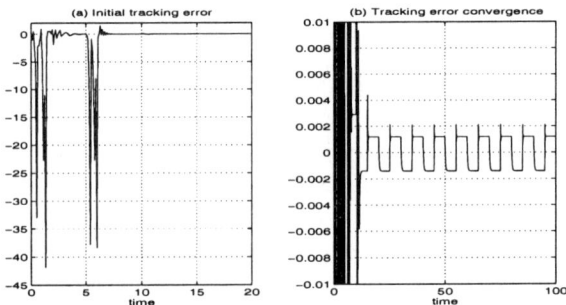

Fig. 5.3. Results of Example 1; (a) Initial tracking error (b) Steady-state tracking error

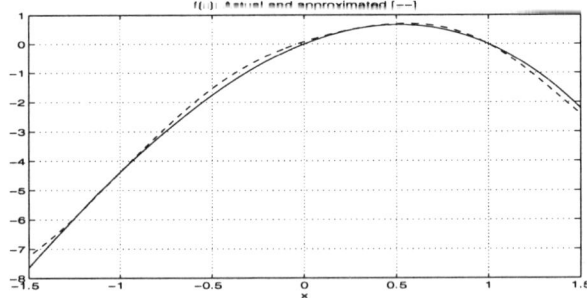

Fig. 5.4. Results of Example 1; Function $f(x)$ (solid) and the network approximation (dashed)

which is shown superimposed as a solid curve. It is important to note that in this scheme, like all adaptive control methods, there is no guarantee that the parameter error converges to zero unless the input is persistently exciting. This explains why the the curves do not match *perfectly* well. Nevertheless the approximation accuracy is impressive and the diagram confirms the learning ability of the neural network.

5.8.2 Example 2

The second example concerns a second order nonlinear plant

$$y(k) = \frac{1.5y(k-1)y(k-2)}{1+y^2(k-1)+y^2(k-2)} + 0.35\sin[y(k-1)+y(k-2)] + u(k-1),$$

where

$$f[\mathbf{x}(k-1)] = \frac{1.5y(k-1)y(k-2)}{1+y^2(k-1)+y^2(k-2)} + 0.35\sin[y(k-1)+y(k-2)]$$

represents the unknown dynamics. The order of the system ($n = 2$) is assumed known. The reference trajectory, $y_d(k)$, is the same as in Example 1. The network approximation region is chosen as $\chi = [-2.5, 2.5] \times [-2.5, 2.5]$. Within this region, an optimal network approximation accuracy $\epsilon_f = 0.005$ is assumed when using a network consisting of 121 basis functions having width parameter of 0.75 and covering a regular mesh of spacing 0.5. The network output layer weights are initialized at zero and the controller design parameters are shown in Table 5.2. With these settings $|e(k)|$ is expected to converge within the range $[0, 0.0056]$.

Table 5.2. Design parameters for Example 2

c_1	β	γ	G
-0.1	0.001	0.001	10000

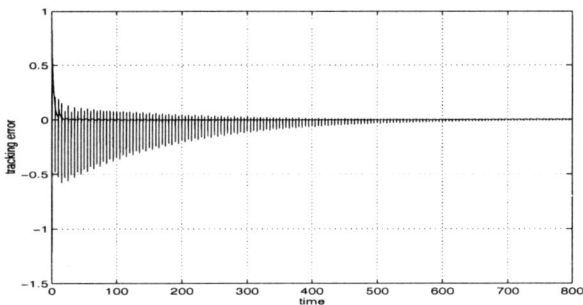

Fig. 5.5. Results of Example 2; Tracking error

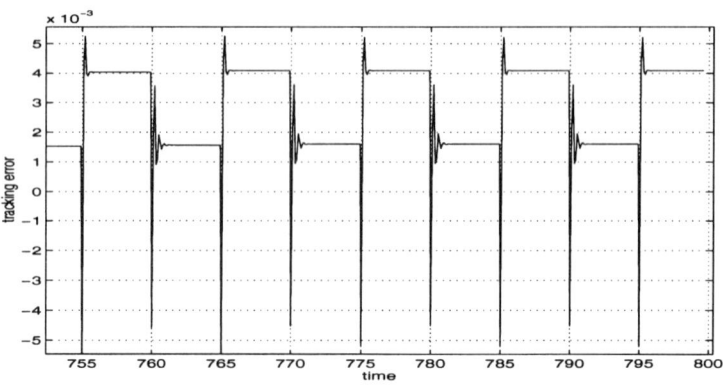

Fig. 5.6. Results of Example 2; Steady-state convergence

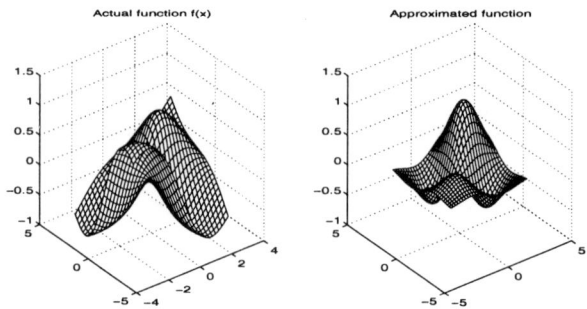

Fig. 5.7. Results of Example 2; Function approximation

Figure 5.5 shows how the tracking error $e(k)$ converges in a stable manner and Figure 5.6 shows only the last few time steps of the tracking error. Note that the predicted error convergence region of ±0.0056 is asymptotically attained. Figure 5.7 shows the actual function $f(\mathbf{x})$ and the network approximation $\hat{f}(\mathbf{x})$ obtained when utilizing the parameters estimated at the end of the simulation. Note that the previous comment regarding the effect of non-persistently exciting inputs on the parameter error is more pronounced in this example, because the approximation is not particularly good. This is not a problem however, because the main concern of a control system is the convergence of the tracking error towards zero, and not the approximation error. As far as function approximation is concerned, the important issue is for the parameter estimates to remain bounded, a feature which is mathematically guaranteed in the proposed scheme and which was also confirmed during the simulation.

5.9 Extension to Adaptive Sliding Mode Control

The theory of sliding mode control for continuous-time systems is well-established [257, 258]. In this case a sliding mode is usually achieved by instantaneous switching of the control law across a suitably defined *sliding manifold* in state-space. In this way, after a finite time, the state trajectories remain confined to the sliding manifold. Sliding mode schemes exhibit robustness to parameter uncertainty and external disturbances, provided the uncertainty or disturbance could be quantified in some way, typically via knowledge of bounds on the value that they could take. The use of instantaneous switching leads to Variable Structure Control (VSC) laws, so called because they are characterized by discontinuous functions. In practice, the inability of actuators to switch instantaneously often leads to the chattering effect, where the state trajectories continuously zigzag across the sliding manifold rather than remain on it.

For the discrete-time case the concept of instantaneous switching is inapplicable because by definition, the controller is sampling inputs and generating outputs only at discrete time intervals [222]. In general, VSC laws in discrete-time do not ensure state confinement to any manifold [143]. Hence the concept of a discrete-time sliding mode, and how to achieve it, is not a straightforward extension of the continuous-time case. In fact a number of different definitions of what constitutes a discrete-time sliding mode have been proposed. These include such concepts as *convergent quasi-sliding mode* [222], *discrete-time sliding mode* [259], *sliding mode for discrete VSC systems* [87] and *sliding sectors* [86]. Some of these ideas still attempt to obtain a discrete-time sliding regime by using VSC control laws. However it has been shown that in discrete-time systems, a sliding regime could be achieved without using discontinuous control [34, 229], which contrasts sharply with the continuous-time case.

5.9.1 Definitions of a Discrete-time Sliding Mode

In this section we review the various sliding mode definitions that have been put forward, with respect to the following discrete-time system:

$$\mathbf{x}(k+1) = \mathbf{f}[\mathbf{x}(k), u(k)]$$
$$y(k) = h[\mathbf{x}(k)]$$

where $\mathbf{x}$, u, y, $\mathbf{f}$, h are defined as usual. If y_d is the reference input and $e(k) = (y(k) - y_d(k))$ denotes the tracking error, the error state vector is defined as

$$\mathbf{e}(k) := [e(k)\ e(k-1) \cdots e(k-n+1)]^T.$$

The sliding manifold S is defined by the equation $s(k) = 0$, where

$$s(k) := \mathbf{c}^T \mathbf{e}(k)$$
$$= e(k) + c_1 e(k-1) + \cdots + c_{n-1} e(k-n+1).$$

Vector $\mathbf{c} := [1 \; c_1 \cdots c_{n-1}]^T$ is chosen such that all the roots of the equation $s(k) = 0$ lie inside the unity circle. Hence if we manage to find a control law ensuring that the condition $s(k) = 0$ is maintained, the trajectories of $\mathbf{e}$ remain on manifold S, and $\mathbf{e}(k)$ asymptotically converges (slides) to zero according to the dynamics of the manifold. The controller must therefore ensure that:

- Starting from any initial conditions, $s(k) = 0$ must be satisfied after a finite time. This is often called the *reaching phase*.
- After "reaching", the condition $s(k) = 0$ must be maintained so that the error state trajectories remain on manifold S. This is often called the *sliding phase*.

Figure 5.8 depicts these ideas for a second order system.

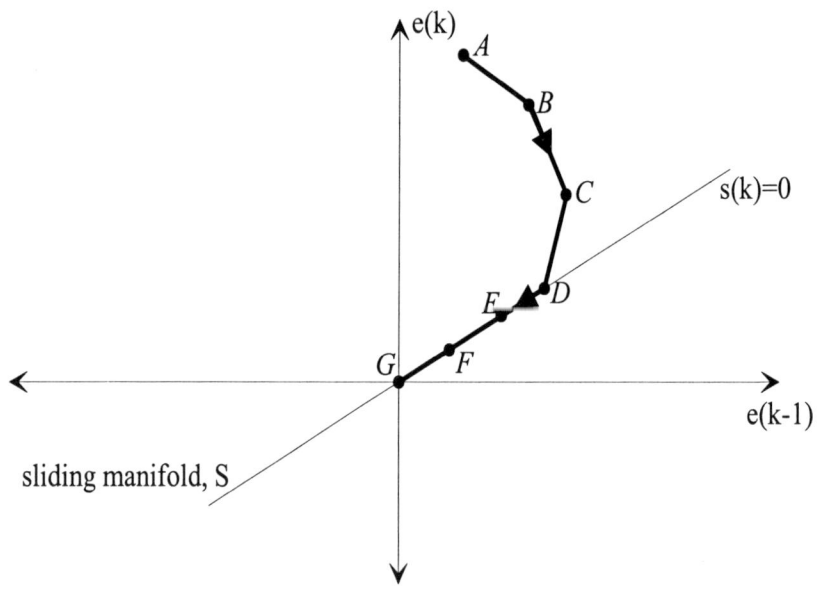

Fig. 5.8. The trajectory of an ideal discrete-time sliding regime in a second order system. Points (A, B, C, D) represent the reaching phase and points (D, E, F, G), the sliding phase.

A number of different conditions that the control law must satisfy to ensure attainment of the reaching and sliding phases have been proposed. These include:

5.9 Extension to Adaptive Sliding Mode Control

1. *The Quasi-Sliding condition* [222]: This is a direct discretization of the continuous-time sliding mode condition $s\dot{s} < 0$. It is given by

   $$s(k)[s(k+1) - s(k)] < 0.$$

   However, on its own, this condition is not enough to ensure proper sliding action. Indeed, the condition could be satisfied with the error state trajectories chattering about S with *increasing* amplitude, leading to instability [229]. It is however a *necessary* (though insufficient) condition for proper sliding action, as shown in the next proposal.

2. *Convergent Quasi-Sliding condition* [222]: This condition, due to Sarptürk et al., does ensure convergence to the sliding manifold and is given by

   $$|s(k+1)| < |s(k)|.$$

   Note that this condition implies that the previous quasi-sliding condition (1) is valid, but note vice-versa.

3. *Furuta's condition* [85]: This is based on Lyapunov-like analysis and is given by

   $$s(k)[s(k+1) - s(k)] < -0.5[s(k+1) - s(k)]^2 \text{ for } s(k) \neq 0.$$

   It ensures that $V(k+1) < V(k)$, where $V(k) := 0.5s^2(k)$ defines a Lyapunov type of function.

As explained in [34] and [87], although the last two conditions ensure that the error state tends to manifold S, they fail to ensure that this occurs in finite time, which is what one normally requires of a sliding mode system. In addition, they lack a proper definition of what really constitutes a discrete-time sliding mode. Hence the following suggestions were put forward to remedy this situation:

4. *Reaching Law Approach condition* [87]: Gao et al. define the attributes of a discrete-time sliding mode regime as follows:
   - Starting from any initial condition, **e** should move monotonically towards S and cross it in finite time.
   - After the first crossing, the trajectory will cross S again during every successive sampling instant, giving rise to a chattering-like trajectory.
   - The amplitude of each successive "chatter" is non-increasing and so the trajectory remains within a boundary layer of S.

   For the above attributes to be satisfied, they propose the following condition

   $$s(k+1) - s(k) = -qT\text{sign}[s(k)] - \epsilon T\text{sign}[s(k)],$$

   where constants ϵ, $q > 0$, $1 - qT > 0$ and T denotes the sampling period. Unfortunately, [87] lacks an explicit proof of the claim that the above condition does indeed satisfy the three specified attributes for obtaining a discrete-time sliding mode.

5. *Discrete-Time Sliding Mode condition*: Utkin and Drakunov [259] introduced the concept of a discrete-time sliding mode system as one whose dynamics are able to reach some manifold in finite time, and subsequently remain confined to it. Bartolini et al. [34] have shown that according to this definition, a sliding regime could be obtained *without* using a discontinuous control law and that a piecewise-constant control is satisfactory. This has the advantage of avoiding chattering problems. They further clarify Utkin and Drakunov's concept by the following definition [34]:

Definition 5.9.1. *A discrete-time sliding mode is obtained in some subset Σ of manifold S if $\mathbf{e}(k+1) \in \Sigma$ for any $\mathbf{e}(k) \in U$, where U is an open n-dimensional neighbourhood of Σ.*

Note that the above condition implies that $s(k+1) = 0$, since Σ is part of manifold S. Hence a discrete-time sliding mode is obtained if we find a control law which ensures that $s(k+1) = 0$, independent of any previous value of $s(k)$. This makes the sliding mode controller very similar to a dead-beat controller, with $s(k)$ taking the role of the usual regulated output variable.

5.9.2 Adaptive Sliding Mode Control

The main attraction of sliding mode control methodology lies in its ability to ensure robustness to modelling imprecision. This assumes that the level of imprecision is known, because it is taken into consideration in the design. On the contrary, if the system parameters or their level of uncertainty are unknown, the sliding regime may not be reached and instability might ensue. This problem was first solved for the continuous-time case by introducing concepts from adaptive control [231], so as to reduce the unknown imprecision, whilst simultaneously preserving as much as possible the appropriate sliding conditions by generating trajectories that lie close to the desired sliding manifold.

For the linear discrete-time case, Furuta [86] developed an adaptive sliding mode scheme based on condition (3) outlined in the previous section. This utilizes a VSC law that switches across a sliding *sector* rather than the usual sliding manifold. The scheme was extended to linear MIMO systems and applied to control an underwater Remotely Operated Vehicle in [59].

Bartolini et al. [34] developed an alternative adaptive sliding mode scheme for linear systems that guarantees the asymptotic satisfaction of the Discrete-Time Sliding Mode condition (5) outlined before. Chan extended the approach to handle plants that are subject to an additive bounded disturbance, by utilizing dead-zone adaptation [49, 50]. In contrast to Furuta's adaptive scheme, these methods avoid VSC laws.

Recently there have been some attempts to apply adaptive sliding mode control to functional uncertain nonlinear systems [161, 181, 240]. All these utilize neural networks, together with VSC control laws. The stability results of these schemes are somewhat limited however. In [181] neither Lyapunov stability nor the condition for a convergent quasi-sliding regime is ensured, whilst [161] only ensures the convergence of a model prediction error rather than the actual plant tracking error. The work of [240] has already been reviewed in Section 2.3.2 whose main limitations, namely the difficulty to correctly choose a number of important design parameters and a strong assumption introduced to obtain a causal control law, were already pointed out.

In this section we therefore re-address the nonlinear functional adaptive sliding mode problem to obtain more complete stability results [130]. The approach avoids VSC laws by utilizing ideas from the scheme of Bartolini et al. [34], but extended to the nonlinear case. We show that a suitable modification of the control law developed in the beginning of this chapter, leads to a novel and more globally stable adaptive sliding mode system when compared with those currently proposed for functional uncertain nonlinear plants. As in the first part of this chapter, error convergence and parameter boundedness are proved by Lyapunov stability techniques. The proposed methodology ensures that the discrete-time sliding mode condition (5) of Utkin and Drakunov is approached asymptotically up to within a boundary layer, whose size mainly depends on the network's optimal approximation accuracy.

5.9.3 Problem Formulation

The plant under consideration is the nth order nonlinear system of Equation (5.1), reproduced here for convenience:

$$y(k+1) = f[\mathbf{x}(k)] + u(k).$$

According to Definition 5.9.1, a discrete-time sliding mode in the sense of Utkin and Drakunov is obtained by a control law ensuring that $\forall k \geq 0$; the sliding condition $s(k+1) = 0$ is satisfied for any initial $\mathbf{e}(0) \in U$. If $f(\mathbf{x})$ were known, it is straightforward to see from the definition of $s(k)$, that this condition is satisfied by the control law

$$u(k) = y_d(k+1) - f[\mathbf{x}(k)] - c_1 e(k) - c_2 e(k-1) - \cdots - c_{n-1} e(k-n+2). \quad (5.29)$$

Note that this control is not of the variable structure type and yet, given any initial condition, the sliding manifold S is reached within one time step and future trajectories remain confined to it.

Since we are interested in the situation where $f(\mathbf{x})$ is unknown, adaptive control techniques are used so as to reduce the impact of this uncertainty, whilst preserving as much as possible the ideal sliding condition. For this purpose we will use the same neural network $\hat{f}[\mathbf{x}(k), \hat{\mathbf{w}}(k)]$ described in Section 5.3, together with the same notation and assumptions.

5.9.4 The Control Law

Consider the control law

$$u(k) = y_d(k+1) - \hat{f}[\mathbf{x}(k), \hat{\mathbf{w}}(k)] + \sigma(k) \tag{5.30}$$

where $\sigma(k) := -c_1 s(k) - c_1 e(k) - \cdots c_{n-1} e(k-n+2)$.

Note that this control law is not simply a certainty equivalent version of Equation (5.29). It was purposely chosen so as to obtain a first order relation between $s(k)$ and $\tilde{f}[\mathbf{x}(k-1)]$ and hence use the same stability analysis as that developed previously for adaptive control of discrete-time systems, with the role of $e(k)$ being replaced by $s(k)$. This is seen by substituting Equation (5.30) into the plant Equation (5.1) and using the definitions for $s(k)$ and $\sigma(k)$, yielding:

$$s(k+1) + c_1 s(k) = \tilde{f}[\mathbf{x}(k)],$$

or in transfer function form:

$$s(k) = \Gamma^{-1}(z^{-1})[\tilde{f}[\mathbf{x}(k-1)]], \tag{5.31}$$

where $\Gamma(z^{-1})$ has been defined before and it is assumed that $-c_1$, the root of $\Gamma(z^{-1})$, lies within the unity circle.

Note that this relation is similar to Equation (5.6) with $s(k)$ replacing $e(k)$. Hence if an augmented error, adaptation law and auxiliary signal similar to Equations (5.7), (5.9) and (5.12) are used, with appropriate replacement for signals expressed in terms of $e(k)$ by $s(k)$, a stable adaptive scheme in terms of $s(k)$ is obtained. These laws are specified in the next section.

5.9.5 The Adaptive System

Since the aim is to ensure that the sliding condition is approached, adaptation is driven by the objective of reducing $s(k)$ to zero, rather than $e(k)$. Hence, on the same principle as in the previous controller, the augmented error $s_1(k)$ is proposed:

$$s_1(k) := \beta(s(k) - \Gamma^{-1}(z^{-1})[v(k)]), \tag{5.32}$$

where the auxiliary signal $v(k)$ is given by

$$v(k) = v_1(k) + v_2(k), \tag{5.33}$$

where $v_1(k) = \dfrac{\beta}{2\gamma c_1^2} \Phi^T[\mathbf{x}(k-1)] \Phi[\mathbf{x}(k-1)] s_1(k)$

and $v_2(k) = G s_1(k)$, $G > 0$ is a constant.

The adaptation law also takes the same form of the non-sliding adaptive scheme, but in $s_1(k)$ rather than $e_1(k)$:

$$\Delta \hat{\mathbf{w}}(k) = \begin{cases} \dfrac{\beta}{\gamma c_1^2} \Phi[\mathbf{x}(k-1)] s_1(k) & \text{if } |s_1(k)| > \epsilon_f/G \\ 0 & \text{if } |s_1(k)| \leq \epsilon_f/G \end{cases} \tag{5.34}$$

5.9 Extension to Adaptive Sliding Mode Control

where, as before, $\Delta\hat{\mathbf{w}}(k) := \hat{\mathbf{w}}(k) - \hat{\mathbf{w}}(k-1)$, γ is a strictly positive constant and c_1 is a non-zero constant with magnitude less than 1.

5.9.6 Stability Analysis

For this scheme, the following Lyapunov function candidate is proposed:

$$V_s(k) = s_1^2(k) + \gamma \tilde{\mathbf{w}}^T(k)\tilde{\mathbf{w}}(k).$$

Define the first difference $\Delta V_s(k) := V_s(k) - V_s(k-1)$. Then using Equations (5.31), (5.32) and adaptation law (5.34) in a similar manner to the non-sliding adaptive case of Section 5.6, we obtain an equation that is the same as (5.11) but having e_1 replaced by s_1. This leads to the following lemma:

Lemma 5.9.1. *Given the previous assumptions, the proposed Lyapunov function candidate $V_s(k)$, the system of Equation (5.1) and control law (5.30); then use of adaptation law (5.34) and auxiliary signal (5.33) result in $\Delta V_s(k) < 0$ for $|s_1(k)| > \epsilon_f/G$.*

Proof. The proof follows exactly the same argumentation as the proof of Lemma 5.6.1 with $s_1(k)$ replacing $e_1(k)$ and V_s, ΔV_s replacing V, ΔV. □

This lemma leads to the following theorem, which is the dual of Theorem 5.6.1:

Theorem 5.9.1. *Given the relevant previously-stated assumptions, when $\|\tilde{\mathbf{w}}(0)\|^2$, $s_1^2(0) < \infty$ and plant (5.1) is subjected to control law (5.30), adaptation law (5.34) and auxiliary signal (5.33); $s_1^2(k)$ and $\|\tilde{\mathbf{w}}(k)\|^2$ are bounded $\forall k$. In addition, $\lim_{k\to\infty} |s_1(k)| \leq \epsilon_f/G$.*

Proof. The proof follows the same argumentation as the proof of Theorem 5.6.1 with $s_1(k)$, $V_s(k)$, $\Delta V_s(k)$ replacing $e_1(k)$, $V(k)$, $\Delta V(k)$, Lemma 5.9.1 replacing Lemma 5.6.1 and set I redefined as $I := \left\{ (\|\tilde{\mathbf{w}}\|, s_1) \,\middle|\, |s_1| \leq \epsilon_f/G \right\}$. □

5.9.7 Sliding and Tracking Error Convergence

As in the non-sliding adaptive case, the convergence of $s_1(k)$ specified in Theorem 5.9.1 leads to the convergence of the sliding error $s(k)$ to a bound that depends on the value of c_1, G and the network approximation accuracy ϵ_f. In fact using the same argumentation as in Section 5.7, with s and s_1 replacing e and e_1 respectively, it follows that when G is chosen to satisfy the condition

$$G \gg \frac{1-|c_1|}{\beta} + \frac{\beta\overline{\Phi}}{2\gamma c_1^2}, \tag{5.35}$$

then

$$\lim_{k \to \infty} |s(k)| \leq \frac{\epsilon_f}{1 - |c_1|}. \tag{5.36}$$

Hence in the steady-state, the ideal sliding manifold $s = 0$ is approached to within a boundary layer of $\pm \epsilon_f/(1 - |c_1|)$.

From the definition of the sliding manifold, the tracking error $e(k)$ is related to the sliding error $s(k)$ according to the stable $(n-1)$th order dynamics:

$$s(k) = e(k) + c_1 e(k-1) + \cdots + c_{n-1}(k-n+1).$$

If these dynamics are chosen to have repeated $(n-1)$ poles at $z = -c_1$, i.e.,

$$s(k) = e(k)(1 + c_1 z^{-1})^{n-1},$$

then $e(k)$ could be considered as the output of a series connection of $(n-1)$ first order filters, each having transfer function $1/(1 + c_1 z^{-1})$. Denoting the output of the first of this series of filters as $e_{i_1}(k)$, it follows that

$$e_{i_1}(k) + c_1 e_{i_1}(k-1) = s(k).$$

This relation is identical in form to Equation (5.23), with α and v replaced by e_{i_1} and s. Hence, by the same reasoning used in deriving Expression (5.25), it follows that

$$\lim_{t \to \infty} |e_{i_1}(k+t)| \leq \frac{\overline{S}}{1 - |c_1|},$$

where $\overline{S}$ is an upper bound on $|s(k+1)|, \cdots, |s(k+t)|$. However, by Condition (5.35) and Expression (5.36), $\overline{S} \to \epsilon_f/(1 - |c_1|)$ for sufficiently large k, which yields:

$$\lim_{k \to \infty} |e_{i_1}(k)| \leq \frac{\epsilon_f}{(1 - |c_1|)^2}.$$

Applying the same reasoning to the rest of the series connection of $(n-1)$ filters, it follows that the tracking error (given by the output of the final filter) satisfies the relation

$$\lim_{k \to \infty} |e(k)| \leq \frac{\epsilon_f}{(1 - |c_1|)^n}.$$

Hence, with this choice of design parameters, the steady-state tracking error converges to a value of $0 \pm \epsilon_f/(1 - |c_1|)^n$.

5.9.8 Simulation Example

The adaptive sliding-mode scheme was tested by simulation of the same nonlinear system of Section 5.8.2. The same neural network and design parameter values were utilized and so $|s(k)|$ is expected to converge within the range $[0, 0.0056]$ and $|e(k)|$ to $[0, 0.0062]$. The results are shown in Figures 5.9, 5.10, 5.11 and 5.12. Figure 5.9 shows $s(k)$ asymptotically converging to the vicinity of the ideal sliding condition $s = 0$. Figure 5.10 shows that, as predicted,

$s(k)$ actually converges to within a boundary layer of ±0.0056 of the sliding condition in the steady-state. Finally, Figures 5.11 and 5.12 show that the tracking error $e(k)$ also reduces consistently and that it converges to the predicted error bound of ±0.0062.

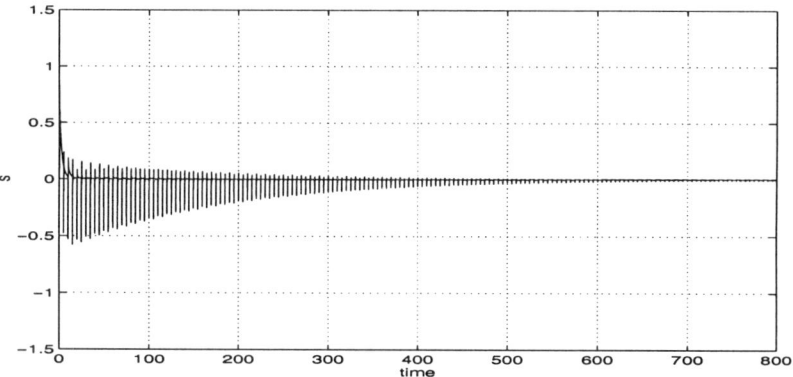

Fig. 5.9. Sliding error $s(k)$

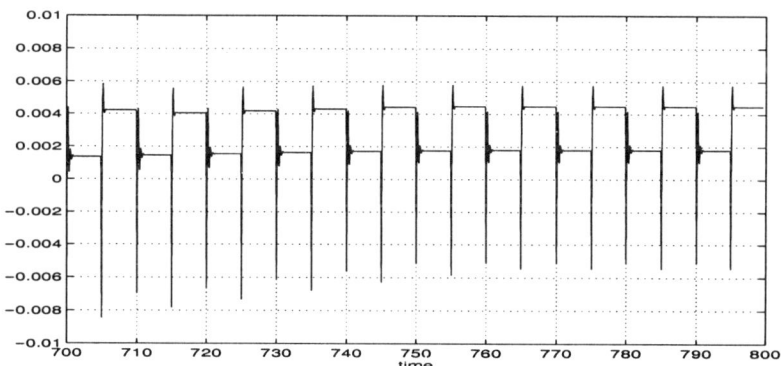

Fig. 5.10. Steady-state convergence of $s(k)$ to the boundary layer

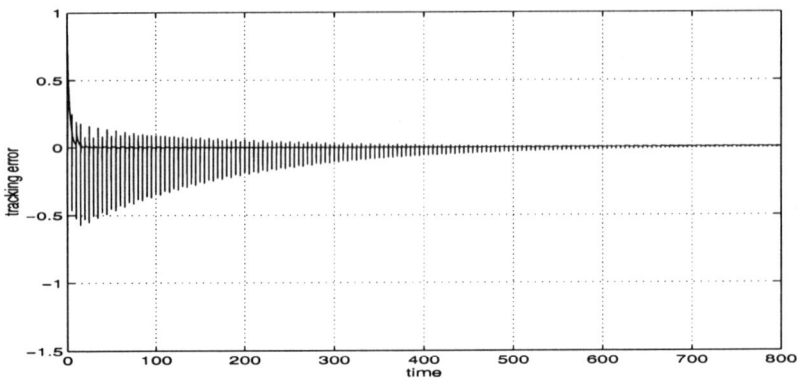

Fig. 5.11. Tracking error $e(k)$

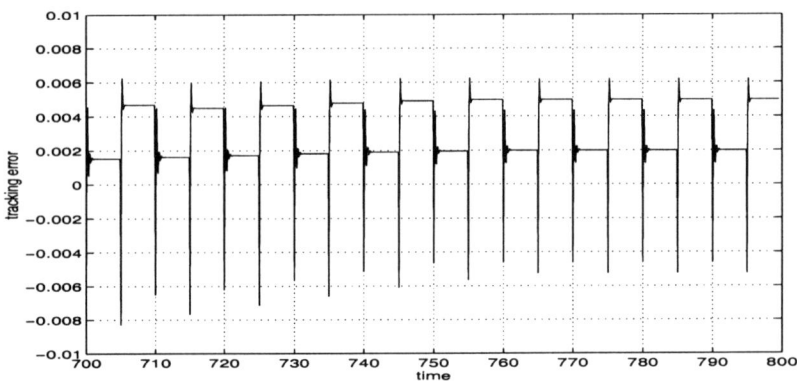

Fig. 5.12. Steady-state convergence of $e(k)$ to the error bound

5.10 Summary

When compared with the continuous-time case, fewer results have been published on functional adaptive control of discrete-time systems, particularly for schemes that are based on Lyapunov stability concepts. In this chapter we develop a novel stable adaptive scheme for a class of functional uncertain, nonlinear discrete-time systems. The scheme uses a GaRBF neural network whose parameters are adjusted in an on-line manner via augmented error techniques. A dead-zone adaptation law is used to take into consideration the possible de-stabilizing effect of the network inherent approximation error. The contribution of the scheme follows because the stability results obtained are more global than those of [53]. Additionally, as opposed to [120], the proposed design does not require any prior knowledge on the unknown optimal network parameters. The convergence and stability properties of the system are proved by Lyapunov stability concepts. An original analysis technique is introduced to prove asymptotic convergence of the tracking error and boundedness of the network parameters.

In this chapter, a number of issues concerning discrete-time sliding mode control have also been discussed. It is shown that the above adaptive scheme could be applied to effect adaptive, discrete-time sliding mode control of the same class of nonlinear systems via a slight modification of the control and adaptation laws. The method avoids the use of VSC laws and results in more complete stability results than other published neural network-based proposals for discrete-time adaptive sliding mode control.

Part III

Stochastic Systems

6. Stochastic Control

6.1 Introduction

In this chapter, some important techniques underpinning stochastic control theory are reviewed. Although the emphasis is on stochastic adaptive methods, *i.e.*, those dealing with systems having unknown parameters, completeness and clarity demand that occasionally the non-adaptive case is also considered. As in most publications on stochastic control, only the discrete-time case is considered because the controller is likely to be implemented as a digital computer programme.

Stochastic control theory deals with systems whose behaviour exhibits the characteristics of a stochastic process, *i.e.*, one whose variables have a certain degree of "chance" associated with the value they actually take and which is modelled in terms of mathematical probability or related concepts. The rationale for modelling systems in this way is threefold [15]:

1. Practical systems are subject to external disturbances or noise whose future values cannot be predicted precisely. Typically we only know something about the statistical properties, such as the mean or the variance or both.
2. The parameters of the system itself may be prone to vary at random, perhaps because of over-sensitivity to ambient temperature or component ageing.
3. It represents a natural way of dealing with uncertainty arising from imprecise knowledge about the system, such as state or parametric uncertainty. Note that this is different from the previous point, because the parameter itself might not be random but the uncertainty stems from our ignorance or lack of proper measurement devices.

The last point has obvious connections with adaptive control, even though most adaptive control schemes do not handle parametric uncertainty in this way. This sounds counterintuitive, but the fact is that very often complexity issues make it impossible to implement an adaptive controller that handles uncertainty in this way, unless a number of approximations are introduced. Nevertheless, a stochastic design for handling both external disturbances and parametric uncertainty should not be ruled out because it can offer distinct advantages in some situations.

All these ideas will be clarified as the chapter unfolds. More detailed descriptions are found in books [17, 114, 236], which give a good coverage of the basic ideas on stochastic control theory. Chapter 7 of [23] covers the basics of stochastic adaptive control. More advanced treatments are found in the seminal work of Fel'dbaum on dual control [76, 77, 78] and Aoki's subsequent extensions [14, 15]. Good tutorial papers include [27, 30, 238], whilst [149, 270, 271] offer good surveys of stochastic adaptive control.

6.2 Fundamental Principles

Due to the complexity associated with its implementation, results on stochastic control theory are rather fragmented, with the consequence that a number of different suboptimal techniques have been developed to handle specific problem classes. A common point of departure however, is to formulate the problem within an optimal control framework based on the minimization of an N-stage Performance Index. This is typically the mathematical expectation of a loss function L, as shown below:

$$J = E\{L(S^N, U^{N-1})\}$$

where

- $E\{\cdot\}$ denotes mathematical expectation taken over all the random variables.
- set $U^N := \{\mathbf{u}(k)\}_{k=0}^{N}$ is the collection of the plant control inputs $\mathbf{u}$ from beginning up to time N.
- Set $S^N := \{\mathbf{s}(k)\}_{k=0}^{N}$ is the collection of some system variable s to be controlled. Typically s is the state vector when using state-space models, or the tracking error when using input-output (IO) models.
- $L(\cdot)$ is a real valued loss function usually having the form

$$\sum_{k=0}^{N-1} c_k(\mathbf{s}(k+1), \mathbf{u}(k)),$$

where c_k is a function that reflects how s and u are required to behave during the N stages of the control horizon.

Due to the stochastic nature of the plant, s is a random variable. This makes the loss function $L(S^N, U^{N-1})$ a random variable as well, which could therefore not be minimized as such. This explains why L is converted to a related non-random quantity via the expectation operator in the Performance Index J.

The problem now reduces to finding the set of control inputs U^{N-1} that minimizes J. This is often called the *control policy* in the jargon of optimal control theory. The solution to the problem however depends on what information is made available to the controller at any one time. Due to

6.2 Fundamental Principles

causality, the "fullest" information that could be provided at any time k consists of all output measurements up to that time, collectively denoted by set $Y^k := \{\mathbf{y}(i)\}_{i=0}^{k}$, and all previous inputs U^{k-1}. The union of these two sets, denoted by $I^k := \{Y^k, U^{k-1}\}$; is called the *information state* and the optimal sequence is obtained by conditioning the expectation upon it at every time step k.

Using the techniques of Dynamic Programming, the minimizing control sequence is obtained by applying Bellman's Principle of Optimality [27, 30, 36, 238]. Briefly this states that at any time k, whatever control inputs were applied before, the overall minimum cost is obtained by using a policy that minimizes the remaining $(N-k)$ stages of the problem with respect to the state at time k. Hence if we start from the terminal point N and move backwards, optimizing at each backward pass, the overall optimal control policy should ensue. In mathematical terms this means that we should first find the optimal value of $\mathbf{u}(N-1)$ that minimizes J conditioned on the information state I^{N-1}, i.e.,

$$\min_{\mathbf{u}(N-1)} E\left\{L(S^N, U^{N-1})|I^{N-1}\right\}.$$

But I^{N-1} depends on $\mathbf{u}(N-2)$. Hence, moving one stage back, we must find $\mathbf{u}(N-2)$ that minimizes the expectation of the above cost, but conditioned on I^{N-2} for causality, i.e.,

$$\min_{\mathbf{u}(N-2)} E\left\{\min_{\mathbf{u}(N-1)} E\left\{L(S^N, U^{N-1})|I^{N-1}\right\}|I^{N-2}\right\}.$$

Repeating this backwards until the first stage, the optimal cost J^* is given as

$$J^* = \min_{\mathbf{u}(0)} E\left\{\min_{\mathbf{u}(1)} E\left\{\min_{\mathbf{u}(2)} E\left\{\min_{\mathbf{u}(3)} \cdots \right.\right.\right.$$
$$\left.\cdots \min_{\mathbf{u}(N-2)} E\left\{\min_{\mathbf{u}(N-1)} E\left\{L(S^N, U^{N-1})|I^{N-1}\right\}|I^{N-2}\right\}\cdots\right.$$
$$\left.\left.\left.\cdots|I^2\right\}|I^1\right\}|\mathbf{y}(0)\right\}$$

It can be shown [17, 30, 238] that when the loss function L is of the following additive type,

$$L = \sum_{k=0}^{N-1} c_k(\mathbf{s}(k), \mathbf{u}(k)) + c_N(\mathbf{s}(N)),$$

the Principle of Optimality leads to a simpler iterative equation for J^* in terms of c_k, called the *Bellman Equation*, which is shown below:

$$J_k^* = \min_{\mathbf{u}(k)} E\left\{c_k + J_{k+1}^*|I^k\right\}. \tag{6.1}$$

The term J_{k+1}^* appearing in the above equation is given as

$$J_{k+1}^* = \min_{\mathbf{u}(k+1)\cdots\mathbf{u}(N-1)} E\left\{c_N + \sum_{t=k+1}^{N-1} c_t(\mathbf{s}(t), \mathbf{u}(t))\right\}$$

and it is called the *optimal cost-to-go* at time $(k+1)$. It represents the optimal cost for the future stages with respect to the stage at time $(k+1)$.

In principle, the solution is found from the Bellman Equation by starting at $k = N-1$, solving $J_{N-1}^* = \min_{\mathbf{u}(N-1)} E\{c_{N-1} + c_N | I^{N-1}\}$ for every possible value of I^{N-1}, and working backwards. Although the Bellman Equation does show how to arrive at the optimal control policy, in practice it is not very helpful [23] because:

1. In general it does not lead to *analytical* closed form solutions, except for a few particular combinations of performance indices and plants.
2. A *numerical* solution is highly computationally and memory intensive, even for the simplest of systems, rendering its implementation practically impossible. This is mainly due to the *curse of dimensionality*, because a numerical solution requires discretization of the space of variables, whose order is typically > 3 even for simple systems.

6.3 Classes of Stochastic Control Problems

Broadly speaking, stochastic control has been applied on two main problem classes: (a) *known* state-space models with inaccessible states and (b) state-space or input-output (IO) models with *unknown* parameters. More details on this are given below.

(a) Incomplete State Information (ISI) problems. These are state-space models whose state vector $\mathbf{x}$ is not accessible for measurement. The stochastic nature of the system is a consequence of the presence of noise signals $\mathbf{v}(k)$ and $\mathbf{w}(k)$ that affect the state and measurement equations respectively, as follows:

$$\mathbf{x}(k+1) = \mathbf{f}(k, \mathbf{x}(k), \mathbf{u}(k), \mathbf{w}(k))$$
$$\mathbf{y}(k) = \mathbf{h}(k, \mathbf{x}(k), \mathbf{v}(k)).$$

Functions $\mathbf{f}$ and $\mathbf{h}$ are assumed known, and output $\mathbf{y}(k)$ is measurable at time k. The N-stage Performance Index to be minimized is typically of quadratic form

$$J_{NSS} = E\left\{\mathbf{x}^T(N)\mathbf{Q}_0\mathbf{x}(N) + \sum_{k=0}^{N-1} \mathbf{x}^T(k)\mathbf{Q}_1\mathbf{x}(k) + \mathbf{u}^T(k)\mathbf{Q}_2\mathbf{u}(k)\right\} \quad (6.2)$$

where weighting matrices $\mathbf{Q}_0, \mathbf{Q}_1$ are symmetric positive semidefinite and matrix $\mathbf{Q}_2$ is positive definite.

When $\mathbf{f}$ and $\mathbf{h}$ are linear, and noise signals $\mathbf{v}$ and $\mathbf{w}$ are mutually independent, zero-mean and Gaussian with known covariance matrices, the problem reduces to the well-known Linear Quadratic Gaussian (LQG) problem. It turns out that this is one of the few exceptions where Equation (6.1) admits a simple closed form solution [17, 37].

(b) Adaptive control problems. In this case, the system equations include some parameter vector θ that is unknown. Both state-space and IO models (ARX and ARMAX) have been considered, as detailed below:

1. *The adaptive ISI problem:* This concerns state-space models and the problem is similar to the ISI problem, in that the state is inaccessible for measurement, but with the additional presence of unknown parameters θ in the equations, *i.e.*:

$$\mathbf{x}(k+1) = \mathbf{f}(k, \mathbf{x}(k), \mathbf{u}(k), \theta, \mathbf{w}(k))$$
$$\mathbf{y}(k) = \mathbf{h}(k, \mathbf{x}(k), \mathbf{u}(k), \theta, \mathbf{v}(k)).$$

The problem could be reformulated as an "extended" ISI problem by defining an *augmented state* $\mathbf{x}_a$ consisting of the state $\mathbf{x}$ and the unknown parameter vector θ, $\mathbf{x}_a := [\mathbf{x}^T \; \theta^T]^T$ [225]. Hence, the original system with unknown parameters becomes

$$\mathbf{x}_a(k+1) = \overline{\mathbf{f}}(k, \mathbf{x}_a(k), \mathbf{u}(k), \mathbf{w}(k))$$
$$\mathbf{y}(k) = \mathbf{h}(k, \mathbf{x}_a(k), \mathbf{u}(k), \mathbf{v}(k)),$$

where

$$\overline{\mathbf{f}}(k, \mathbf{x}_a(k), \mathbf{u}(k), \mathbf{w}(k)) = \begin{vmatrix} \mathbf{f}(k, \mathbf{x}(k), \mathbf{u}(k), \theta(k), \mathbf{w}(k)) \\ \theta(k) \end{vmatrix}$$

and the Performance Index J_{NSS} of Equation (6.2) is reformulated in terms of $\mathbf{x}_a$ rather than $\mathbf{x}$, with $\mathbf{Q}_0$ replaced by $\overline{\mathbf{Q}}_0 = \begin{vmatrix} \mathbf{Q}_0 & 0 \\ 0 & 0 \end{vmatrix}$ and $\mathbf{Q}_1$ by $\overline{\mathbf{Q}}_1 = \begin{vmatrix} \mathbf{Q}_1 & 0 \\ 0 & 0 \end{vmatrix}$. Hence the system can now be treated as an ISI problem defined in terms of the augmented state vector $\mathbf{x}_a$.

2. *The adaptive IO problem:* This concerns stochastic IO models. Typically the linear ARMAX model is used:

$$\begin{aligned} y(k) = {} & a_1 y(k-1) + \cdots + a_n y(k-n) + \\ & b_1 u(k-d) + \cdots + b_p u(k-p-d+1) + \\ & e(k) + c_1 e(k-1) + \cdots + c_l e(k-l) \end{aligned} \quad (6.3)$$

where y, u and e respectively denote the system output, input and noise. The coefficients are assumed unknown and collected together as a parameter vector

$$\theta := [b_1 \cdots b_p \; a_1 \ldots a_n \; c_1 \cdots c_l]^T.$$

The noise is assumed to be zero-mean Gaussian identically, independently distributed (iid), with known variance.

The N-stage Performance Index is expressed in terms of the tracking error as

$$J_{NIO} = E\left\{\sum_{k=0}^{N-1}[y(k+d) - y_d(k+d)]^2 + q(k)u^2(k)\right\}, \qquad (6.4)$$

where y_d is the desired output and $q(k)$ is a weighting factor. Different approaches for handling this type of problem are explained in more detail in Section 6.4.2.

6.4 Dual Control

As pointed out before, in almost all situations, it is impossible to implement the ideal solution to the stochastic control problem as given by the Bellman Equation (6.1). The complexity arises as a result of the multistage optimization reflected by the N-stage summation in the Performance Index, and the presence of some unknown variables that require estimation, namely the state in the ISI problem and also the parameters in the adaptive problem. However, N-stage criteria of this type should not be ruled out because they lead to *dual control* schemes that take into consideration the interaction between estimation and control, and manage to strike a balance between these two conflicting demands. This yields some desirable control characteristics.

The interaction between estimation and control was first studied by Fel'dbaum who developed the concept of *dual control* [76, 77, 78]. This is a realization of the fact that the control signal can affect not only the value of the state or the tracking error, but also the accuracy of the estimation process. Additionally, the accuracy of the estimates affects the quality of the control performance. This leads to two important considerations that could be summed up as follows:

1. The control law typically depends directly on the estimates of the unknown variables. Hence the uncertainty of the current estimates directly affects the regulation or tracking performance. High uncertainty could lead to bad control action and large output errors unless due consideration is given to it when calculating the control input. This desirable feature is called *caution*.
2. The control input could actively act so as to render the estimation process more efficient and thus help in effecting accurate tracking or regulation even faster. In general, "richer" inputs help to reduce the estimation uncertainty more rapidly. Hence it would be desirable to plan the control input in a way that the estimator actively gathers richer information about the unknown variables and thereby reduce the uncertainty more rapidly. This attractive feature of the control law is called *probing*.

Dual control manages to strike a balance between caution and probing, so as to obtain an optimal response.

6.4.1 Degrees of Interaction

In a few specific situations, the interaction between estimation and control is absent. One such example is the LQG problem, which explains why the Bellman Equation reduces to a simple closed form solution for that case. To characterize explicitly the different levels of interaction (or lack of it) that might exist in a system, the following definitions have been proposed [29, 114, 238]:

Definition 6.4.1. NEUTRALITY: *A system is said to be neutral if the rate of reduction of uncertainty about the estimates is independent of the control input. Hence there is no reason for introducing probing in a neutral system.*

E.g., in the LQG problem, the Riccati equation propagating the estimation error covariance matrix $\mathbf{P}$ is independent of the control input.

Definition 6.4.2. SEPARABILITY: *A system is said to be separable if the N-stage optimal control law depends only on the value of the estimate and not on terms reflecting its accuracy or uncertainty. Hence no caution is required to optimally control a separable system.*

E.g., in the LQG problem, the estimation error covariance matrix $\mathbf{P}$ does not appear in the control law, only the state estimate $\hat{\mathbf{x}}$ does.

Definition 6.4.3. CERTAINTY EQUIVALENCE: *A certainty equivalent system is one that is (1) separable and (2) its optimal control law is identical to the optimal control law of the equivalent deterministic/known system, with the estimates directly replacing any unknown variables appearing in the deterministic/known control law.*

By an equivalent deterministic/known system we mean that for the case of the ISI problem there is no noise and the state is completely measurable, whilst for the adaptive control problem there are no unknown parameters. Note that certainty equivalence is a special case of separability. *E.g.*, the LQG control law is identical to the optimal control law of the deterministic linear quadratic problem with complete state information, but having state $\mathbf{x}$ replaced by the estimate $\hat{\mathbf{x}}$.

Of all these properties, Certainty Equivalence (CE) implies the least interaction between estimation and control. Note that although CE implies separability, the converse is not true. The LQG problem satisfies all three properties and that is why its optimal N-stage solution could be found in a straightforward way and is easy to implement. Reference [30] discusses a few other cases where, like the LQG case, CE holds.

6.4.2 Solutions to the Implementation Problem

In those situations where estimation and control do interact, a dual control scheme is desirable because caution and probing would enhance the performance of the system. Such situations include the ISI problem with nonlinear dynamics, the adaptive ISI problem (including the linear case) and the adaptive IO problem. However due to the complexity associated with implementing the optimal dual scheme by direct use of the Bellman Equation, a number of suboptimal solutions have been proposed. These could be grouped in two [270]:

1. *Non-dual solutions*; these disregard partially or completely the interaction between estimation and control, and lack either probing or caution or both.
2. *Suboptimal dual solutions*; these include approximations which are thought out in such a way that the resulting control law retains, to a certain extent, the desirable properties of dual control: caution and probing.

Non-dual Solutions. Non-dual schemes represent the crudest way of implementing a suboptimal solution to the ideal dual control scheme. There are two categories of non-dual solutions: (a) Heuristic Certainty Equivalent Control and (b) Cautious Control.

Heuristic Certainty Equivalent (HCE) Control. This represents the most straightforward non-dual solution by assuming heuristically that the system is certainty equivalent, even when this is not the case. Hence, an optimal control problem is first solved for the equivalent deterministic/known system and then the unknown variables are simply replaced by their estimates, calculated by some estimation algorithm. Such a control law lacks both caution and probing because the equivalent system from which the control law is derived does not require either feature. Hence if the actual plant is neither neutral nor separable, a HCE solution is likely to perform unsatisfactorily.

Despite this, the class of stochastic adaptive controllers that has been most widely used in industry - *the self-tuning regulator* - is of the HCE type [18, 22, 23, 94]. Self-tuning regulators tackle the adaptive IO problem for the ARMAX model of Equation (6.3). The control objective is specified in terms of the *Minimum Variance Criterion*, given by

$$J_{MV} = E\left\{[y(k+d) - y_d(k+d)]^2\right\}. \quad (6.5)$$

This could be considered as a simplified, one-stage version of Performance Index J_{NIO} of Equation (6.4). When the plant parameters θ are known, the solution to the problem is known as the *Minimum Variance Control Law* [17]. In its adaptive form, *i.e.*, the self-tuning regulator, a recursive extended least squares (ELS) algorithm is used to obtain an estimate $\hat{\theta}$ of the unknown parameter vector θ, which is then used on a HCE basis inside the Minimum Variance Control Law that is derived by minimizing Equation (6.5) on the assumption of known parameters. Åström and Wittenmark showed that *provided $\hat{\theta}$ converges*, the HCE controller would indeed optimize J_{MV} asymptotically [22]. This probably accounts for the popularity of self-tuning regulators in industry. However, establishing whether the parameter estimates really converge or not, is a crucial problem.

To overcome this, Goodwin *et al.* [91] obtained global convergence results by using a recursive stochastic approximation algorithm for parameter estimation instead of ELS. However, least squares estimators are usually faster to converge and so Sin and Goodwin [228] proposed a modified version of the least squares algorithm, to ensure global convergence of the self-tuning system. Recently, Guo and Chen [97, 98] have proved a number of important convergence properties on the self-tuning regulator as originally formulated, based on ELS estimation. An important extension of self-tuning control is the *Generalized Minimum Variance (GMV) Controller* of Clarke and Gawthrop [55, 56]. Unlike the self-tuning regulator, the GMV could also be used on non-minimum phase systems and it allows greater flexibility on the choice of the weighting coefficients in its performance index, which also penalizes the control input. More recent developments are concentrating on self-tuning type schemes for nonlinear stochastic systems subjected to parametric uncertainty [273].

A more sophisticated class of self-tuning schemes is the *Adaptive Predictive Control* methodology [23, 57, 58]. This is based on a receding-horizon version of performance index J_{NIO}, rather than the one-stage version of Equation (6.5). However, the philosophy remains of the HCE type because the parameter estimates are directly substituted inside a control law derived on the assumption of a known system.

Cautious Control. In contrast to their HCE counterpart, cautious controllers take the uncertainty of the estimates into consideration when calculating the control. However they do not actively plan any probing signals to reduce the future estimation uncertainty. Hence they are still classified as non-dual schemes. In brief, a cautious controller applies caution but lacks probing. This comes about from the simplifications introduced when minimizing the ideal performance index.

For the adaptive IO case, the simplification is applied straightaway on performance index J_{NIO} by reducing it to a one-stage function, just like the Minimum Variance Criterion J_{MV} of Equation (6.5). However, contrary to the self-tuning methodology, certainty equivalence is not assumed and the control law is derived by minimization of J_{MV}, but with the uncertainty of the unknown parameters taken into consideration by treating them as random variables. Instead of working on the ARMAX model, cautious control is typically applied on the less general ARX model, that takes form:

$$y(k) = a_1 y(k-1) + \cdots + a_n y(k-n) + b_1 u(k-1) + \cdots + b_p u(k-p) + e(k). \quad (6.6)$$

The unknown parameter vector

$$\theta := [b_1 \cdots b_p \; a_1 \cdots a_n]^T$$

is assumed to have a prior Gaussian distribution with known mean and covariance matrix, and the noise $e(k)$ is assumed to be zero mean Gaussian. This permits the use of a Kalman filter to calculate the mean $\hat{\theta}(k+1)$, and covariance $\mathbf{P}(k+1)$, of θ conditioned on the information state I^k. This same mean is taken to be the predictive estimate of θ, and the covariance indicates the uncertainty of the estimate. Both terms appear in the control law that results from minimization of performance index J_{MV}, because θ is now treated as a random variable. It is the appearance of $\mathbf{P}(k+1)$ inside the control law which leads to the desirable feature of caution. Hence when the estimates are poor, matrix $\mathbf{P}$ has large elements, which typically leads to small control signals $u(k)$. This helps to eliminate the typical large overshoot that would have resulted when using an HCE control scheme. However it also has the disadvantage of reducing the richness of the input, with the consequence that $\mathbf{P}$ might always remain appreciably large and the controller may get caught up in a vicious circle of a low control encouraging large covariance, which in turn leads to lower control. This phenomenon is called *turn-off* and is the obvious consequence of applying caution without any probing. Further details on cautious control for adaptive IO systems are discussed in [21, 23, 269].

For the ISI state-space problem, the equivalent of cautious control is usually called the *Open-Loop Optimal Feedback (OLOF)* policy. This is based around a simplification in the procedure used for optimizing the N-stage Performance Index J_{NSS} of Equation (6.2), introduced in order to obtain a practical suboptimal solution. The optimal solution given by the Bellman Equation (6.1) is called the *Closed-Loop (CL)* policy because it takes into consideration the availability of the information state I^k during *all* stages of the control sequence. This explains why the inner iterate J_{k+1}^* of the Bellman Equation has a future value with respect to the subject of the equation J_k^*, and why the Bellman Equation has to be solved by starting from the final stage and then working backwards. By contrast, the suboptimal OLOF policy ignores the fact that the information state is going to be available during future stages. Essentially, at every time step k, the sequence $\mathbf{u}(k) \cdots \mathbf{u}(N-1)$ that minimizes an expression similar to J_k^* (the cost-to-go at time k) *but* conditioned on I^k is calculated, *i.e.*:

$$\arg\min_{\mathbf{u}(k)\cdots\mathbf{u}(N-1)} E\left\{c_N + \sum_{t=k}^{N-1} c_t(\mathbf{s}(t), \mathbf{u}(t))|I^k\right\}. \tag{6.7}$$

The conditioning on I^k shows that the minimization is based only on whatever information has been available up to time k; future values of the information state are ignored. That is why the minimization is "open-loop" with respect to the information from time $(k+1)$ onwards. This fact and the absence of iteration, renders a much simpler determination of the sequence $\mathbf{u}(k) \cdots \mathbf{u}(N-1)$ than would have been the case from the optimal Bellman Equation. The first element of this sequence, $\mathbf{u}(k)$, is applied as control at time k and the procedure is repeated for subsequent time steps.

The OLOF policy algorithm is shown in the following pseudo-code listing:

FOR $k = 0 \cdots N-1$:

Calculate the OLOF policy: $\arg \min\limits_{\mathbf{u}(k)\cdots\mathbf{u}(N-1)} E\left\{c_N + \sum\limits_{t=k}^{N-1} c_t | I^k\right\}$

Apply $\mathbf{u}(k)$ as input to the plant

REPEAT

Note that this procedure is *not* equivalent to the iterations of the Bellman Equation and its policy is a suboptimal solution. As in the cautious control scheme for adaptive IO models, since the future state is assumed unavailable at every time step, the OLOF policy lacks probing and only exhibits caution. Further details on the OLOF policy are discussed in [247].

Suboptimal-dual Solutions. In contrast with non-dual solutions, suboptimal dual schemes seek to obtain a control law that is easy to implement, but which still retains the desirable properties of dual control: caution and probing. There are two distinct ways of accomplishing this task [79, 270]:

- *Implicit solutions*, which try to find approximate ways of solving the Bellman Equation without losing the dual characteristics of the ideal control law.
- *Explicit solutions*, which modify the one-stage cost function of cautious control by explicitly including a term related to parameter estimation, so as to induce a "probing-like" effect. This approach has been applied on adaptive IO models only.

Implicit solutions. The problems associated with obtaining an exact numerical solution of Equation (6.1) are not only related to the curse of dimensionality, but also to the calculation of expectations. This requires the availability of appropriate *a posteriori* probability density functions which, in general, are not available in analytical closed form. Fel'dbaum [78] and Aoki [14, 15] develop recursive methods for generating these density functions, but the process remains quite involved and computationally intensive, with the exception of some very simple problems [238]. In an attempt to mitigate this situation, Alspach [6] proposed to approximate the density functions by Gaussian sums and developed suboptimal control laws therefrom. The approach is still computationally demanding however, both in terms of calculating the approximate densities and also because the curse of dimensionality problem is still not addressed.

Tse *et al.* developed a suboptimal scheme for the nonlinear ISI problem which they called *wide-sense adaptive dual control* [250]. The technique was also applied on a linear adaptive ISI problem [248]. The approximations introduced consist of the following:

1. The information state I^k is approximated by what they call the wide-sense information state, defined as

    $P^k := \{\hat{\mathbf{s}}(k|k), \mathbf{P}(k|k)\},$

where $\hat{\mathbf{s}}(k|k)$ and $\mathbf{P}(k|k)$ are the conditional mean and covariance of state $\mathbf{s}(k)$, assumed to be calculated by some *approximate* method such as the Extended Kalman filter (EKF) or a second order filter.

2. At each time k, an approximation to the optimal cost-to-go J^*_{k+1} in Equation (6.1) is found by first expanding up to second order terms the expression

$$c_N + \sum_{t=k+1}^{N-1} c_t(\mathbf{s}(t), \mathbf{u}(t))$$

about a *nominal* trajectory $\{\mathbf{s}_0(k+2), \cdots, \mathbf{s}_0(N)\}$. This trajectory is the state sequence that the equivalent *deterministic* plant would follow as a result of applying some *nominal* control sequence

$$\{\mathbf{u}_0(k+1), \cdots, \mathbf{u}_0(N-1)\};$$

using the conditional prediction of the state $\hat{\mathbf{s}}(k+1|k)$ from the EKF as initial condition $\mathbf{s}_0(k+1)$. The nominal control sequence is typically the solution from the HCE or OLOF non-dual control scheme.

Then the expectation of this second-order expansion, conditioned on the wide-sense information state, is minimized with respect to

$$\{\mathbf{u}(k+1) \cdots \mathbf{u}(N-1)\}$$

in a closed-loop fashion, yielding an approximation $J^*_{0_{k+1}}$ to J^*_{k+1}, the optimal cost-to-go. Being a second-order expansion about a nominal, this task is much less complex than if the actual cost-to-go were found.

3. Finally the following problem is solved

$$\min_{\mathbf{u}(k)} E\left\{c_k + J^*_{0_{k+1}} | P^k\right\}.$$

This minimization must be done by using a search procedure, such as quadratic interpolation, because $\mathbf{u}(k)$ appears nonlinearly in the equation. This value of $\mathbf{u}(k)$ is then applied to the system and the procedure repeated at every time instant.

This suboptimal scheme retains both caution and probing, the latter being due to the fact that the calculation of the approximate cost-to-go is performed on a closed-loop basis. However the scheme is still rather complicated and computationally involved, because at every time step one must determine the nominal trajectory, perform a second-order expansion about it, evaluate the approximate cost-to-go and use a search procedure to effect the final minimization in terms of the approximate cost-to-go. Dersin et al. [64] report some interesting analytical comparisons between the wide-sense dual control and the optimal (CL) control policy for a very simple linear adaptive ISI problem. They show that the former is a first-order approximation of the latter if the parameter covariances are small. On the other hand, the HCE control is only a zeroth-order approximation, and hence inferior to the wide-sense dual policy.

Recently, a wide-sense adaptive control approach has been developed for adaptive IO systems (ARX models) by Pronzato et al. [148, 209]. The main difference from the original work is that rather than using an EKF to calculate the wide-sense information state for future values i.e.,

$$\{\hat{\mathbf{s}}(k+j|k), \mathbf{P}(k+j|k)\}, \ j > 0,$$

they assume that $\hat{\mathbf{s}}(k+j|k) = \hat{\mathbf{s}}(k|k)$ and $\mathbf{P}(k+j|k)$ is approximated by the inverse of the information matrix calculated at time k.

Explicit solutions. As noted in the previous section, implicit suboptimal solutions remain somewhat computationally intensive. Explicit solutions avoid this by starting off with the simplified performance index of cautious control and then extend it by the addition of some terms that enhance the quality of the estimation. A control law is then derived by minimizing this extended criterion. This approach has been applied on the adaptive IO problem, typically using ARX models. Various extensions to the cautious performance index of Equation (6.5) have been proposed, a few of which are discussed next.

One idea depends on the addition of a term that encourages $\mathbf{P}$ (the conditional covariance matrix of the parameters) to reduce, leading to a performance index of the form

$$J = \min_{u(k)} E\left\{[y(k+1) - y_d(k+1)]^2 + \lambda f | I^k\right\}$$

where $\lambda > 0$ is a constant design parameter and f is the additional term, for which various forms have been proposed. Goodwin and Payne's Actively Adaptive Controller (AAC) [92] sets it as

$$f = -\frac{|\mathbf{P}(k+1)|}{|\mathbf{P}(k+2)|},$$

whilst Wittenmark and Elevitch's Actively Suboptimal Dual Controller (ASOD) [272] sets

$$f = \sigma^2 \frac{\mathbf{P}_{b_1}(k+2)}{\mathbf{P}_{b_1}(k+1)},$$

where $\mathbf{P}_{b_1}$ is the $[1,1]$ element of the covariance matrix $\mathbf{P}$, corresponding to the conditional variance of parameter b_1 in Equation (6.6). This parameter is the coefficient of the most recent input to affect the output, and hence its role is considered as being crucial for the control law. However, in ASOD, the minimization requires a numerical search method because the loss function is not convex but includes two local minima. Allison et al. [5] extended this control law to cover ARX systems that include an input-output delay > 1 and applied it to control a thermomechanical pulping refiner.

Instead of concentrating on the covariance matrix, Milito et al. [176] suggest an additional term that utilizes the innovations

$$\epsilon(k+1) := y(k+1) - \hat{\theta}^T(k+1)\phi(k),$$

where $\phi(k) := [u(k)\ u(k-1)\cdots u(k-p+1)\ y(k)\cdots y(k-n+1)]^T$ is the regressor vector and $\hat{\theta}(k+1)$ is the conditional mean of θ derived from a Kalman filter. The aim is to encourage $\epsilon^2(k+1)$ to remain high so that the Kalman filter updates of $\hat{\theta}$, which depend on $\epsilon(k+1)$, are driven by richer information. Hence they choose

$$f = -\epsilon^2(k+1),$$

with $0 \le \lambda \le 1$. In the case when $\lambda = 0$, Milito et al.'s Innovations Dual Control (IDC) law reduces to the usual cautious controller and when $\lambda = -1$ it is the same as the HCE controller. For values of λ in between, a good compromise is reached between the cautious case, which lacks probing and often leads to turn-off, and the HCE case, which does not take into consideration the estimate uncertainty and often leads to excessive transient overshoot. Ishihara et al. [112] extended the IDC law to cover an ARX model having delay > 1 and Chan and Zarrop [51] introduced concepts from IDC into the Generalized Minimum Variance framework of Clarke and Gawthrop's HCE self-tuning control for ARMAX systems. Radenković [211] analyzed the convergence properties of Chan and Zarrops's controller, concluding that, subject to the passivity of two time-varying operators, the system is globally stable.

Padilla et al. [195] choose the additional term f as a combination of two sensitivity functions (i.e., the partial derivative of the output with respect to the unknown parameters). The logic behind the scheme is that the sensitivity functions reflect the amount of information on the parameters carried by the output.

Some other explicit approaches address the suboptimal dual problem in a different manner. For example, Filatov et al.'s Bicriterial Approach [79, 80] depends on two separate performance indices: (i) J_k^c that involves minimization of the expected value of the tracking error and which is therefore similar to the cautious performance index, and (ii) J_k^a that involves the minimization of the expected value of an innovations term, similar to the choice of f in the IDC scheme. However, in the Bicriterial Approach the minimizations are kept separate, in the sense that first the cautious control is calculated on the basis of J_k^c and then J_k^a is minimized such that u remains inside a domain Ω that is symmetrically distributed around the cautious control. The control from J_k^a therefore produces a perturbation around the cautious control to enhance the innovations and, as a consequence, the estimation process. The size of Ω is made proportional to the trace of the covariance matrix $\mathbf{P}$, so that the larger the uncertainty, the greater is the excitation signal due to J_k^a. In a different scheme, Alster and Belanger [8] minimize the usual cautious performance index under the constraint that the trace of the information matrix $\mathbf{P}^{-1}$ should remain greater than some lower limit. Maitelli and Yoneyama [167] suggest extending the cautious performance index from one to two stages, i.e., $E\left\{\sum_{i=1}^{2}[y(k+i) - y_d(k+i)]^2 | I^k\right\}$. The resulting control law is more complex than cautious control, but is not as "short-sighted" and accounts for an efficient reduction of future uncertainty.

A third approach for deriving explicit suboptimal dual solutions involves direct modification of the cautious control signal. Jacobs and Patchell [115] and Wieslander and Wittenmark [269] suggest the addition of a white noise or a pseudo random binary sequence onto the cautious control when the covariance exceeds some limit. Hughes and Jacobs [105] suggest lower bounding the cautious control signal such that its magnitude is never less than some threshold value, thereby avoiding turn-off.

6.5 Conclusions

The principles of stochastic control suggest an elegant framework for dealing with uncertainty. Imprecise system knowledge is handled by treating the unknowns as random variables that are characterized by probability distributions. The control task is usually specified as an optimal control problem that requires minimization of some suitable N-stage Performance Index via Dynamic Programming. This results in a *dual controller* that possesses the interesting property of taking into consideration the interaction that exists between the estimation of unknown information and control. Hence the dual controller actively seeks to learn better the uncertainty of the system, for the purpose of effecting better control. Unfortunately, in most cases it is impossible to find an analytical closed form solution to the optimization problem, and realistic computational limitations prohibit the implementation of a numerical solution. Hence a number of suboptimal schemes are used in practice. Some of these, *e.g.* self-tuning regulators, ignore completely the attractive properties of dual control. Other schemes, the so-called suboptimal dual controllers, attempt to preserve the dual properties as much as possible, according to the constraints dictated by realistic implementation issues.

Although the ideal dual control solution cannot be practically implemented, the majority of suboptimal dual schemes are certainly feasible. This is particularly true for explicit solutions applied to the adaptive IO problem, which easily outperform the non-dual class of self-tuning regulators, particularly when the control horizon is short, the initial parameter uncertainty is large or the parameters are changing rapidly [23]. Dual control has been applied on a number of practical problems where improved performance has been reported. A few examples include optimization of economic systems [31], chip refiner control in the pulp industry [5] and roll angle control of a vertical-takeoff pilot aircraft [79]. The active learning features of dual control suggest interesting possibilities for intelligent control schemes, where the levels of uncertainty are typically large and the benefits of dual control should therefore be more pronounced. These ideas are investigated in the following chapters.

7. Dual Adaptive Control of Nonlinear Systems

7.1 Introduction

In this chapter, two suboptimal dual adaptive control schemes for a stochastic class of functional uncertain nonlinear systems are developed. The two schemes are based on GaRBF and sigmoidal MLP neural networks respectively. The idea of applying dual control principles within a functional adaptive context first appeared in [72]. Most other approaches typically adopt an HCE procedure that often leads to an inadequate transient response because the initial uncertainty of the unknown network parameters is large. Some of the neural network control schemes that have been put forward avoid this by performing intensive off-line training to identify the plant in open-loop and reduce the prior uncertainty of the unknown parameters [53, 193, 215]. Only later is an adaptive control phase started, with the initial network parameters set to the pre-trained values that are already substantially close to the optimal. In a certain sense this procedure defeats the main objective of adaptive control because the off-line training phase reduces most of the uncertainty existing prior to application of the control.

In the dual control scheme proposed here, the pre-control neural network training phase is avoided by taking into consideration the parameter uncertainty and its effect on tracking, at the same time that control is being effected. This is more in tune with the features expected from adaptive control, and also more efficient and economical. In practice, performing an off-line training scheme is usually time consuming and expensive. The proposed scheme is of the explicit suboptimal dual type. It is based around the Innovations Dual Control (IDC) law of Milito et al. [176] originally developed for parametric uncertain ARX systems. Here it is extended to cover the case of functional uncertain nonlinear systems and functional adaptive control by neural networks. The proposed performance index includes an "innovations-squared" term as in the IDC, and also an additional cost term to penalize the control input. In the GaRBF controller, a Kalman filter is used to estimate the parameters and their uncertainty. The MLP controller is more complicated because the unknown parameters do not appear linearly in the equations. Hence additional assumptions have to be introduced, and an Extended Kalman filter (EKF) is used for parameter estimation. The main contributions of this chapter could therefore be summarized as follows:

148   7. Dual Adaptive Control of Nonlinear Systems

1. The derivation of a suboptimal dual control scheme for a class of functional uncertain, stochastic nonlinear systems. This represents a generalization of traditional suboptimal dual control schemes developed for the adaptive IO problem, which have been concerned only with linear ARX models.
2. The use of stochastic adaptive control techniques within a functional adaptive control paradigm, utilizing both GaRBF and MLP neural networks.
3. The performance improvement over HCE neural-adaptive schemes, particularly the avoidance of an off-line neural network training phase preceding control operation.
4. The extension of the IDC suboptimal dual scheme for systems that are nonlinear in the parameters, as represented by the proposed MLP dual controller.

7.2 Problem Formulation

The objective is to control a stochastic, single-input single-output, affine class of nonlinear, discrete-time systems having the general form:

$$y(k) = f[\mathbf{x}(k-1)] + g[\mathbf{x}(k-1)]u(k-1) + e(k) \tag{7.1}$$

where $y(k)$ is the output, $u(k)$ is the control input, $\mathbf{x}(k-1) = [y(k-n)\ldots y(k-1)\quad u(k-1-p)\ldots u(k-2)]^T$ is the system state vector, $f[\mathbf{x}(k-1)], g[\mathbf{x}(k-1)] : \Re^{n+p} \mapsto \Re$ are unknown nonlinear functions of the state and $e(k)$ is an additive noise signal. The output is required to track a reference input $y_d(k)$. The following conditions are also assumed to hold:

Assumption 7.2.1 *The noise $e(k)$ is independent and has a zero-mean Gaussian distribution of variance σ^2.*

Assumption 7.2.2 *The state vector's dimensionality parameters p and n, and the noise variance σ^2 are known.*

Assumption 7.2.3 *The reference input is known at least one time step ahead and is bounded.*

Assumption 7.2.4 *The system is minimum phase and $g(\mathbf{x})$ is bounded away from zero.*

The problem is addressed from a stochastic adaptive perspective. This means that we have to find an *admissible* control sequence that minimizes some specified Performance Index. Admissibility implies that $u(k)$ must be restricted to depend upon the information state I^k, which consists of all the outputs measured up to the present time, Y^k, and all the previous inputs, U^{k-1}. This ensures that the control law derived from the optimization procedure is causal. Using the terminology from probability theory, when $u(k)$ is some function of (k, I^k) we say that $u(k)$ is I^k-*measurable*.

If the nonlinear functions $f[\mathbf{x}(k-1)]$, $g[\mathbf{x}(k-1)]$ were known, and the system were noiseless, then control law

$$u(k) = \frac{y_d(k+1) - f[\mathbf{x}(k)]}{g[\mathbf{x}(k)]} \quad (7.2)$$

yields $y(k+1) = y_d(k+1)$, which does indeed minimize the desirable N-stage performance index J_{NIO} of Equation (6.4) with $d = 1$ and $q(k) = 0$. Equation (7.2) therefore represents the optimal control law of the deterministic/known system. Note that Assumption 7.2.4 ensures that this control law does not give rise to unbounded $u(k)$. Even in the presence of independent noise $e(k)$, the same control law results in $y(k+1) - y_d(k+1) = e(k+1)$, which is also optimal because the term in the summation of J_{NIO} would then be $e^2(k+1)$, which is independent of $u(k)$ and could not be minimized further [23].

For the case of unknown functions, however, the problem reduces to the *adaptive* IO form, whose optimal N-stage solution is known to require a dual controller. In addition, the nonlinear nature of the unknown functions suggests the use of neural networks as one possible way of estimating and representing them.

7.3 Dual Controller Design

The dual controller based on GaRBF networks is derived first, because its design is more straightforward. This is then followed by the more involved sigmoidal MLP-based controller.

7.3.1 GaRBF Dual Controller

Two GaRBF neural networks are used to approximate the nonlinear functions $f[\mathbf{x}(k-1)]$ and $g[\mathbf{x}(k-1)]$ within a compact set $\chi \subset \Re^{n+p}$, that represents the network approximation region. The outputs of the two neural networks approximating functions f and g are respectively given by

$$\begin{aligned} \hat{f}_r[\mathbf{x}, \hat{\mathbf{w}}_\mathbf{f}] &= \hat{\mathbf{w}}_\mathbf{f}^T \Phi_\mathbf{f}[\mathbf{x}] \\ \hat{g}_r[\mathbf{x}, \hat{\mathbf{w}}_\mathbf{g}] &= \hat{\mathbf{w}}_\mathbf{g}^T \Phi_\mathbf{g}[\mathbf{x}] \end{aligned} \quad (7.3)$$

where $\hat{\mathbf{w}}_\mathbf{f}, \hat{\mathbf{w}}_\mathbf{g}$ are the network output layer parameter vectors and $\Phi_\mathbf{f}, \Phi_\mathbf{g}$ are the Gaussian basis function vectors. The following conditions are assumed to hold for the neural networks:

Assumption 7.3.1 *The system state is always confined within a bounded region that is also enclosed by χ. Note that χ could be chosen arbitrarily large by the designer.*

Assumption 7.3.2 *There exist some optimal basis function centres, width parameters and output layer parameters ensuring that within χ, the magnitude of the network approximation errors are bounded by a negligibly small value. It is assumed that these basis function centres and width parameters are known a priori.*

This assumption is justified by the Universal Approximation Property of neural networks and, as explained before, there exist systematic ways of choosing appropriate basis function centres and width parameters *a priori*. This leaves the *optimal* output layer parameter vectors of the two networks, denoted by $\mathbf{w}_\mathbf{f}^*$ and $\mathbf{w}_\mathbf{g}^*$, as the only unknown variables. They will therefore be estimated by recursive adjustment of $\hat{\mathbf{w}}_\mathbf{f}$ and $\hat{\mathbf{w}}_\mathbf{g}$.

Parameter Estimation. By Assumptions 7.3.1, 7.3.2 and Equations (7.1) and (7.3), it follows that within region χ, the system could be represented in the following state-space form:

$$\begin{aligned}\mathbf{w}^*(k+1) &= \mathbf{w}^*(k) \\ y(k) &= \mathbf{w}^{*T}(k)\Phi[\mathbf{x}(k-1)] + e(k)\end{aligned} \quad (7.4)$$

where $\mathbf{w}^*(k) = [\mathbf{w}_\mathbf{f}^{*T}(k) \vdots \mathbf{w}_\mathbf{g}^{*T}(k)]^T$ and $\Phi[\mathbf{x}(k-1)] = [\Phi_\mathbf{f}^T[\mathbf{x}(k-1)] \vdots \Phi_\mathbf{g}^T[\mathbf{x}(k-1)]u(k-1)]^T$.

Since the optimal parameters requiring estimation appear linearly in the above measurement equation, the well established techniques of Kalman filtering [17, 121] could be used if the following additional assumption on the optimal output layer parameters is imposed:

Assumption 7.3.3 *The optimal network parameter vector $\mathbf{w}^*(k)$ is a random variable whose initial value $\mathbf{w}^*(0)$ has a Gaussian distribution of mean $\mathbf{m}$ and covariance matrix $\mathbf{R}$.*

Note that in practice the significance of $\mathbf{m}$ and $\mathbf{R}$ is that the former reflects the initial estimate of the parameters based on *a priori* knowledge, whilst the latter reflects the accuracy of this estimate; larger values indicating great uncertainty, and hence less confidence in the accuracy of the initial estimate [163]. We are now in a position to state the following lemma on parameter estimation.

Lemma 7.3.1. *Subject to Assumptions 7.2.1, 7.2.2, 7.3.1, 7.3.2 and 7.3.3, the distribution of $\mathbf{w}^*(k+1)$ conditioned on information state I^k is Gaussian and the optimal minimum mean-square (predictive) estimate of $\mathbf{w}^*$ conditioned on I^k, is given by the mean:*

$$\hat{\mathbf{w}}(k+1) := E\left\{\mathbf{w}^*(k+1)|I^k\right\}.$$

The conditional covariance of $\mathbf{w}^*(k+1)$, defined as

$$\mathbf{P}(k+1) := E\left\{[\mathbf{w}^*(k+1) - \hat{\mathbf{w}}(k+1)][\mathbf{w}^*(k+1) - \hat{\mathbf{w}}(k+1)]^T|I^k\right\},$$

7.3 Dual Controller Design

and the mean $\hat{\mathbf{w}}(k+1)$, satisfy the following recursive Kalman filter equations:

$$\mathbf{K}(k) = \frac{\mathbf{P}(k)\Phi[\mathbf{x}(k-1)]}{\sigma^2 + \Phi^T[\mathbf{x}(k-1)]\mathbf{P}(k)\Phi[\mathbf{x}(k-1)]}$$
$$\hat{\mathbf{w}}(k+1) = \hat{\mathbf{w}}(k) + \mathbf{K}(k)\left\{y(k) - \hat{\mathbf{w}}^T(k)\Phi[\mathbf{x}(k-1)]\right\} \quad (7.5)$$
$$\mathbf{P}(k+1) = \left\{\mathbf{I} - \mathbf{K}(k)\Phi^T[\mathbf{x}(k-1)]\right\}\mathbf{P}(k)$$

with initial conditions $\hat{\mathbf{w}}(0) = \mathbf{m}$, $\mathbf{P}(0) = \mathbf{R}$.

Additionally, the distribution of $y(k+1)$ conditioned on I^k is also Gaussian with mean $\hat{\mathbf{w}}^T(k+1)\Phi[\mathbf{x}(k)]$ and variance $\Phi^T[\mathbf{x}(k)]\mathbf{P}(k+1)\Phi[\mathbf{x}(k)] + \sigma^2$.

Proof. The proof follows directly by applying a standard predictive type Kalman filter on the system represented by state space Equations (7.4), which are a valid representation of plant (7.1) by Assumptions 7.3.1 and 7.3.2. □

Note that the mean and variance of the density $p(y(k+1)|I^k)$ depend on $u(k)$ through $\Phi[\mathbf{x}(k)]$. Since I^k does not include $u(k)$, this might appear as an inconsistency. However this is not the case because $u(k)$ is I^k-measurable and any dependence on $u(k)$ could be re-expressed as a function of (k, I^k).

The Kalman filter Equations (7.5) therefore represent the adaptation law for this dual control scheme. Note that although the Kalman filter might incur a significant computational burden when estimating the parameters of large networks due to a large $\mathbf{P}$, its use is nevertheless fundamental for updating the conditional probability distribution of y. This information is essential in dual control because the uncertainty of the estimates is not ignored. If the computational burden could not be met in practice, some *ad hoc* techniques might be introduced to relieve the situation, such as treating the covariance matrix $\mathbf{P}$ as a diagonal matrix and storing only the diagonal terms.

The Control Law. The control law depends upon an explicit-type, suboptimal dual performance index based on the innovations dual controller developed by Milito et al. [176] for linear systems. This performance index explicitly includes a term concerning the *innovations* ϵ at time $(k+1)$, defined as

$$\epsilon(k+1) := y(k+1) - \hat{\mathbf{w}}^T(k+1)\Phi[\mathbf{x}(k)].$$

The idea is to reward performance that encourages the size of the innovations to remain high, so that the parameter updating in Equation (7.5) (which depends directly on the innovations $\epsilon(k)$) is driven by richer information. Hence the performance index takes the form:

$$J_{inn} = E\{[y(k+1) - y_d(k+1)]^2 + qu^2(k) + r\epsilon^2(k+1)|I^k\} \quad (7.6)$$

where $E\{\cdot|I^k\}$ denotes mathematical expectation conditioned on the information state I^k, and design parameters q and r are scalar weighting factors chosen within the range $q \geq 0$, $-1 \leq r \leq 0$.

152   7. Dual Adaptive Control of Nonlinear Systems

The difference between this performance index and that originally proposed in [176], is the inclusion of a penalty term for $u(k)$. Higher q induces a penalty on large control signals, reflecting that in practice the control amplitude needs to be constrained. Factor r affects the innovations and is used to induce a dual-like effect. This is explained in detail further on. Minimization of the explicit suboptimal dual performance index (7.6) leads to the control law specified in the following theorem.

Theorem 7.3.1. *The control law minimizing performance index J_{inn} of Equation (7.6) subject to the system of Equation (7.1) and all previously-mentioned assumptions, is given by*

$$u^*(k) = \frac{\left(y_d(k+1) - \hat{f}_r[\cdot]\right)\hat{g}_r[\cdot] - (1+r)v_{gf}}{\hat{g}_r^2[\cdot] + q + (1+r)v_{gg}} \quad (7.7)$$

where the arguments $[\cdot]$ of $\hat{f}_r$ and $\hat{g}_r$ are $[\mathbf{x}(k), \hat{\mathbf{w}}_\mathbf{f}(k+1)]$ and $[\mathbf{x}(k), \hat{\mathbf{w}}_\mathbf{g}(k+1)]$ respectively, and

$$v_{gf} := \Phi_\mathbf{g}^T[\mathbf{x}(k)]\mathbf{P}_{\mathbf{gf}}(k+1)\Phi_\mathbf{f}[\mathbf{x}(k)]$$
$$v_{gg} := \Phi_\mathbf{g}^T[\mathbf{x}(k)]\mathbf{P}_{\mathbf{gg}}(k+1)\Phi_\mathbf{g}[\mathbf{x}(k)],$$

and matrix $\mathbf{P}(k+1)$ has been repartitioned as:

$$\mathbf{P}(k+1) = \begin{vmatrix} \mathbf{P}_{\mathbf{ff}}(k+1) & \vdots & \mathbf{P}_{\mathbf{gf}}^T(k+1) \\ \cdots & \cdots & \cdots \\ \mathbf{P}_{\mathbf{gf}}(k+1) & \vdots & \mathbf{P}_{\mathbf{gg}}(k+1) \end{vmatrix}$$

where $\mathbf{P}_{\mathbf{ff}}, \mathbf{P}_{\mathbf{gg}}$ are square $(n_{rf} \times n_{rf}), (n_{rg} \times n_{rg})$ sub matrices with n_{rf}, n_{rg} denoting the number of basis functions in the $\hat{f}_r, \hat{g}_r$ networks respectively.

Proof. By the Gaussian distribution of $y(k+1)$ conditioned on I^k as specified in Lemma 7.3.1, and using the general result that for a Gaussian random variable ξ, $E\{\xi^2\} = [E\{\xi\}]^2 + \text{variance}\{\xi\}$, it follows that

$$E\left\{[y(k+1) - y_d(k+1)]^2 | I^k\right\} = \{\hat{\mathbf{w}}^T(k+1)\Phi[\mathbf{x}(k)] - y_d(k+1)\}^2 + \Phi^T[\mathbf{x}(k)]\mathbf{P}(k+1)\Phi[\mathbf{x}(k)] + \sigma^2$$

and

$$E\left\{\epsilon^2(k+1)|I^k\right\} = \Phi^T[\mathbf{x}(k)]\mathbf{P}(k+1)\Phi[\mathbf{x}(k)] + \sigma^2.$$

Hence the performance index J_{inn} could be re-written as

$$J_{inn} = (r+1)(\Phi^T[\mathbf{x}(k)]\mathbf{P}(k+1)\Phi[\mathbf{x}(k)] + \sigma^2) + qu^2(k) + \{\hat{\mathbf{w}}^T(k+1)\Phi[\mathbf{x}(k)] - y_d(k+1)\}^2.$$

Re-expressing $\Phi[\mathbf{x}(k)]$ in terms of its originally defined components, $\Phi_\mathbf{f}[\mathbf{x}(k)]$ and $\Phi_\mathbf{g}[\mathbf{x}(k)]u(k)$, yields,

$$\begin{aligned}
J_{inn} = (1+r) &\{\Phi_{\mathbf{f}}^T[\mathbf{x}(k)]\mathbf{P}_{\mathbf{ff}}(k+1)\Phi_{\mathbf{f}}[\mathbf{x}(k)] \\
&+ 2\Phi_{\mathbf{g}}^T[\mathbf{x}(k)]\mathbf{P}_{\mathbf{gf}}(k+1)\Phi_{\mathbf{f}}[\mathbf{x}(k)]u(k) \\
&+ \Phi_{\mathbf{g}}^T[\mathbf{x}(k)]\mathbf{P}_{\mathbf{gg}}(k+1)\Phi_{\mathbf{g}}[\mathbf{x}(k)]u^2(k) + \sigma^2\} + qu^2(k) \\
&+ \{\hat{\mathbf{w}}_{\mathbf{f}}^T(k+1)\Phi_{\mathbf{f}}[\mathbf{x}(k)] + \hat{\mathbf{w}}_{\mathbf{g}}^T(k+1)\Phi_{\mathbf{g}}[\mathbf{x}(k)]u(k) - y_d(k+1)\}^2.
\end{aligned}$$

Note that this equation is quadratic in $u(k)$ and that the coefficients of $u^2(k)$ are all positive. This means that J_{inn} has a unique minimum with respect to $u(k)$, which is found directly by differentiation with respect to $u(k)$ and equating to zero, leading to control law (7.7). □

7.3.2 Sigmoidal MLP Dual Controller

As opposed to the GaRBF network, the sigmoidal MLP network does not preserve the advantage of linearity in the unknown parameters. Hence its parameter adjustment rules tend to be more complex. However, because the support of its basis functions is not localized, a relatively smaller number of units is usually required to achieve similar levels of function approximation accuracy. This consideration is particularly important for cases of high dimensional state, where RBF networks are hindered by the *curse of dimensionality* problem. Hence the use of MLP networks ought to be given significant attention.

The dual controller developed in this section utilizes two sigmoidal MLP networks, each having one hidden layer and one summing output node, to approximate the unknown functions $f[\mathbf{x}(k-1)]$, $g[\mathbf{x}(k-1)]$ inside a compact set χ of state space, where the state is known to be contained. The output of the two neural networks is respectively given by

$$\begin{aligned}
\hat{f}_s[\mathbf{x}, \hat{\mathbf{c}}_{\mathbf{f}}, \hat{\mathbf{W}}_{\mathbf{f}}] &= \hat{\mathbf{c}}_{\mathbf{f}}^T \Phi_{\mathbf{f}}[\mathbf{x}, \hat{\mathbf{W}}_{\mathbf{f}}] \\
\hat{g}_s[\mathbf{x}, \hat{\mathbf{c}}_{\mathbf{g}}, \hat{\mathbf{W}}_{\mathbf{g}}] &= \hat{\mathbf{c}}_{\mathbf{g}}^T \Phi_{\mathbf{g}}[\mathbf{x}, \hat{\mathbf{W}}_{\mathbf{g}}]
\end{aligned} \quad (7.8)$$

where $\hat{\mathbf{c}}_{\mathbf{f}}, \hat{\mathbf{c}}_{\mathbf{g}}$ are vectors containing the parameters (weights) of the output layer unit, $\hat{\mathbf{W}}_{\mathbf{f}}$, $\hat{\mathbf{W}}_{\mathbf{g}}$ are matrices whose columns are the parameter vectors of the hidden units, and $\Phi_{\mathbf{f}}$, $\Phi_{\mathbf{g}}$ are the sigmoidal activation function vectors. These represent the basis functions of the units in the hidden layer, whose ith element is given by:

$$\Phi_{f_i} = \frac{1}{1 + \exp(-\hat{\mathbf{w}}_{\mathbf{f}_i}^T \mathbf{x_a})}$$

$$\Phi_{g_i} = \frac{1}{1 + \exp(-\hat{\mathbf{w}}_{\mathbf{g}_i}^T \mathbf{x_a})},$$

where $\hat{\mathbf{w}}_{\mathbf{f}_i}^T$, $\hat{\mathbf{w}}_{\mathbf{g}_i}^T$ are the parameter vectors of the ith basis function in the hidden layer. These are also the columns of $\hat{\mathbf{W}}_{\mathbf{f}}$, $\hat{\mathbf{W}}_{\mathbf{g}}$. Vector $\mathbf{x_a} := [\mathbf{x}^T \vdots 1]^T$ denotes the system state vector augmented by an additional constant input serving as a bias term. The number of basis functions in the hidden layers of the $\hat{f}_s$ and $\hat{g}_s$ networks is denoted by n_{sf} and n_{sg} respectively.

As in the GaRBF case, we assume the following conditions on the neural networks:

Assumption 7.3.4 *The state is always confined within a compact region that is enclosed by χ and there exist some optimal network parameters ensuring that within χ, the magnitude of the network approximation errors are bounded by a negligibly small value.*

The values of the optimal network parameters are unknown and hence require estimation. For convenience, the network parameters are grouped in a single vector $\hat{\mathbf{w}}$, defined as follows:

$$\hat{\mathbf{w}} := [\hat{\mathbf{c}}_\mathbf{f}^T \; \hat{\mathbf{w}}_{\mathbf{f}_1}^T \; \ldots \; \hat{\mathbf{w}}_{\mathbf{f}_i}^T \; \ldots \; \hat{\mathbf{c}}_\mathbf{g}^T \; \hat{\mathbf{w}}_{\mathbf{g}_1}^T \; \ldots \; \hat{\mathbf{w}}_{\mathbf{g}_i}^T \; \ldots]^T.$$

In contrast with the RBF network, not all of these appear linearly in Equations (7.8), because the parameters of the hidden layer neurons are also assumed unknown.

Parameter Estimation. Denoting the unknown *optimal* network parameters by vector

$$\mathbf{w}^* := [\mathbf{c}_\mathbf{f}^{*T} \; \mathbf{w}_{\mathbf{f}_1}^{*T} \; \ldots \; \mathbf{w}_{\mathbf{f}_i}^{*T} \; \ldots \; \mathbf{c}_\mathbf{g}^{*T} \; \mathbf{w}_{\mathbf{g}_1}^{*T} \; \ldots \; \mathbf{w}_{\mathbf{g}_i}^{*T} \; \ldots]^T,$$

and using Assumption 7.3.4 and Equations (7.1) and (7.8), the plant could be represented by the following equations inside χ:

$$\begin{aligned} \mathbf{w}^*(k+1) &= \mathbf{w}^*(k) \\ y(k) &= h(\mathbf{w}^*, \mathbf{x}(k-1), u(k-1)) + e(k) \end{aligned} \quad (7.9)$$

where

$$\begin{aligned} h(\mathbf{w}^*, \mathbf{x}(k-1), u(k-1)) := \; & \mathbf{c}_\mathbf{f}^{*T} \Phi_\mathbf{f}[\mathbf{x}(k-1), \mathbf{W}_\mathbf{f}^*] + \\ & \mathbf{c}_\mathbf{g}^{*T} \Phi_\mathbf{g}[\mathbf{x}(k-1), \mathbf{W}_\mathbf{g}^*] u(k-1) \end{aligned} \quad (7.10)$$

is a nonlinear function of the unknown optimal parameters $\mathbf{w}^*$, and $\mathbf{W}_\mathbf{f}^*$, $\mathbf{W}_\mathbf{g}^*$ are matrices whose columns are the optimal hidden unit parameter vectors $\mathbf{w}_{\mathbf{f}_i}^*$, $\mathbf{w}_{\mathbf{g}_i}^*$.

Since the parameters to be estimated do not appear linearly in the system model, nonlinear estimation techniques have to be used. The Extended Kalman filter (EKF) [9, 114, 121] is the most widely used nonlinear estimator and for our case it also represents a natural progression from the (linear) Kalman filter used in the RBF network case. The idea of using an EKF for weight estimation of neural networks is not new. For example, it has been applied in system identification by MLP networks [140, 263] and shown to give better results than the back-propagation training algorithm, and also for sequential learning with GaRBF networks [136].

The EKF applied to the model of Equation (7.9) gives the following recursive equations

7.3 Dual Controller Design

$$\mathbf{K}(k) = \frac{\mathbf{P}(k)\nabla_{\mathbf{h}}^{T}(k)}{\sigma^2 + \nabla_{\mathbf{h}}(k)\mathbf{P}(t)\nabla_{\mathbf{h}}^{T}(k)}$$

$$\hat{\mathbf{w}}(k+1) = \hat{\mathbf{w}}(k) + \mathbf{K}(k)\left\{y(k) - h(\hat{\mathbf{w}}(k), \mathbf{x}(k-1), u(k-1))\right\} \quad (7.11)$$

$$\mathbf{P}(k+1) = \left\{\mathbf{I} - \mathbf{K}(k)\nabla_{\mathbf{h}}(k)\right\}\mathbf{P}(k)$$

with initial conditions $\hat{\mathbf{w}}(0) = \mathbf{m}, \mathbf{P}(0) = \mathbf{R}$. As in the GaRBF case, these initial conditions respectively reflect the prior estimate of the unknown optimal parameter vector and its uncertainty.

$\nabla_{\mathbf{h}}(k)$ denotes the gradient of $h(\mathbf{w}^*, \mathbf{x}(k-1), u(k-1))$ with respect to $\mathbf{w}^*$ evaluated at $\mathbf{w}^* = \hat{\mathbf{w}}(k)$, i.e.,:

$$\nabla_{\mathbf{h}}(k) = \left[\frac{\partial h}{\partial w_1^*} \frac{\partial h}{\partial w_2^*} \cdots \frac{\partial h}{\partial w_i^*} \cdots\right]_{|\mathbf{w}^* = \hat{\mathbf{w}}(k)}$$

where w_i^* denotes the ith element of $\mathbf{w}^*$. From Equation (7.10), this could be expressed in closed form as follows:

$$\nabla_{\mathbf{h}}(k) = [\nabla_{\mathbf{h_f}}(k) \vdots \nabla_{\mathbf{h_g}}(k)u(k-1)]$$

where

$$\nabla_{\mathbf{h_f}}(k) = [\Phi_{\mathbf{f}}^T \ldots \hat{c}_{f_i} \exp(-\hat{\mathbf{w}}_{f_i}^T \mathbf{x_a})(\Phi_{f_i})^2 \mathbf{x_a}^T \ldots], \quad i = 1 \ldots n_{sf}$$

$$\nabla_{\mathbf{h_g}}(k) = [\Phi_{\mathbf{g}}^T \ldots \hat{c}_{g_i} \exp(-\hat{\mathbf{w}}_{g_i}^T \mathbf{x_a})(\Phi_{g_i})^2 \mathbf{x_a}^T \ldots], \quad i = 1 \ldots n_{sg}$$

and $\hat{c}_{f_i}$ and $\hat{c}_{g_i}$ denote the ith element of vectors $\hat{\mathbf{c}}_\mathbf{f}, \hat{\mathbf{c}}_\mathbf{g}$ respectively. Note that in the above equations, all parameter estimates are indexed in terms of k whilst the state is indexed in terms of $(k-1)$.

To be able to proceed in a similar manner as in the GaRBF controller, the following approximation is introduced:

$$p(\mathbf{w}^*(k+1)|I^k) \approx N(\hat{\mathbf{w}}(k+1), \mathbf{P}(k+1)). \quad (7.12)$$

This means that we are approximating the conditional distribution of $\mathbf{w}^*(k+1)$ by a Gaussian of mean $\hat{\mathbf{w}}(k+1)$ and covariance matrix $\mathbf{P}(k+1)$ as calculated by the EKF Equations (7.11). It should be emphasized that this represents an *approximation* and does not follow naturally as in Lemma 7.3.1 for the GaRBF controller, even if the prior distribution of the parameters were assumed Gaussian, because now the plant representation of Equations (7.9) is nonlinear. By contrast, the linearity in the parameters of Equations (7.4) in the GaRBF case, preserves the Gaussian propagation of all random variables.

Despite the above approximation regarding the conditional distribution of $\mathbf{w}^*(k+1)$, there are still no straightforward equations for the distribution of $y(k+1)$ conditioned on I^k, due to the nonlinear relationship between y and $\mathbf{w}^*$ in Equation (7.9). Hence a second approximation is introduced by expanding $y(k+1)$ as a first order Taylor series around $\mathbf{w}^* = \hat{\mathbf{w}}(k+1)$ in Equation (7.9), to yield the following approximate linear relation

$$y(k+1) \approx h(\hat{\mathbf{w}}(k+1), \mathbf{x}(k), u(k)) +$$
$$\nabla_{\mathbf{h}}(k+1)(\mathbf{w}^*(k+1) - \hat{\mathbf{w}}(k+1)) + e(k+1), \qquad (7.13)$$

where $\nabla_{\mathbf{h}}(k+1)$ is the same as $\nabla_{\mathbf{h}}(k)$ but evaluated at $\hat{\mathbf{w}}(k+1)$, $u(k)$ and $\mathbf{x}(k)$. These two approximations lead to the following lemma.

Lemma 7.3.2. *On the basis of approximations (7.12) and (7.13), the conditional distribution of $y(k+1)$ given I^k is also approximately Gaussian with mean $h(\hat{\mathbf{w}}(k+1), \mathbf{x}(k), u(k))$ and variance $\nabla_{\mathbf{h}}(k+1)\mathbf{P}(k+1)\nabla_{\mathbf{h}}^T(k+1) + \sigma^2$.*

Proof. The proof follows directly from the linearity of Equation (7.13), the Gaussian distribution of noise e and the approximate conditional distribution of $\mathbf{w}^*$ given by Equation 7.12. □

The Control Law. Using the same performance index J_{inn} as for the GaRBF controller, with the innovations defined as $\epsilon(k+1) := y(k+1) - h(\hat{\mathbf{w}}(k+1), \mathbf{x}(k), u(k))$, together with the approximations leading to Lemma 7.3.2, we obtain the optimal control law specified in the following theorem.

Theorem 7.3.2. *The control law minimizing performance index J_{inn} of Equation (7.6) subject to system (7.1), Assumptions 7.2.1, 7.2.2, 7.2.3, 7.3.4 and the approximations of Equations (7.12) and (7.13), is given by*

$$u^*(k) = \frac{(y_d(k+1) - \hat{f}_s[\cdot])\hat{g}_s[\cdot] - (1+r)\mu_{gf}}{\hat{g}_s^2[\cdot] + q + (1+r)\mu_{gg}}, \qquad (7.14)$$

where the arguments $[\cdot]$ of $\hat{f}_s$ and $\hat{g}_s$ are $[\mathbf{x}(k), \hat{\mathbf{c}}_\mathbf{f}(k+1), \hat{\mathbf{W}}_\mathbf{f}(k+1)]$ and $[\mathbf{x}(k), \hat{\mathbf{c}}_\mathbf{g}(k+1), \hat{\mathbf{W}}_\mathbf{g}(k+1)]$ respectively,

$$\mu_{gf} := \nabla_{\mathbf{h}_\mathbf{g}}(k+1)\mathbf{P}_{\mathbf{gf}}(k+1)\nabla_{\mathbf{h}_\mathbf{f}}^T(k+1)$$
$$\mu_{gg} := \nabla_{\mathbf{h}_\mathbf{g}}(k+1)\mathbf{P}_{\mathbf{gg}}(k+1)\nabla_{\mathbf{h}_\mathbf{g}}^T(k+1),$$

and covariance matrix $\mathbf{P}(k+1)$ has been re-partitioned as before, but in this case $\mathbf{P}_{\mathbf{ff}}$, $\mathbf{P}_{\mathbf{gg}}$ are square $(n_{sf}(n+p+2) \times n_{sf}(n+p+2))$, $(n_{sg}(n+p+2) \times n_{sg}(n+p+2))$ sub-matrices respectively. The parameter estimates and the covariance matrix required in the above equations are calculated by the EKF Equations (7.11).

Proof. From the approximate Gaussian distribution of $y(k+1)$ conditioned on I^k as given in Lemma 7.3.2, and proceeding as for the proof of Theorem 7.3.1, it follows that

$$J_{inn} \approx (r+1)(\nabla_{\mathbf{h}}(k+1)\mathbf{P}(k+1)\nabla_{\mathbf{h}}^T(k+1) + \sigma^2) + qu^2(k) +$$
$$\{h(\hat{\mathbf{w}}(k+1), \mathbf{x}(k), u(k)) - y_d(k+1)\}^2.$$

Re-expressing $\nabla_{\mathbf{h}}(k+1)$ in terms of its components $\nabla_{\mathbf{h}_\mathbf{f}}(k+1), \nabla_{\mathbf{h}_\mathbf{g}}(k+1)u(k)$ and minimizing with respect to $u(k)$ leads to control law (7.14). □

7.3.3 Analysis of the Control Laws

The two control laws (7.7) and (7.14) for the GaRBF and MLP controller have a very similar structure. A closer look reveals that both controllers take into consideration the uncertainty of the parameter estimates via inclusion of the variance-related terms v_{gf}, v_{gg} in one case, and μ_{gf}, μ_{gg} in the other. Design parameter r acts as a weighting factor where at one extreme, with r set to -1, the controller completely ignores the parameter uncertainty terms and at the other extreme, with r set to 0, it gives maximum attention to them. For intermediate settings, $-1 < r < 0$, a balance is struck between these two extremes.

The case $r = 0$, $q = 0$ reduces to cautious control, because it makes J_{inn} equivalent to the cautious performance index. This carries with it the well-known disadvantages of turn-off and slowness of response, because too strong an emphasis is placed on the uncertainty of the parameter estimates. In fact, from the control law Equations (7.7) and (7.14), it follows directly that very small control signals will result when the uncertainty terms v_{gg} and μ_{gg} are large.

On the other hand, the case $r = -1$, $q = 0$ corresponds to HCE control. In fact, substitution of these values in the control laws yields an equation that is identical to replacing the actual nonlinear system functions f, g with the network approximations $\hat{f}, \hat{g}$ inside control law (7.2) of the corresponding deterministic/known plant. Hence function uncertainty is ignored completely, which results in excessively high peak overshoot and bad tracking, especially during the transient part of the response when the parameter estimates are highly inaccurate.

The case $-1 < r < 0$ represents a compromise between the above two extremes: the control is neither too cautious, thereby avoiding a sluggish response, nor too bold, thereby avoiding crude tracking performance. This is the philosophy behind innovations dual control, where the level of caution could be tuned within a range that varies from no caution at all, to full caution.

Apart from the practical consideration of restraining the energy of the control signal, the second design parameter q, also serves to prevent numerical ill-conditioning of the control law in case its denominator tends to zero. As seen in Equations (7.7) and (7.14), since the rest of the terms in the denominator are zero or positive, a value of $q > 0$ guarantees that u never tends to infinity and prevents division by zero.

Comparing the control laws of the two controllers, it is interesting to note that the Gaussian basis function vectors appearing in the uncertainty terms of the GaRBF controller are replaced by the corresponding gradient of h evaluated at $\hat{w}$ in the MLP controller. This reflects the principle behind the EKF technique that was used for parameter estimation in the MLP case; namely that a nonlinear system is linearized about the most recent estimate at every time instant. Consequently the statistical properties of the estimates, particularly the uncertainty terms μ_{gf} and μ_{gg}, are calculated on this basis. Although this technique is based on an approximation, the EKF provides us with a practical solution for handling nonlinearity.

7.4 Simulation Examples and Performance Evaluation

In the section we demonstrate the performance and benefits of the system by simulating the dual control schemes on two different plants that satisfy the general affine form and assumptions outlined before. Note that in all simulations, the neural networks were never subjected to an initial off-line training phase. Closed loop control was activated immediately, with the initial parameter estimates selected at random from a uniform distribution within the range $[-0.1, 0.1]$.

7.4.1 Example 1

The dynamic equations of the plant in this first example are:

$$y(k+1) = \sin(x(k)) + \cos(3x(k)) + (2 + \cos(x(k)))u(k) + e(k+1)$$

where state $x(k) = y(k)$ and the noise variance $\sigma^2 = 0.001$. The functions $f(x) = \sin(x(k)) + \cos(3x(k))$ and $g(x) = 2 + \cos(x(k))$ represent the unknown nonlinear dynamics.

The control reference input $y_d(k)$ is obtained by sampling a unit amplitude, $0.1Hz$ square wave filtered through a network of transfer function $1/(s+1)$. The sampling frequency is $10Hz$. A GaRBF controller is implemented in this example. The network approximation region is chosen as $\chi = [-2, 2]$. The $\hat{f}_r$ network is structured with nine Gaussian basis functions of width parameter 1, placed on a mesh of spacing 0.5. The $\hat{g}_r$ network consists of three basis functions having a width parameter of 1.9 and placed on a mesh of spacing 2. The initial parameter covariance of the Kalman filter is set to $\mathbf{P}(0) = 1000\mathbf{I}$.

For comparison purposes, trials were conducted using three different design-parameter settings corresponding to HCE ($r = -1$), cautious ($r = 0$) and innovations dual ($r = -0.7$) control. The same noise sequence, initial conditions, control penalty $q = 0.0001$, and reference input were used in each case. A typical output is shown in Figure 7.1. As expected, the figure shows that the HCE controller initially responds crudely, exhibiting large transient overshoot because it is not taking into consideration the inaccuracy of the parameter estimates. Only after the initial period, when the parameters converge, does the control assume good tracking. On the contrary the cautious controller (part (b) of the figure) does not react so hastily during the initial period, knowing that the parameter estimates are still inaccurate. But although there is no large overshoot, the controller is practically inactive during the first 2 seconds. The innovations dual controller (part (c) of the figure)

7.4 Simulation Examples and Performance Evaluation 159

strikes a compromise between these two extremes, clearly showing no particularly unacceptable overshoot whilst tracking the reference input at least one second earlier than the cautious controller. Hence, even qualitatively, it is clear that the performance of the innovations dual controller is the better one.

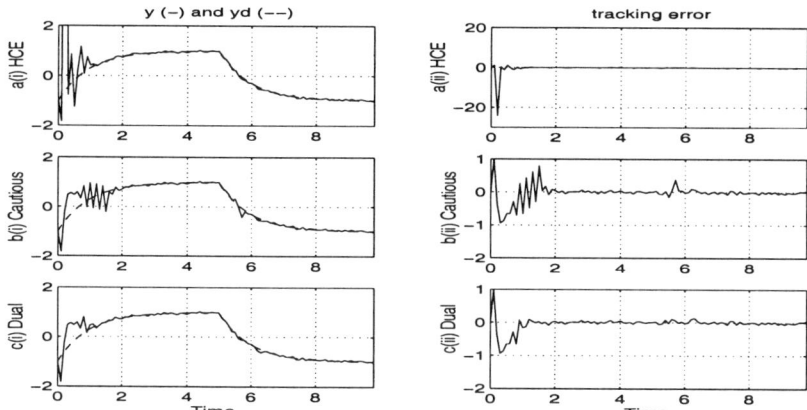

Fig. 7.1. Output and Tracking error; (a) HCE (b) Cautious (c) Dual. N.B: In plot a(i), the y-axis is truncated to enable clear visualization of the steady-state tracking. The actual amplitude during the initial period of the response could be seen in plot a(ii), which is purposely drawn at a different scale from the rest.

To quantify the performance objectively, a Monte Carlo analysis involving 500 trials was performed. The accumulated cost

$$V(T) = \sum_{k=0}^{T}(y_d(k) - y(k))^2$$

was calculated over the whole simulation interval time T after each trial. A fresh realization of the noise sequence was generated at each trial. The results, showing $V(T)$ for each trial are depicted in Figure 7.2. The average of the accumulated cost and the variance over 500 trials are shown in Table 7.1, where the lower values of mean and variance confirm, in quantitative terms, that the cautious and dual controllers are much better than the HCE scheme. Additionally the dual controller shows a slight improvement over cautious control, mainly due to its faster speed of response.

In order to reduce the overshoot of the HCE controller, it is tempting to increase the performance index control weight q. Although this helps to reduce overshoot, it also causes a general deterioration of the tracking capabilities in the steady-state, as shown in Figure 7.3, where q was set to 1 for the HCE controller and 0.0001 for the other two. The HCE accumulated cost

160   7. Dual Adaptive Control of Nonlinear Systems

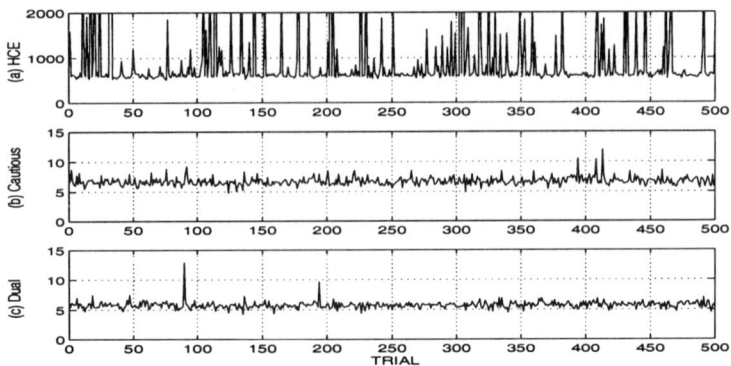

Fig. 7.2. Accumulated cost; (a) HCE (b) Cautious (c) Dual

Table 7.1. Example 1; Average cost and variance from Monte Carlo analysis

	HCE	CAUTIOUS	DUAL
Average cost	1434.0	6.7	5.7
Variance	1.03×10^8	0.516	0.369

is reduced drastically to around 15 but it is still higher than 6, the order of magnitude of the cautious and dual controllers. The reason is that q tends to limit the amplitude of the control at all times and not only during those periods when parameter uncertainty is large. Hence there is a larger tracking error in the steady state, which leads to a higher accumulated cost.

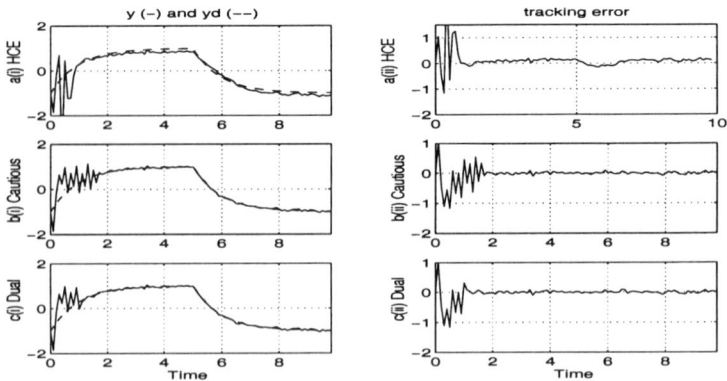

Fig. 7.3. Effect of q; (a) HCE ($q = 1$) (b) Cautious ($q = 0.0001$) (c) Dual ($q = 0.0001$)

7.4.2 Example 2

The plant of the second example is given by the equations

$$y(k+1) = \frac{1.5y(k)y(k-1)}{1+y^2(k)+y^2(k-1)} + 0.35\sin(y(k)+y(k-1)) + 1.2u(k) + e(k+1)$$

where the state vector $\mathbf{x}(k) = [y(k-1)\ y(k)]^T$. The unknown nonlinear dynamics are given by

$$g(\mathbf{x}) = 1.2$$
$$f(\mathbf{x}) = \frac{1.5y(k)y(k-1)}{1+y^2(k)+y^2(k-1)} + 0.35\sin(y(k)+y(k-1)).$$

The noise $e(k)$ has variance $\sigma^2 = 0.05$. The reference input is the same as in Example 1.

A MLP neural controller is tested on this plant where the $\hat{f}_s$ and $\hat{g}_s$ networks are structured with 10 and 5 neurons in the hidden layer respectively. The initial parameter estimates are chosen at random and the initial covariance matrix $\mathbf{P}(0)$ has a diagonal structure with the terms corresponding to $\hat{f}_s$ and $\hat{g}_s$ set to 50 and 10 respectively. As before, trials were conducted with the three different control schemes: HCE, cautious and dual ($r = -0.5$). The control penalty weight q was set to 0.0001 in all cases.

A typical output is shown in Figure 7.4. Note that the same comments as before also apply in this case, with the innovations dual controller performing better in terms of overshoot and speed of response. This is corroborated by Figure 7.5 which shows the accumulated costs of the three controllers from a Monte Carlo analysis of 100 trials, whose average and variance are shown in Table 7.2. The means show that the innovations dual controller has the

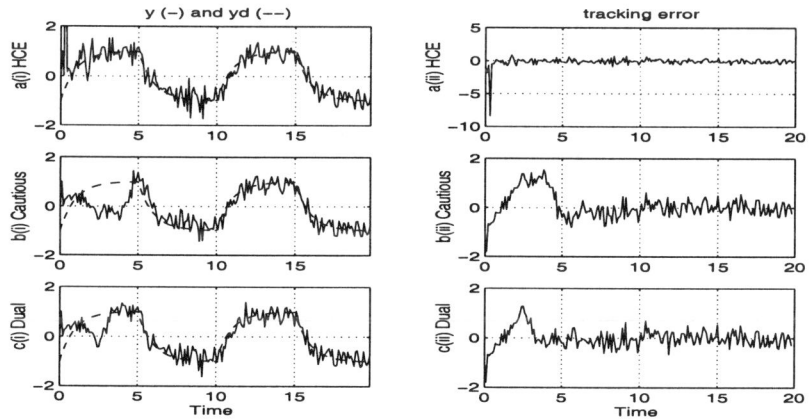

Fig. 7.4. Output and Tracking error; (a) HCE (b) Cautious (c) Dual

better performance, despite the slightly higher variance over the cautious case, which was mainly due to one atypical cost at trial 60, as can be seen in Figure 7.5(c).

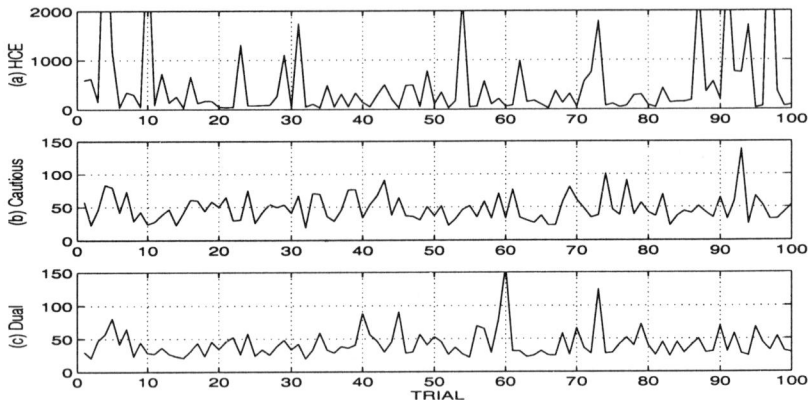

Fig. 7.5. Accumulated cost; (a) HCE (b) Cautious (c) Dual

Table 7.2. Example 2; Average cost and variance for MLP controller

	HCE	CAUTIOUS	DUAL
Average cost	497.0	48.3	42.0
Variance	7.3×10^5	404.1	469.1

A GaRBF controller was also tested on the same plant and subjected to the same noise conditions. Once again, the results confirm the superiority of dual control as shown in Table 7.3 and Figures 7.6 and 7.7. Note that the cautious and dual GaRBF controllers yield a lower mean cost than the corresponding MLP controller, and vice-versa for the HCE case.

Table 7.3. Example 2; Average cost and variance for GaRBF controller

	HCE	CAUTIOUS	DUAL
Average cost	1235.0	26.1	18.3
Variance	1.5×10^7	88.1	17.93

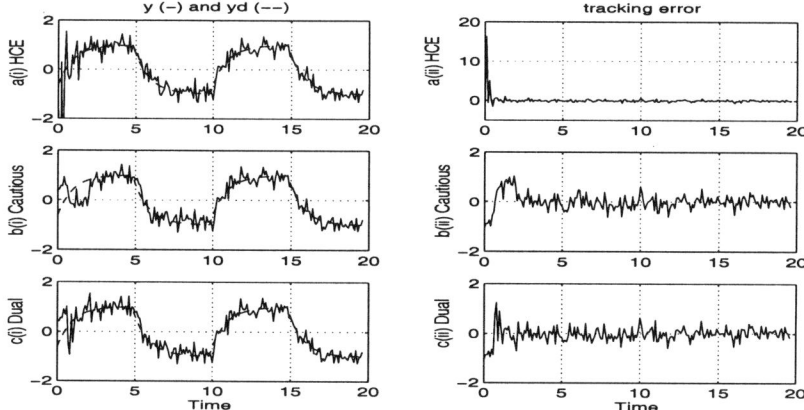

Fig. 7.6. Output and Tracking error; (a) HCE (b) Cautious (c) Dual

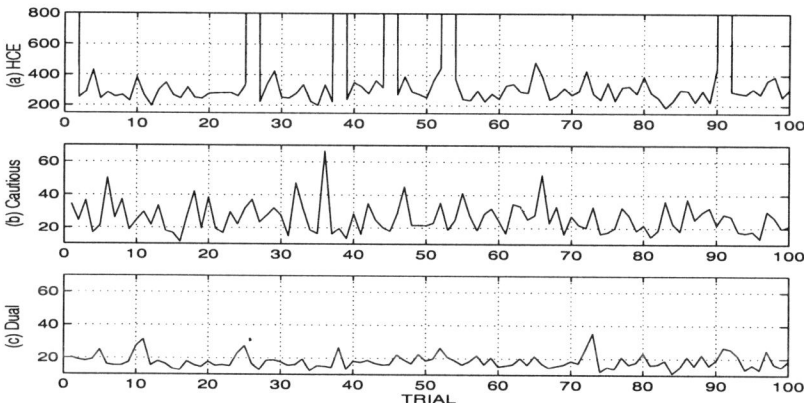

Fig. 7.7. Accumulated cost; (a) HCE (b) Cautious (c) Dual

7.5 Summary

In this chapter, two novel adaptive control schemes for a class of discrete-time, functional uncertain nonlinear systems have been presented. The novelty comprises the introduction of dual control concepts within a neural network-based methodology. The two schemes, one for GaRBF and the other for MLP networks, exhibit great improvement over non-dual adaptive schemes, especially in the transient response. This was confirmed by simulation and Monte Carlo analysis. The control law takes into consideration the neural network parameter uncertainty, so that control and estimation could be performed simultaneously from the outset without the need of a prior, off-line system identification exercise. At the same time, the level of caution could be adjusted so as to avoid an excessively slow transient response and induce a probing-like effect.

The proposed method is a generalization of the IDC scheme originally developed for parametric uncertain linear systems [176]. The generalizations permit consideration of functional uncertain systems, as well as nonlinear parameterized models for the case of MLP networks. The approach is important for intelligent control because the caution and probing-like properties allow the inclusion of higher *a priori* uncertainty without compromising on performance. The price to pay for this is a slightly more complex algorithm based on Kalman filtering, whose implementation however is not impracticable.

8. Multiple Model Approaches

8.1 Introduction

During the mid-to-late 1960's there emerged a new state estimation and control methodology for handling the adaptive Incomplete State Information (ISI) problem [150, 166]. This methodology is known as *Multiple Model Adaptive Estimation/Control* (MMAE/C) or *Partitioned Adaptive Filtering/Control* (PAF/C). It originally appeared as a response to the fact that the reformulation of the adaptive ISI problem in terms of an augmented state (as explained in Section 6.3) yields a set of nonlinear equations, even if the original system were linear. Although this technique seems attractive, because it enables the uncertain parameters to be treated as part of the augmented state vector, estimation and control of nonlinear equations is not a simple task. This was explained in Chapter 6 when it was noted that in general, *suboptimal* solutions of the nonlinear ISI problem still remain computationally intensive.

The MMAE/C approach avoids state augmentation altogether and uses a different technique that yields simpler and more practical algorithms. Although the issue of dual control is not usually addressed, MMAE/C is still highly appealing because it was found to be useful for other problems and applications. These include the identification of unknown noise statistics and system model order, tracking of fast-moving targets, and control of systems whose parameters switch value suddenly in time (jump systems), leading to fault-tolerant control schemes [1, 28, 150, 153, 262]. This renders the technique quite versatile and worthy of consideration in intelligent control, particularly for dealing with multimodal complexity.

8.2 Basic Formulation

The underlying assumption of multiple model (MM) techniques is that although the model of the physical system is unknown, it is a member of a *known* finite set of models. In other words, each member of the set has the potential of representing the system dynamics, but only one of them is actually correct. The problem therefore reduces to finding which member of the set best represents the actual system.

Suppose that the model of the true system consists of the following linear, stochastic state-space equations:

$$\mathbf{x}(k+1) = \mathbf{F}\mathbf{x}(k) + \mathbf{G}\mathbf{u}(k) + \mathbf{B}\mathbf{w}(k)$$
$$\mathbf{y} = \mathbf{H}\mathbf{x}(k) + \mathbf{v}(k)$$

where the constant matrices $\mathbf{F}, \mathbf{G}, \mathbf{B}, \mathbf{H}$ are unknown. Signals $\mathbf{w}, \mathbf{v}$ are Gaussian noise having statistics $\mathbf{w} \sim N(\mathbf{0}, \mathbf{Q})$, $\mathbf{v} \sim N(\mathbf{0}, \mathbf{R})$. For notational convenience, matrices $\mathbf{F}, \mathbf{G}, \mathbf{B}, \mathbf{H}$ will be collectively represented by the symbol θ and referred to as the "model". Although θ is itself unknown, it is a member of a *known* finite set $\mathcal{M}$ of H distinct *candidate models*

$$\mathcal{M} := \{\theta_1, \theta_2, \cdots, \theta_H\}.$$

If the candidate model that represents the actual plant is θ_m (*i.e.*, $\theta = \theta_m$), then model uncertainty reduces to the condition that the value of m is unknown, even though it is known that $m \in \{1, 2, \cdots, H\}$. This is coupled with the problem of estimating state $\mathbf{x}$, which is assumed unmeasurable.

To handle model uncertainty, a Bayesian technique is used to sequentially calculate the posterior probability that a candidate model θ_i does indeed represent the actual plant dynamics. This is the same as the probability of the event that the unknown index m is equal to i, which will be denoted by symbol M_i. The posterior probability of this event is therefore $\Pr(M_i|Y^k)$, where Y^k denotes the measurement sequence $\{\mathbf{y}(i)\}_{i=0}^{k}$. From Bayes' rule it follows that

$$\Pr(M_i|Y^k) = \Pr(M_i|\mathbf{y}(k), Y^{k-1})$$
$$= \frac{p(\mathbf{y}(k)|M_i, Y^{k-1})\Pr(M_i|Y^{k-1})}{\sum_{j=1}^{H} p(\mathbf{y}(k)|M_j, Y^{k-1})\Pr(M_j|Y^{k-1})}. \quad (8.1)$$

The initial prior probabilities $\Pr(M_j|Y^0)$ are assumed known for all candidate models. Very often, these are all set equal to $1/H$, which means that the initial prior probabilities are uniformly distributed. The term $p(\mathbf{y}(k)|M_i, Y^{k-1})$ denotes the conditional probability density of $\mathbf{y}(k)$ and is sometimes called the likelihood function.

Remarks 8.2.1

1. The conditioning on Y^k is introduced to denote explicitly that the probabilities are updated by utilizing all the measurement information available at each time instant k.
2. Since the plant model is linear and the noise is Gaussian, it turns out that the likelihood $p(\mathbf{y}(k)|M_i, Y^{k-1})$ is a Gaussian density with mean $\mathbf{H}_i\hat{\mathbf{x}}_i(k|k-1)$ and covariance $\mathbf{R} + \mathbf{H}_i\mathbf{P}_i(k|k-1)\mathbf{H}_i^T$, where $\mathbf{H}_i$ denotes the $\mathbf{H}$ matrix from candidate model θ_i. $\hat{\mathbf{x}}_i(k|k-1)$ denotes the conditional expectation of state $\mathbf{x}(k)$, calculated on the assumption that θ_i is

the actual plant mode and $\mathbf{P}_i(k|k-1)$ denotes the corresponding state estimation error covariance.
3. Since all the candidate models are linear and the noise is Gaussian, $\hat{\mathbf{x}}_i(k|k-1)$ and $\mathbf{P}_i(k|k-1)$ are obtained directly from a standard Kalman filter matched to model i, i.e., a Kalman filter with parameters corresponding to the matrices of θ_i.

The above implies that a bank of H Kalman filters is required to evaluate Equation (8.1), each matched to one of the H candidate models $\theta_1, \cdots, \theta_H$. This enables calculation of all the likelihood terms in the denominator of Equation (8.1). Each of these Kalman filters is effectively calculating the mean square estimate of the state and the covariance of the estimation error, conditioned on the assumption that the matrices of the actual system are equal to the corresponding candidate model. Equation (8.1) then calculates a probability measure for the validity of each of these assumptions.

But what about the optimum estimate of the state? In general the minimum mean square estimate of the state, denoted by $\hat{\mathbf{x}}(k|k)$, is equal to the expectation of $\mathbf{x}(k)$ conditioned on Y^k. In the MM case, this is given as a combination of the individual state estimates from the bank of Kalman filters. This follows from Lainiotis' Partitioning Theorem [150, 151, 262], which leads to the equation

$$\hat{\mathbf{x}}(k|k) = \sum_{j=1}^{H} \hat{\mathbf{x}}_j(k|k) \Pr(M_j|Y^k). \tag{8.2}$$

The covariance matrix of the state estimation error, denoted as $\mathbf{P}(k|k)$ and defined as $E\{[\mathbf{x}(k) - \hat{\mathbf{x}}(k|k)][\mathbf{x}(k) - \hat{\mathbf{x}}(k|k)]^T\}$, is similarly given as:

$$\mathbf{P}(k|k) = \sum_{j=1}^{H} [\mathbf{P}_j(k|k) + (\hat{\mathbf{x}}_j(k|k) - \hat{\mathbf{x}}(k|k))(\hat{\mathbf{x}}_j(k|k) - \hat{\mathbf{x}}(k|k))^T] \Pr(M_j|Y^k).$$

The terms $\hat{\mathbf{x}}_j(k|k)$ and $\mathbf{P}_j(k|k)$ required in the above equations are obtained from the same bank of Kalman filters that generated $\hat{\mathbf{x}}_j(k|k-1)$ and $\mathbf{P}_j(k|k-1)$, which were used to calculate the likelihoods. The former are filtered estimates, and the latter predictive estimates.

The basis of the MM state estimation scheme for unknown, time-invariant state-space systems was originally proposed by Magill in 1965 [166]. Magill's algorithm, however, calculates the posteriors $\Pr(M_i|Y^k)$ differently and is more memory and computationally demanding than the form shown here. Research on several aspects of this basic MM formulation is still ongoing. Recent contributions include those of Maybeck and Hanlon [170], who suggest a number of techniques for enhancing the performance of the MM estimation scheme, and Li and Bar-Shalom [157] who discuss the problem of selecting an appropriate set $\mathcal{M}$ of candidate models and develop a strategy for making it variable.

8.2.1 Multiple Model Adaptive Control

Having just described the MM approach for state estimation, we will now discuss its extension to the control problem. The Performance Index is the same as J_{NSS} of Equation (6.2), repeated here for convenience,

$$J_{NSS} = E\{\mathbf{x}^T(N)\mathbf{Q}_0\mathbf{x}(N) + \sum_{k=0}^{N-1} \mathbf{x}^T(k)\mathbf{Q}_1\mathbf{x}(k) + \mathbf{u}^T(k)\mathbf{Q}_2\mathbf{u}(k)\}.$$

Although the system is linear and Gaussian, the uncertainty of the parameters leads to an interaction between control and estimation. Hence unlike the LQG problem, where the parameters are known, the optimal solution for this case requires a dual control scheme. As usual however, because of complexity issues, the ideal solution cannot be implemented in practice.

The MM approach suggests a suboptimal *non-dual* solution, which turns out to be computationally efficient, reliable and much simpler than non-MM schemes such as the augmented state method. Several MM-based control solutions have been proposed for the adaptive ISI problem [65, 153, 261]. These differ in the approximations introduced to simplify the problem. The DUL algorithm by Deshpande, Upadhyay and Lainiotis [65, 151] is the most popular because it is the least computationally demanding. It starts from the idea of utilizing an OLOF policy to minimize J_{NSS}. In this case, the cost-to-go at time k is given by Equation (6.7) as

$$J_k^{OLOF} = \min_{\mathbf{u}(k)\cdots\mathbf{u}(N-1)} E\{\mathbf{x}^T(N)\mathbf{Q}_0\mathbf{x}(N) + \sum_{t=k}^{N-1} \mathbf{x}^T(t)\mathbf{Q}_1\mathbf{x}(t) + \mathbf{u}^T(t)\mathbf{Q}_2\mathbf{u}(t)|I^k\}$$

Even though this represents an OLOF policy, its minimization is not easy because of the unknown system parameters. Using the Smoothing Property of Conditional Expectations, J_k^{OLOF} could be written as

$$J_k^{OLOF} = \min_{\mathbf{u}(k)\cdots\mathbf{u}(N-1)} E\{E\{\mathbf{x}^T(N)\mathbf{Q}_0\mathbf{x}(N) + \sum_{t=k}^{N-1} \mathbf{x}^T(t)\mathbf{Q}_1\mathbf{x}(t) + \mathbf{u}^T(t)\mathbf{Q}_2\mathbf{u}(t)|M_j, I^k\}|I^k\}.$$

Even in this form, the minimization is still not feasible. Hence to simplify matters, the DUL algorithm suggests to interchange the minimization and the first expectation. Naturally, this leads to a different value for J_k^{OLOF} because it represents an approximation. In essence, it is assumed that:

$$J_k^{OLOF} \approx$$
$$E\{\min_{\mathbf{u}(k)\cdots\mathbf{u}(N-1)} E\{\mathbf{x}^T(N)\mathbf{Q}_0\mathbf{x}(N) +$$
$$\sum_{t=k}^{N-1} \mathbf{x}^T(t)\mathbf{Q}_1\mathbf{x}(t) + \mathbf{u}^T(t)\mathbf{Q}_2\mathbf{u}(t)|M_j, I^k\}|I^k\}.$$

The inner minimization now represents an LQG problem for model j, due to the conditioning on M_j in the inner expectation. This part of the expression could therefore be easily calculated for all candidate models, each utilizing the state estimate from the corresponding Kalman filter inside the LQG control laws. Let us denote this LQG control at time k for model j by $\hat{\mathbf{u}}(k|M_j)$. This way, the inner minimization reduces to a finite set of H discrete control values. Hence, the outer expectation is calculated as a probability weighted sum of these individual values, where the probability weights are given by the posteriors from Equation (8.1), i.e., $\{\Pr(M_i|Y^k)\}_{i=1}^H$. The DUL algorithm interprets this as indicating that a reasonable control law is given by

$$\mathbf{u}(k) = \sum_{j=1}^{H} \hat{\mathbf{u}}(k|M_j) \Pr(M_j|Y^k).$$

In other words, the control is calculated as a probability weighted average of the LQG controls from every candidate model, as shown in Figure 8.1. Alternative MM control schemes include those of Saridis and Dao [221], whose performance was shown to be inferior to the DUL algorithm [65], and the

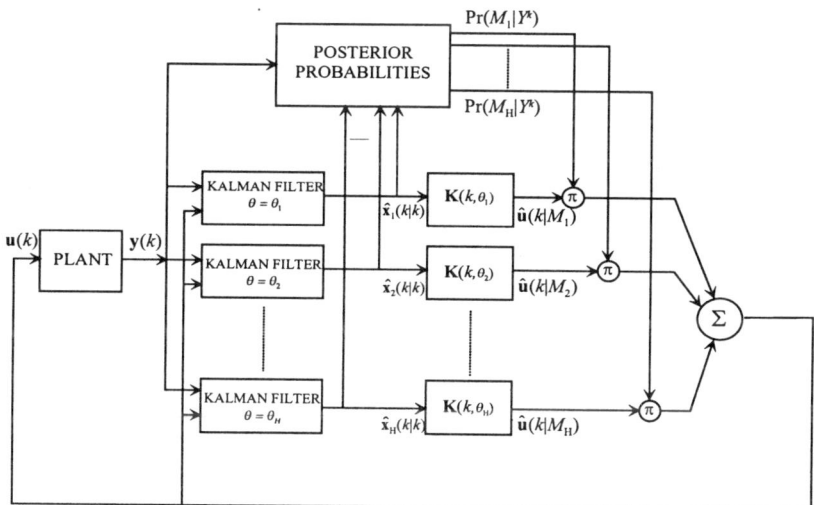

Fig. 8.1. The DUL multiple model control scheme

algorithm of Lee and Simms [153] which is more computationally intensive than DUL. Watanabe's method [261] claims better performance than DUL by utilizing more fully the information from the individual optimal controls $\hat{u}(k|M_j)$ in the Kalman filter update equations. An interesting development due to Watanabe and Tzafestas [265] extends the MM scheme to a distributed control scenario, which they call Hierarchical MMAC. Some interesting real-world applications of MMAC are described in [24, 99, 172].

8.2.2 Jump Systems

The term *jump systems* refers to the case where the system parameters are not constant but prone to change value abruptly. This feature is used to model situations such as plant component failure or environmental changes, tracking of moving objects that suddenly change course (*e.g.*, aircraft manoeuvres) or sudden changes in a patient's heart rhythm (cardiac arrhythmia). Jump systems are an example of temporal multimodality, as explained in Section 1.3.1, because there is no direct information that specifies the mode activity.

The plant equations of a jump system are given as

$$\mathbf{x}(k+1) = \mathbf{F}_{m(k)}\mathbf{x}(k) + \mathbf{G}_{m(k)}\mathbf{u}(k) + \mathbf{B}_{m(k)}\mathbf{w}(k)$$
$$\mathbf{y}(k) = \mathbf{H}_{m(k)}\mathbf{x}(k) + \mathbf{v}(k)$$

where $\mathbf{F}_{m(k)}, \mathbf{G}_{m(k)}, \mathbf{B}_{m(k)}, \mathbf{H}_{m(k)}$, which are collectively represented by the symbol $\theta_{m(k)}$, represent the unknown system matrices. In this case, the matrices could switch in time from one model to another, as indexed by $m(k)$. As in the non-jump case, it is assumed that $\theta_{m(k)}$ is a member of a set $\mathcal{M}$ that consists of H *known* candidate models

$$\mathcal{M} = \{\theta_1, \theta_2, \cdots, \theta_H\}.$$

Hence index $m(k) \in \{1, 2, \cdots, H\}$ is the real unknown variable. In contrast to the non-jump case, the index is now time-varying. In this representation, the candidate models specify the possible modes of operation of the system.

Let $M_i(k)$ denote the event that the plant dynamics at time k correspond to model θ_i, and $S_j(k)$ denote one specific *sequence* of such events from start till time k. (*e.g.*, $S_3(k) = \{M_1(0), M_2(1), M_2(2), M_1(3), \cdots, M_1(k)\}$ denotes one arbitrarily selected particular sequence). Since every element of this sequence has H possibilities, there exist H^k different possible sequences at time k in all, *i.e.*, $S_1(k), S_2(k), \cdots, S_{H^k}(k)$. Naturally, only one of these H^k sequences has actually taken place, namely

$$\{M_{m(0)}(0), M_{m(1)}(1), \cdots, M_{m(k)}(k)\}.$$

However it is not known which sequence it is, because $m(k)$ is unknown for all k. Hence in this case, partitioning theory suggests to consider *all* possible sequences, leading to the following expressions for the minimum mean square estimate of the state $\hat{\mathbf{x}}(k|k)$ and the estimation error covariance $\mathbf{P}(k|k)$:

$$\hat{\mathbf{x}}(k|k) = \sum_{j=1}^{H^k} \hat{\mathbf{x}}_j(k|k) \Pr(S_j(k)|Y^k) \tag{8.3}$$

$$\mathbf{P}(k|k) = \sum_{j=1}^{H^k} [\mathbf{P}_j(k|k) +$$

$$(\hat{\mathbf{x}}_j(k|k) - \hat{\mathbf{x}}(k|k))(\hat{\mathbf{x}}_j(k|k) - \hat{\mathbf{x}}(k|k))^T] \Pr(S_j(k)|Y^k) \tag{8.4}$$

where $\hat{\mathbf{x}}_j(k|k)$ is the state estimate calculated on the assumption that the actual mode sequence is $S_j(k)$, and $\mathbf{P}_j(k|k)$ is the corresponding estimation error covariance. These are given by a Kalman filter matched to sequence $S_j(k)$.

The term $\Pr(S_j(k)|Y^k)$ denotes the posterior probability of the event that the actual mode sequence is $S_j(k)$. This posterior is given by Bayes' rule as

$$\Pr(S_i(k)|Y^k) = \Pr(S_i(k)|\mathbf{y}(k), Y^{k-1})$$
$$= \frac{p(\mathbf{y}(k)|S_i(k), Y^{k-1}) \Pr(S_i(k)|Y^{k-1})}{\sum_{j=1}^{H^k} p(\mathbf{y}(k)|S_j(k), Y^{k-1}) \Pr(S_j(k)|Y^{k-1})}. \tag{8.5}$$

The likelihood terms $p(\mathbf{y}(k)|S_j(k), Y^{k-1})$ are Gaussian distributions with mean $\mathbf{H}_j \hat{\mathbf{x}}_j(k|k-1)$ and covariance $\mathbf{R} + \mathbf{H}_j \mathbf{P}_j(k|k-1) \mathbf{H}_j^T$, where $\mathbf{H}_j$ is the $\mathbf{H}$ matrix from θ corresponding to the kth event of sequence $S_j(k)$. The predictive state estimate and error covariance $\hat{\mathbf{x}}_j(k|k-1)$ and $\mathbf{P}_j(k|k-1)$, are also calculated from the Kalman filter matched to sequence $S_j(k)$.

The prior terms $\Pr(S_j(k)|Y^{k-1})$ appearing on the right hand side of Equation (8.5) are calculated as follows. Let us first split up sequence $S_j(k)$ as

$$S_j(k) = \{S_{js}(k-1), M_{jm}(k)\},$$

where $M_{jm}(k)$ denotes the last entry of $S_j(k)$ and $S_{js}(k-1)$ denotes the sequence of all events up to the one before the last. Then by the Chain Rule,

$$\Pr(S_j(k)|Y^{k-1}) =$$
$$\Pr(M_{jm}(k)|S_{js}(k-1), Y^{k-1}) \Pr(S_{js}(k-1)|Y^{k-1}) \tag{8.6}$$

where $\Pr(S_{js}(k-1)|Y^{k-1})$ is given by the posteriors from Equation (8.5) that were evaluated during the previous time step.

Although the formulation of Equations (8.3), (8.5) and (8.6) is optimal, the procedure is hindered by two complications:

1. Evaluation of the term $\Pr(M_{jm}(k)|S_{js}(k-1), Y^{k-1})$ in Equation (8.6).
2. At time k, H^k Kalman filters are required for evaluating $\hat{\mathbf{x}}_j$ and $\mathbf{P}_j$ for all possible mode sequences. The number of filters required therefore increases exponentially in time! Any computer system would quickly run out of memory when trying to meet such a fast growing computational demand.

There are two principal ways of dealing with these problems:

1. The first, rather *ad hoc* method of handling both problems simultaneously, is to ignore the fact that the parameters are time-varying and use Equations (8.1) and (8.2) in place of the theoretically correct Equations (8.5) and (8.3). This way, only H Kalman filters are required at any one time and Equation (8.6) is not required at all. The approach is obviously an approximation and suffers from the problem that Equation (8.1) is insensitive to parameter switches because it was derived for the non-jump case. Inspection of this equation shows that if at some time k the posteriors $\Pr(M_i|Y^k)$ were equal to 1 for some $i = l$ and 0 for all $i \neq l$, then at time $(k+1)$ and all subsequent instances, $\Pr(M_i|Y^{k+1})$ would remain clipped at zero for all $i \neq l$, even if there is a mode jump from model l. To circumvent this problem, the posterior probabilities are usually forcefully lower bounded by some small positive value. Despite the approximations introduced, this method seems to work reasonably well in practice [24, 172].

2. An alternative and more elegant solution is to establish a relationship for $\Pr(M_{jm}(k)|S_{js}(k-1), Y^{k-1})$ of Equation (8.6), by assuming that the parameter switching mechanism is a finite state Markov Chain. This means that the probability of a current event is independent of the previous information sequence Y^{k-1} and all the previous events, except for the last one. Hence for our case this implies,

$$\Pr(M_{jm}(k)|S_{js}(k-1), Y^{k-1}) = \Pr(M_{jm}(k)|M_{jsm}(k-1))$$

where $M_{jsm}(k-1)$ denotes the latest event of sequence $S_{js}(k-1)$, *i.e.*, $S_{js}(k-1) = \{S_{jss}(k-2), M_{jsm}(k-1)\}$. If in addition it is assumed that the transition probability matrix $\Pi = [\pi_{i,j}]_{H \times H}$ of the Markov Chain is known, where by definition

$$\pi_{i,j} = \Pr(M_j(k)|M_i(k-1)),$$

then

$$\Pr(M_{jm}(k)|S_{js}(k-1), Y^{k-1}) = \pi_{jsm,jm}.$$

Hence under this assumption, and using Equation (8.6), the posteriors Equation (8.5) becomes

$$\Pr(S_i(k)|Y^k) = \frac{p(\mathbf{y}(k)|S_i(k), Y^{k-1})\pi_{ism,im}\Pr(S_{is}(k-1)|Y^{k-1})}{\sum_{j=1}^{H^k} p(\mathbf{y}(k)|S_j(k), Y^{k-1})\pi_{jsm,jm}\Pr(S_{js}(k-1)|Y^{k-1})}. \quad (8.7)$$

Of course, the last expression still does not solve the problem of requiring an exponentially increasing number of Kalman filters. The only way to deal with this problem is to resort to suboptimal approximations, several of which have been proposed.

8.2 Basic Formulation

Suboptimal Techniques. The suboptimal techniques that have been put forward for dealing with the problem of the amount of Kalman filters could be grouped into 4 classes:

- Generalized Pseudo-Bayes (GPB) methods [1, 52, 119, 153, 264].
- The Interacting Multiple Model (IMM) algorithm [40, 173].
- The Random Sampling Algorithm (RSA) [2].
- The Detection-Estimation Algorithm (DEA) [253].

We will limit ourselves to discussing only the first two because their principles are closely related. The result of implementing the full optimal algorithm as given by Equations (8.3), (8.4) and (8.7) for a 2 model problem (*i.e.*, $H = 2$), is represented in the evolution diagram of Figure 8.2. Note how the

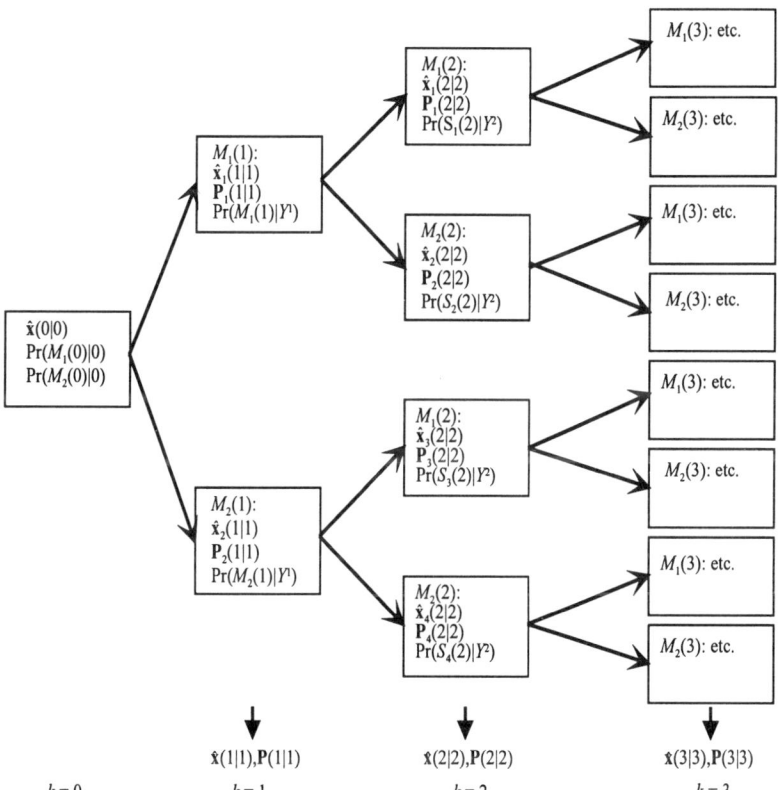

Fig. 8.2. Evolution diagram for the optimal scheme with $H = 2$. Each block represents one Kalman filter, the event $M_j(k)$ to which it is matched and the corresponding state estimate, covariance and probability updates. The arrow links show how information is propagated from one time instant to the next along a particular sequence.

174   8. Multiple Model Approaches

number of Kalman filters required to calculate the optimal minimum mean square estimate of the state increases exponentially in time. Each linked path represents one specific mode sequence and the state estimates along one such path are related via the propagation of information through the Kalman filter equations. The distribution of **x** conditioned on the information state *and* the modes specified along one such linked path is Gaussian. However the distribution of **x** conditioned uniquely on the information state Y^k is given by the probability weighted sum of the individual Gaussian distributions along every path and hence is not Gaussian. In mathematical terms,

$$p(\mathbf{x}(k)|Y^k) = \sum_{j=1}^{H^k} p(\mathbf{x}(k)|S_j(k), Y^k)\Pr(S_j(k)|Y^k)$$

where $p(\mathbf{x}(k)|S_j(k), Y^k)$ is Gaussian.

In the GPB methods, the increase in the number of Kalman filters is curbed by *pruning* the growing tree of linked paths down to H after every d time steps. This is shown in Figure 8.3 for the case of $H = 2$ and $d = 2$, where the 4 filters at $k = 2$ are reset back to 2 "new" filters at $k = 3$. Each of the new

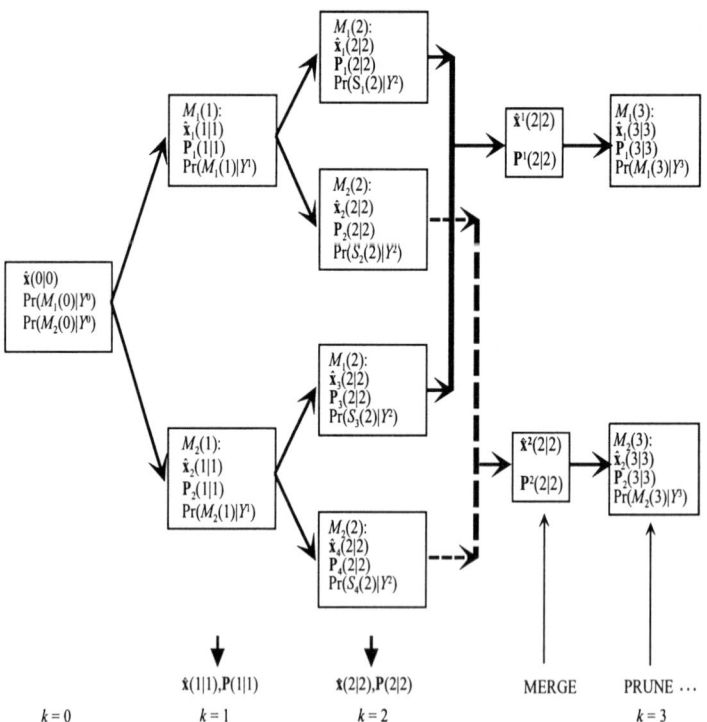

Fig. 8.3. Evolution diagram for GPB with $d = 2$

filters is matched to the event of a specific mode being active at that particular time. However pruning introduces the issue of what previous state estimation information is to be propagated to the new filters as prior statistics. Should it be the mean $\hat{\mathbf{x}}_i$ and covariance $\mathbf{P}_i$ from one particular path existent just prior to the prune, or should it be a probability weighted mixture of these statistics from various paths? The process of determining this information is called *merging*. In GPB methods, merging is performed by propagating a probability weighted sum of the means and covariances from the previous sequences ending in event M_1 to the new Kalman filter matched to M_1, and similarly for the rest of the filters. This is clearly shown in the example of Figure 8.3, where the merged statistics are denoted as $\hat{\mathbf{x}}^i(k|k)$, $\mathbf{P}^i(k|k)$; $i = 1, 2$. Note that the conditional distribution of the state from the merged sequences is not Gaussian, because it is given by the probability-weighted mixture of all the Gaussians corresponding to each individual sequence. However, when propagating the information to the new Kalman filters in the cycle that follows pruning, this mixture distribution is effectively being approximated by a single Gaussian having mean and covariance equal to the merged statistics.

For the special case of GPB with $d = 1$, since merging and pruning is done at every single time step, there is only one previous event matched to each specific mode. This is shown in Figure 8.4. In this case, it is not

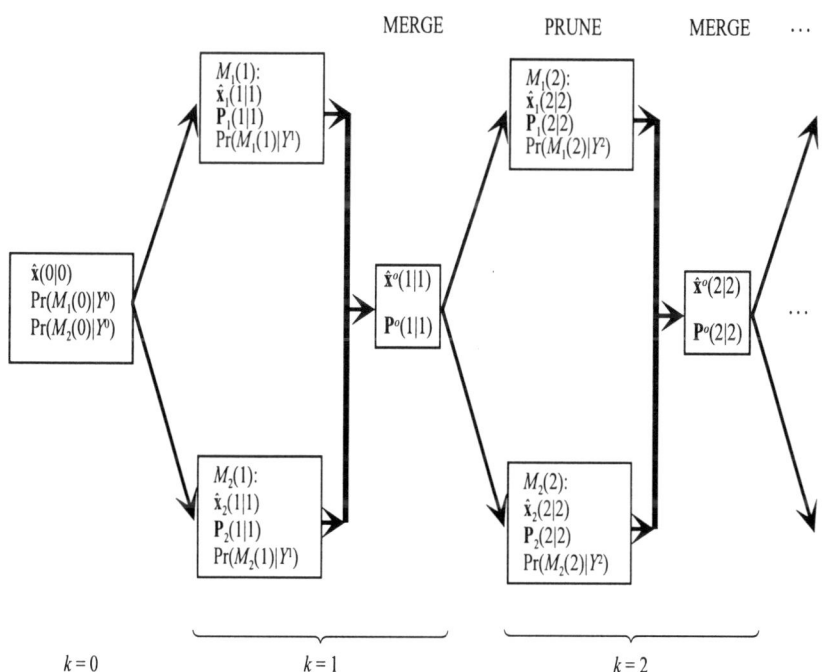

Fig. 8.4. Evolution diagram for GPB with $d = 1$

176   8. Multiple Model Approaches

possible to merge the statistics from the set of sequences ending in the same event, simply because each of these sets has only one element. Rather than avoiding merging altogether however, it is preferred to merge the statistics from all previous events, even though they are matched to different modes [1]. In contrast to the $d > 1$ case, this means that the *same* prior information is passed to all the new Kalman filters, as shown in Figure 8.4. This leads to a poorer propagation of information that discriminates between previous events and it could give rise to worse state estimates. However the $d = 1$ case is attractive because it uses less Kalman filters and is therefore not very computationally demanding.

It is precisely this issue that the IMM algorithm seeks to address. Essentially it strikes a compromise between GPB with $d = 1$ and GPB with $d > 1$, by pruning after every time step but merging differently. The merging process for the mean in GPB with $d = 1$ is done according to the equation

$$\hat{\mathbf{x}}^o(k|k) = \sum_{j=1}^{H} \hat{\mathbf{x}}_j(k|k) \Pr(M_j(k)|Y^k)$$

where $\hat{\mathbf{x}}^o$ denotes the merged statistic for the mean. In the IMM this is modified by making use of the transition probabilities to obtain a set of H

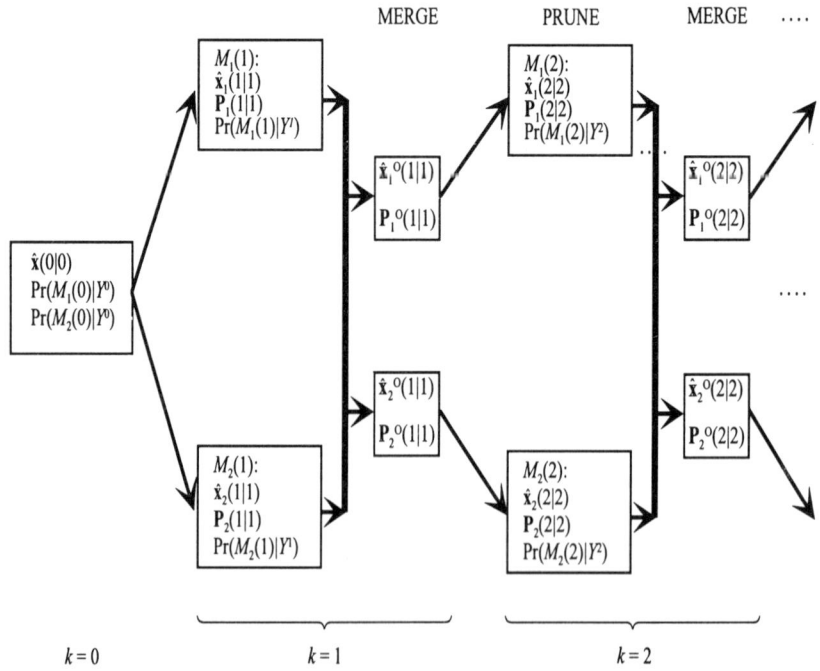

Fig. 8.5. Evolution diagram for IMM algorithm

different merged estimates $\{\hat{\mathbf{x}}_j^o\}_{j=1}^H$, one for every previous event. The IMM merging equations for $j = 1, \cdots, H$ are:

$$\hat{\mathbf{x}}_j^o(k|k) = \sum_{l=1}^H \hat{\mathbf{x}}_l(k|k) \Pr(M_l(k)|Y^k) \left\{ \frac{\pi_{l,j}}{\sum_{q=1}^H \pi_{q,j} \Pr(M_q(k)|Y^k)} \right\}.$$

A similar relation is used for the merged estimation error covariance matrices $\mathbf{P}_j^o$. The merged statistics $(\hat{\mathbf{x}}_j^o, \mathbf{P}_j^o)$ are then propagated as prior information to the new Kalman filter matched to event M_j, for all H events. This is shown in Figure 8.5 for $H = 2$. Simulations reveal that the performance of IMM is similar to that of GPB with $d = 2$, yet utilizing a maximum of 2 Kalman filters as opposed to 4 [40].

Since these suboptimal techniques are based on pruning, the posteriors Equation (8.7) which is based on H^k Kalman filters, requires modification. The form it takes depends on the choice of d, *i.e.*, the number of time steps after which pruning is applied. This form is substantially different according to whether $d = 1$ or $d > 1$. Further detail is beyond the scope of this book and could be found in the relevant references [1, 52, 119, 262, 264].

Control of Jump ISI Systems. As for the non-jump case, although the optimal control scheme has a dual nature in the sense of Fel'dbaum, implementation issues have led to various non-dual suboptimal solutions. Most of these schemes concentrate on the jump adaptive ISI problem characterized by Markovian switching. The earlier publications concentrated on problems where the jump parameters are restricted to the measurement equation only [3, 83, 252]. For jump parameters in both state and measurement equations, Lee and Simms [153] combine their suboptimal control algorithm for the non-switching case with a state estimator that is similar in concept to GPB with $d = 1$. Watanabe and Tzafestas [264] extend the DUL algorithm for non-jump systems to the jump case and use a GPB algorithm with $d > 1$ for state estimation. They update the estimates in a way to ensure a low demand on computational storage. Recently Campo *et al.* [47] developed a new control scheme, the Parallel Control Algorithm, that utilizes the IMM for state estimation. They claim better performance and less computational load than the scheme of [252]. Griffiths and Loparo [95] proposed a different algorithm that attempts to probe the system for a limited number of future stages, thus leading to a suboptimal dual control scheme. A few other MM control schemes for jump systems are based on the technique of lower bounding the posterior probabilities, instead of assuming a Markov switching sequence. Athans et al. [24] and Maybeck and Stevens [172] describe such control methods for aircraft systems. Gustafson and Maybeck [99] also utilize this type of scheme to control oscillations in large flexible space structures and describe a Moving-Bank algorithm for easing the computational demand when the number of possible modes H is very large.

8.3 Adaptive IO Models

As indicated in the previous section, most of the work on MM adaptive estimation and control deals with the adaptive ISI problem and fewer works have been published on the adaptive IO problem. These usually concentrate on ARX systems having form

$$y(k) = \theta_{m(k)}^T \phi(k-1) + e(k) \tag{8.8}$$

where e is zero-mean Gaussian noise, $\phi = [u(k-1) \cdots u(k-p)\ y(k-1) \cdots y(k-n)]^T$ is the regressor vector and $\theta_{m(k)}$ is the parameter vector. Being a multiple model method, $\theta_{m(k)}$ is a member of a finite set $\mathcal{M}$ of H candidate parameter vectors

$$\mathcal{M} = \{\theta_1, \theta_2, \cdots, \theta_H\}$$

as indexed by the mode variable $m(k) \in \{1, 2, \cdots, H\}$. In the non-jump case, the mode variable is constant. For the adaptive IO case, a few schemes consider the case with all H candidate parameters θ_i known, while others perform on-line estimation of θ_i.

Since part of the research reported in this book concerns IO models of nonlinear stochastic form, the theoretical details for handling IO models by MM techniques are given in the appropriate chapters. In this section we only give a review of previously published results.

Narendra and Balakrishnan [187] suggest a multiple model adaptive control scheme for non-jump linear IO models as a way of improving the transient response of conventional (single model) indirect adaptive control. Narendra et al. [188] also address control of jump nonlinear IO systems by a multiple model adaptive scheme. However both these methods are purely deterministic and hence not of direct interest to this book.

Wenk and Bar-Shalom [266] consider a non-jump ARX system with members of set $\mathcal{M}$ assumed known. Hence, as in the ISI problem, the uncertainty concerns the value of the mode variable m, which is constant in this case. The problem is to find a control policy that minimizes an N-step Performance Index of similar form to Equation (6.4). As explained before, since the actual parameter vector is uncertain, this constitutes a dual control problem. The main contribution of this work is the determination of a suboptimal dual control algorithm of the implicit type, which the authors call Model Adaptive Control (MAD). The principles behind MAD are similar to the wide-sense adaptive dual scheme of [248] described previously for the ISI problem, and involve a second order expansion of the cost-to-go about a nominal trajectory. Apart from the work reported in the following chapters of this book, the MAD algorithm seems to be the only attempt at incorporating *dual control* principles within a MM scheme for adaptive IO models.

Millnert [177] considers the system identification problem for an ARX model having jump parameters. In this work, in addition to the mode variable $m(k)$, the candidate parameters θ_i of set $\mathcal{M}$ are also assumed unknown.

The problem is to sequentially estimate the different mode parameters $\theta_{m(k)}$ which the plant experiences during its operation. This requires some means of (a) detecting a change in $m(k)$ and (b) estimating the mode parameters θ_i as separate vectors for every distinct value of $m(k)$ that was detected. It is assumed that $m(k)$ switches according to a Markov chain with known transition probabilities and the number of parameter estimators utilized is equal to the number of modes. The convergence properties of the algorithm are heuristically discussed and this motivates the determination of different methods for detecting changes in $m(k)$.

Kadirkamanathan [129] addresses identification of nonlinear systems whose dynamics are both unknown and subject to jumps. GaRBF neural networks are used to identify the different mode dynamics and the unknown mode transitions are characterized as a Markov chain having known transition probabilities. A set of Kalman filters, equal in amount to the number of modes, is used as the basis for estimating the GaRBF output layer parameters, which form the elements of set $\mathcal{M}$. Two parameter estimation algorithms are suggested: hard and soft competition. The former restricts parameter adjustment only to the model whose posterior probability dominates over the rest, and the latter adjusts the parameters of all models after weighting by a gain that is proportional to the posterior probability of the corresponding mode being active.

Andersson [11] addresses ARX models with unknown, time-varying parameters from a somewhat different perspective than the previous MM schemes. In this work the multiple models are not the consequence of the set $\mathcal{M}$ of estimated mode parameters, but the components of a Gaussian Sum approximation to the posterior distribution of the parameters θ conditioned on Y^{k-1}. θ is therefore treated as a continuous random variable having non-Gaussian distribution, but which is decomposed as a sum of weighted Gaussians. Unfortunately this approach still does not solve the problem of requiring an exponentially increasing number of estimators to solve the optimal estimation problem, because an optimal propagation of the posteriors involves "splitting" each component of the Gaussian sum into two at every time step. This leads to $H2^k$ components at time k, where H is the *initial* number of components. Andersson suggests a pruning scheme for maintaining only H components at all times and combines this with H Kalman filters to derive a sequential algorithm for estimating the parameters of the time-varying (not necessarily jump) ARX system.

8.3.1 Scheduled Mode Transitions

Up to this point we have only considered systems whose parameters are either constant but unkown, or else are subject to jump or switch value suddenly in time. The switching case is an example of *temporal multimodality*, because the onset of a mode transition is an arbitrary event and no measurable signal is available to specify which mode is active at any time. In such a situation, a mode switch could only be detected by its effect on the system output which, in effect, represents a closed-loop or *feedback* process. This explains why the posterior probabilities of Equation (8.5), which reflect the mode activity, are conditioned on the measurement sequence Y^k.

It is not rare to have a situation where the mode transitions depend directly on some auxiliary *scheduling variable* z, that is available for measurement. This case is representative of *spatial multimodality* because the mode switch depends on the value of variable z that exists inside a measurable space. Under these conditions, a mode switch could be detected by monitoring the value of z, rather than the plant output. This therefore represents an *open-loop* method of detection. This approach forms the basis of *Gain Scheduled Control* [23, 217, 226, 227, 243], where different control gains are activated according to the plant mode as indicated by the scheduling variable. The scheduling variable usually consists of the system states, the inputs or some other relevant external variables that are accessible for measurement.

Hence, as opposed to temporal multimodal systems, the mode transitions are not arbitrary but scheduled. This suggests the possibility of utilizing both open and closed-loop techniques to gather information on mode activity, *i.e.*, to use the information provided by *both* the scheduling variable, as well as the effect of a mode transition on the output.

A model of an ARX system that is subject to spatial multimodality is given by the following equations:

$$y(k) = \theta_{m(k)}^T \phi(k-1) + e(k) \tag{8.9}$$
$$m(k) = \Psi(\mathbf{z}) \tag{8.10}$$

where $e, \theta_{m(k)}, \phi, m(k)$ are defined as usual. The difference from Equation (8.8) is the presence of function $\Psi(\cdot)$ that maps from z to $m(k)$. Usually, function Ψ segments the space of z into a number of non-overlapping partitions P_i (sometimes called *regimes*), according to the way that the modes m are scheduled by z. A partition P_i denotes the zones corresponding to mode $m(k) = i$. An example is shown in Figure 8.6. For values of z within some given partition, plant output $y(k)$ is affected only by one particular candidate parameter from set $\mathcal{M}$.

Since $m(k)$ is a discrete variable, it is usually more convenient to define a set of H *indicator variables* $\gamma_i(\mathbf{z})$; $i = 1, \cdots, H$ such that

$$\gamma_i(\mathbf{z}) = \begin{cases} 1 & \text{if } m(k) = i \\ 0 & \text{otherwise.} \end{cases} \tag{8.11}$$

Hence $\gamma_i(\mathbf{z})$ is equal to 1 only when z falls inside the partition where mode i is active, and otherwise it is zero. This way, Equation (8.9) could be equivalently represented as

$$y(k) = \sum_{j=1}^{H} \gamma_j(\mathbf{z}) \theta_j^T \phi(k-1) + e(k). \tag{8.12}$$

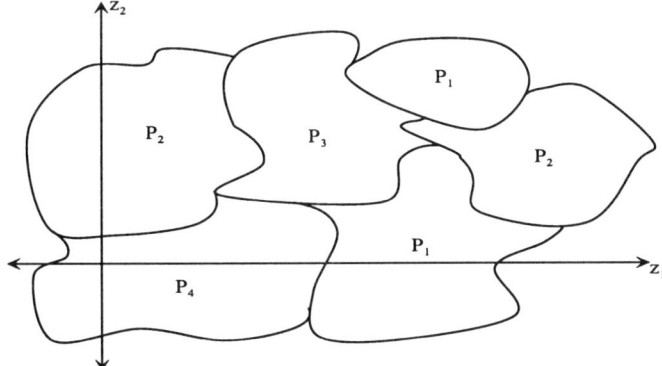

Fig. 8.6. Partitioning of the scheduling space. Note that in general a partition need not be continuous, like P_1 and P_2 in the diagram.

The plant equations as described so far, satisfy the property that at any given time only *one* parameter vector of the candidate models in set $\mathcal{M}$ affects the output. These are often referred to as *hard partitioned, mode switched, piecewise* or *hybrid* dynamical systems [26, 38, 42, 124, 201, 212, 230, 237]. In the statistics and machine learning literature, similar models have been proposed for regression, classification and function approximation [43, 44, 81]. Hard partitioning implies that when z traverses from one partition to another, the parameter vector of Equation (8.12) switches value immediately.

In practice there may be situations where a mode switch is not so sharply defined and instead the dynamics of two neighbouring modes are smoothly blended together while a transition is taking place. In such cases, it is more accurate to utilize a *soft-partitioned* model to represent the plant dynamics. This means that the model output is allowed to depend on the parameter vector of more than one candidate model at a time, particularly in the vicinity of the boundary separating two partitions. Hence when z traverses from some partition P_i, corresponding to mode i, to a neighbouring partition P_j, the output y is smoothed out by mixing the individual outputs from the two modes. Away from the boundary, the influence of the neighbouring mode is gradually reduced to nil. This model is obtained by simply replacing the binary valued indicator variables of Equation (8.11) with continuous functions $\hat{\gamma}_i(\mathbf{z})$, often called the *validity, interpolation* or *gating functions* that satisfy the following three conditions:

$$0 \leq \hat{\gamma}_i \leq 1$$
$$\sum_{i=1}^{H} \hat{\gamma}_i(\mathbf{z}) = 1 \tag{8.13}$$
$$\hat{\gamma}_i \to \begin{cases} 1 \text{ if } & \mathbf{z} \in P_i \\ 0 \text{ if } & \mathbf{z} \notin P_i. \end{cases}$$

The last condition means that $\hat{\gamma}_i$ is very close to 1 while $\mathbf{z} \in P_i$ and smoothly reduces to zero as $\mathbf{z}$ goes outside P_i. The plant output is then represented by the equation

$$y(k) = \sum_{j=1}^{H} \hat{\gamma}_j(\mathbf{z}) \theta_j^T \phi(k-1) + e(k). \tag{8.14}$$

The soft-partitioned approach is inherent in a wide variety of modelling techniques based upon interpolation methods [45, 90, 106, 122, 123, 124, 125] and also the fuzzy modelling approach [242, 278]. Soft-partitioning is particularly effective for modelling nonlinear functions by using the validity functions to *smoothly* interpolate between a set of linear local models, each of which is valid only over a small local operating region. The smooth interpolation serves to blend together the approximations from the local models that lie close to the boundary of an operating region, where the representation of an individual local model starts becoming inaccurate.

In this book we call the type of modelling architecture represented by Equation (8.14), based upon an interconnection of all the outputs $\theta_j^T \phi(k-1)$ from the multiple candidate models and weighted by a set of validity functions $\hat{\gamma}_j$, as a *modular network*. A graphical representation is shown in Figure 8.7. In general, the candidate models need not have a linear ARX structure

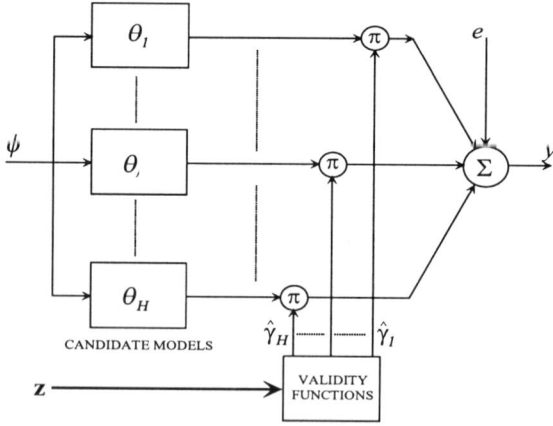

Fig. 8.7. A modular network representation.

as in Equation (8.9), but could take a more complex linear form, such as ARMAX [122], or even a nonlinear structure [125]. Additionally they need not all have the same structure, which leads to heterogeneous candidate model sets. Modular network models exhibit several advantages over alternative methods that attempt to capture nonlinearities by one complex, homogeneous nonlinear architecture such as a conventional neural network [118]. For example, the simple (usually linear) structure of the candidate models leads to fast learning and requires less complex estimation algorithms. Additionally, in principle, superior generalization could be obtained by matching the structure of each candidate model to the features of its own partition. Very often the partitions could also be given a useful physical interpretation.

Probabilistic Approaches to Modular Network Modelling. In this book we are particularly interested in probabilistic approaches towards modular network modelling of spatially complex dynamic systems. The aim is to investigate their use inside a stochastic adaptive control scheme when the plant dynamics are nonlinear and unknown. The basis of most probabilistic approaches to modular networks is the *Mixture Modelling* density estimation technique [175, 246] from statistics. A mixture model is a general tool for representing conditional probability density functions $p(\mathbf{y}|\mathbf{x})$ in parameterized form. This is achieved through a weighted sum of density kernels $\beta_j(\mathbf{y}|\mathbf{x})$ in the following form:

$$p(\mathbf{y}|\mathbf{x}) = \sum_{j=1}^{M} \alpha_j(\mathbf{x})\beta_j(\mathbf{y}|\mathbf{x}).$$

The parameters α_j are called *mixing coefficients* and the density kernels β_j are usually chosen as Gaussian.

The *Mixture of Experts* (ME) architecture of Jordan and Jacobs [117] is one example of a soft-partitioned modular network model derived from probabilistic considerations. Its original formulation was specified in terms of static systems for solving regression problems, however it could easily be extended to the dynamic case. When applied to the plant representation of Equation (8.14), the starting point is the probability that the output $y(k)$ is generated by parameter vector θ_j, given the known input $\phi(k-1)$. Using the usual notation, this probability is denoted as $\Pr(M_j(k)|\phi(k-1))$. This term is related to the probability density function $p(y(k)|\theta_j, \phi(k-1))$ of the jth candidate model[1] and the conditional probability density $p(y(k)|\phi(k-1))$, as represented by the following Mixture Model:

$$p(y(k)|\phi(k-1)) = \sum_{j=1}^{H} \Pr(M_j(k)|\phi(k-1)) p(y(k)|\theta_j, \phi(k-1)).$$

Note that $p(y(k)|\theta_j, \phi(k-1))$ takes the role of the density kernels β_j and $\Pr(M_j(k)|\phi(k-1))$ reflects the mixing coefficients. Since the system noise $e(k) \sim N(0, \sigma^2)$, it follows that

$$p(y(k)|\theta_j, \phi(k-1)) \sim N(\theta_j^T \phi(k-1), \sigma^2) \qquad (8.15)$$

which is easily evaluated in closed form by the Gaussian equation.

In the ME architecture it is assumed that the scheduling variable $\mathbf{z}$ consists of the regressor itself $\phi(k-1)$. Hence the validity function $\hat{\gamma}_j(\mathbf{z})$ could

---
[1] also called Expert Network in ME terminology

be interpreted as an estimate of the probability $\Pr(M_j(k)|\phi(k-1))$ because $\hat{\gamma}_j$ depends on $\phi(k-1)$ and satisfies conditions (8.13), which are compatible with the constraints of mathematical probability. Therefore, the conditional probability density of the output is given by

$$p(y(k)|\phi(k-1)) = \sum_{j=1}^{H} \hat{\gamma}_j(\phi(k-1)) p(y(k)|\theta_j, \phi(k-1)). \quad (8.16)$$

The estimate of the output $\hat{y}$, is taken to be the mean of the above conditional probability density which, from Equations (8.15) and (8.16), is easily calculated as

$$\hat{y} = E\{y(k)|\phi(k-1)\} = \sum_{j=1}^{H} \hat{\gamma}_j \theta_j^T \phi(k-1). \quad (8.17)$$

This shows that the estimate of the output is given as the sum of the individual expert network outputs, weighted by the corresponding validity function. As expected, this is similar to the actual plant Equation (8.14), but with the absence of the noise term $e(k)$ which is unknown and unmeasurable.

When using an ME architecture to identify some nonlinear dynamics, both the validity functions and the expert network parameters are initially unknown. Hence the validity functions are parameterized in terms of the softmax function, which satisfies the three conditions (8.13) and is given as

$$\hat{\gamma}_j(\phi(k-1), \mathbf{h}_1, \cdots, \mathbf{h}_H) = \frac{\exp(\mathbf{h}_j^T \phi(k-1))}{\sum_{i=1}^{H} \exp(\mathbf{h}_i^T \phi(k-1))}, \quad (8.18)$$

where $\{\mathbf{h}_i\}_{i=1}^{H}$ are the parameter vectors of the softmax functions. These require estimation together with the expert network parameters $\{\theta_i\}_{i=1}^{H}$. The subsystem that generates the validity functions according to Equation (8.18) is called the *Gating Network*. All the unknown parameters are calculated using batch estimation methods on a training data set. The ME network has also been extended to a *hierarchical* architecture of experts models [125], where the parameter estimation procedure is formulated in terms of the *Expectation-Maximization* (EM) algorithm [63]. Some interesting convergence results on the ME architecture are given in [126]. Modular network schemes based on sequential (on-line) parameter estimation procedures are of particular interest to adaptive control. A sequential version of the batch estimation schemes of EM was also proposed in [125], which is based on a recursive least squares algorithm with exponential forgetting.

Kadirkamanathan and Kadirkamanathan [133, 134] proposed an ME-like architecture for nonlinear candidate models whose parameter estimation is sequential and based on Kalman filters. The expert models consist of GaRBF neural networks and they propose a different gating network structure that utilizes the same Gaussian basis functions of the expert models. In on-line schemes, mode transitions have to be detected on the fly during parameter estimation. This makes the estimation procedure more challenging than for the batch estimation methods. The mode-switching detection algorithm of [133, 134] combines information from both temporal and spatial characteristics, *i.e.*, the effect of a mode switch on the output, as well as the open-loop information from the value of the scheduling variable.

Skeppstedt *et al.* [230] proposed a different sequential scheme for identifying piecewise linear systems. The parameters of the candidate models are estimated by Andersson's algorithm [11], which was briefly described in Section 8.3. A hard partitioning scheme is implemented, which consists of a set of discriminant functions whose parameters are estimated by a recursive least squares procedure normally used in pattern recognition. They also combine both temporal and spatial information to detect and classify mode switches.

8.4 Summary

Intelligent control systems are expected to deal with situations of multimodal complexity. Multiple model techniques are a promising way of handling such problems. In this chapter we have reviewed and categorized the main multiple model approaches to control and estimation that are based on probabilistic techniques. The methodology was originally developed to effect control and state estimation of adaptive ISI systems, however it was subsequently also used to handle uncertain IO models.

A distinction has been drawn between temporal and spatial multimodal problems. The former are characterized by arbitrary mode jumps whose activity does not depend on any measurable signals, whilst the latter assume the availability of some scheduling variables that reflect the mode activity. The optimal multiple model solution of temporal multimodal systems demands the use of an exponentially increasing number of models. Hence various suboptimal solutions have been proposed to derive estimation and control schemes that could be implemented in practice. For the spatial multimodal case, multiple model techniques typically utilize modular networks to represent the system and effect estimation and control. The resulting control schemes are closely related to gain scheduling.

This chapter has covered the basic theoretical issues concerning multiple model techniques. A number of important publications on probabilistic approaches to multiple model estimation and control have also been described. These will serve as a basis and motivation for the new schemes developed in the following two chapters which deal with handling temporal and spatial multimodality in the presence of uncertain and nonlinear mode dynamics.

9. Multiple Model Dual Adaptive Control of Jump Nonlinear Systems

9.1 Introduction

This chapter considers control of a class of nonlinear, stochastic, multimodal systems whose various mode dynamics are unknown and subject to unscheduled jumps. The practical significance of such systems includes fault-tolerant control or plants working in an unpredictable environment. The task of controlling such systems is challenging because of the presence of both mode jumps, which makes it a temporal multimodal problem, as well as the dynamic uncertainty of the modes. The latter takes into account that in a realistic situation it is difficult to formulate prior accurate models for all modes, particularly when a mode corresponds to some fault condition. The solution adopted combines ideas taken from adaptive control to handle dynamic uncertainty, and multiple model techniques to handle the multimodality. The use of multiple models also serves to make the system more "intelligent" because, as explained in Chapter 1, it furnishes it with characteristics of *learning* (through memorization) and not just adaptation.

The problem under consideration differs from the adaptive ISI case covered by classical MMAC because of the following reasons:

- The mode dynamics are in affine-nonlinear IO form instead of state-space form.
- The uncertainty concerns the unknown mode activity *and* the different mode dynamics. In MMAC, the state-space matrices of all modes are assumed known and the uncertainty is due to the unknown mode activity and unmeasurable state vector only.
- The presence of unknown mode dynamics makes it more difficult to detect mode jumps than for classical MMAC.
- The system must estimate a *set* of distinct nonlinear functions corresponding to the different mode dynamics, rather than a *single* state vector.

Nevertheless, although not exactly similar, the two problems are clearly related: whilst the classical MMAC problem is concerned with state estimation and control of *known* jump linear systems, the problem addressed in this chapter concentrates on function estimation and control of *unknown* jump nonlinear systems.

As explained in Chapter 8, stochastic multiple model schemes for system identification of uncertain jump systems in IO form have been applied before, both for the linear [11, 177] and nonlinear cases [129]. The latter is particularly relevant to this chapter because it addresses the multiple model estimation problem within the context of neural networks. The estimation aspects of the adaptive control scheme proposed in this chapter are based on [129], but here we have included a rigorous theoretical analysis and justification of the approximations introduced. More significantly, the multiple model estimation ideas are extended to derive original multiple model dual adaptive control laws for uncertain, temporal multimodal nonlinear systems. The problem of controlling jump systems in nonlinear IO form was also considered by Narendra et al. [188], who proposed a scheme that combines multiple model and parameter adaptive methods within a NN framework. That scheme however, covers only deterministic systems and so the Bayesian methodology bequeathed by classical MMAC theory cannot be utilized. In this chapter we therefore re-address the stochastic version of this problem.

In general, stochastic models are more representative of realistic situations and naturally lead to the use of stochastic estimation and control algorithms. In the proposed scheme, all the principal tasks of mode detection, estimation of mode dynamics and control signal generation utilize the probabilistic information from a set of Kalman filters in a co-ordinated manner. Rather than using dynamic back propagation, the parameters of the neural networks are adjusted by Kalman filters and the control law is not of the HCE type, but an extension of the suboptimal dual scheme of Chapter 7 to the multiple model case. Additionally, to handle the situation when the number of modes that the plant can assume is unknown, a self-organizing scheme that automatically evolves the set of candidate models in real-time is also developed. The main contributions of this chapter could therefore be summarized as follows:

1. The design of a multiple model adaptive control scheme for a class of multimodal, affine-nonlinear, stochastic plants whose mode dynamics are subject to functional uncertainty and which can jump arbitrarily in time.
2. The development and analysis of a mode detection and estimation scheme within a unified probabilistic framework, similar in spirit to classical MMAC but modified to handle nonlinear functional uncertainty instead of state uncertainty.
3. The development of novel suboptimal dual adaptive control laws for temporal multimodal systems.
4. A self-organizing technique for autonomously growing and assigning the set of candidate models in real-time, when the number of plant modes is not known *a priori*.

9.2 Problem Formulation

The task is to control a stochastic, single-input single-output affine class of jump nonlinear, discrete-time systems having the general form:

$$y(k) = f_{m(k)}[\mathbf{x}(k-1)] + g_{m(k)}[\mathbf{x}(k-1)]u(k-1) + e(k) \tag{9.1}$$

where $y(k)$ is the output, $u(k)$ is the control input, $\mathbf{x}(k-1) := [y(k-n) \ldots y(k-1) \ u(k-1-p) \ldots u(k-2)]^T$ is the system state vector and $e(k)$ is an additive noise signal.

The smooth, nonlinear functions $f_{m(k)}[\mathbf{x}(k-1)]$, $g_{m(k)}[\mathbf{x}(k-1)] : \Re^{n+p} \mapsto \Re$, could switch form at an *arbitrary* instant in time, taking on any of the function pairs in the set

$$\{(f_1, g_1), (f_2, g_2), \ldots (f_H, g_H)\},$$

as indexed by $m(k) \in \{1, \cdots, H\}$. These represent the various modes of operation that the system can assume during the course of time. In this chapter we consider the case where both the mode variable $m(k)$ and the functions (f_i, g_i) for the different modes are unknown.

The system output is required to track a reference input $y_d(k)$ under the following assumptions:

Assumption 9.2.1 *The noise $e(k)$ is independent and has a zero-mean Gaussian distribution of variance σ^2.*

Assumption 9.2.2 *The state vector's dimensionality parameters p and n, and the noise variance σ^2 are known.*

Assumption 9.2.3 *The reference input is known at least one time step ahead and is bounded.*

Assumption 9.2.4 *The dynamics of every mode are minimum phase and functions g_i are bounded away from zero.*

A multiple model approach based on neural networks fits naturally within this scenario. H *local* neural network models, one per mode, are used to identify the nonlinear mode dynamics and to control the system via an indirect adaptive technique. The methodology we adopt is to detect which particular mode is active at a given time, and force a particular local model to learn the dynamics of that mode by on-line adjustment of its neural network weights. Considering that the mode dynamics are not known *a priori*, the mode detection task is not trivial. The local model's approximations are then used inside a control law derived from suboptimal dual techniques. Following allocation to a particular mode and adjustment of its network weights, a local model therefore acts as an "expert" on the dynamics of that mode.

In the scheme being proposed, each local model is intended to capture the dynamics of one particular mode of behaviour. Since every mode depends on a pair of smooth nonlinear functions (f_i, g_i), two GaRBF networks are used per local model to approximate the corresponding functions within a compact set $\chi \subset \Re^{n+p}$ that corresponds to the network approximation region. The outputs of the two neural networks of local model i, denoted by $\hat{f}_i, \hat{g}_i$, are given as follows:

$$\begin{aligned} \hat{f}_i[\mathbf{x}, \hat{\mathbf{w}}_{f_i}] &= \hat{\mathbf{w}}_{f_i}^T \Phi_{f_i}[\mathbf{x}] \\ \hat{g}_i[\mathbf{x}, \hat{\mathbf{w}}_{g_i}] &= \hat{\mathbf{w}}_{g_i}^T \Phi_{g_i}[\mathbf{x}] \end{aligned} \tag{9.2}$$

where $\hat{\mathbf{w}}_{f_i}, \hat{\mathbf{w}}_{g_i}$ denote the network output layer parameter vectors and Φ_{f_i}, Φ_{g_i} denote the Gaussian basis function vectors. The following conditions, connected with the neural networks, are assumed to hold:

Assumption 9.2.5 *The state* $\mathbf{x}(k)$ *is always confined within a bounded region of state space that is enclosed by* χ. *Note that* χ *could be chosen arbitrarily large by the designer.*

Assumption 9.2.6 *For every local model* i, *there exist some optimal basis function centres, width parameters and output layer parameters* $\mathbf{w}_{f_i}^*$, $\mathbf{w}_{g_i}^*$ *ensuring that within region* χ, *networks* $\hat{f}_i, \hat{g}_i$ *approximate the mode dynamics* (f, g) *captured by model* i *arbitrarily well.*

Assumption 9.2.7 *The optimal basis function centres and width parameters are known a priori and they are the same in each individual series of* $\hat{f}_i$ *and* $\hat{g}_i$ *networks, i.e.:*

$$\begin{aligned} \Phi_{f_1} &= \Phi_{f_2} = \cdots = \Phi_{f_H} := \Phi_f \\ \Phi_{g_1} &= \Phi_{g_2} = \cdots = \Phi_{g_H} := \Phi_g. \end{aligned}$$

Assumption 9.2.6 follows from the Universal Approximation Property of GaRBF neural networks. Interpreted in terms of the theory in [219], the last part of Assumption 9.2.7 implies that the f_i functions have a similar degree of smoothness across all modes, and similarly for functions g_i.

Hence, in this scheme, the unknown variables consist of the *optimal* output layer parameters of the networks in all local models, *i.e.*, $\mathbf{w}_{f_i}^*$, $\mathbf{w}_{g_i}^*$; $i = 1 \cdots H$. By Assumptions 9.2.5, 9.2.6, 9.2.7 and Equations (9.1) and (9.2), it follows that *during activity of the mode captured by local model* i, as long as the state lies within region χ, the system dynamics could be represented in the following state space form:

$$\begin{aligned} \mathbf{w}_i^*(k+1) &= \mathbf{w}_i^*(k) \\ y(k) &= \mathbf{w}_i^{*T}(k)\Phi[\mathbf{x}(k-1)] + e(k) \end{aligned} \tag{9.3}$$

where $\mathbf{w}_i^{*T}(k) := [\mathbf{w}_{f_i}^{*T}(k) \; \mathbf{w}_{g_i}^{*T}(k)]$ and $\Phi^T[\mathbf{x}(k-1)] := [\Phi_f^T[\mathbf{x}(k-1)] \; \Phi_g^T[\mathbf{x}(k-1)]u(k-1)]$.

Vector $\mathbf{w}_i^*$ therefore represents the *optimal* value of the output layer parameter vectors of the two neural networks in local model i. Being unknown, it is estimated by recursive adjustment of $\hat{\mathbf{w}}_{f_i}$, $\hat{\mathbf{w}}_{g_i}$, which are collectively denoted by vector $\hat{\mathbf{w}}_i := [\hat{\mathbf{w}}_{f_i}^T \; \hat{\mathbf{w}}_{g_i}^T]^T$. This applies to all local models

$i = 1, \cdots, H$. Hence the estimate of the plant output from local model i is given by

$$\hat{y}_i(k) = \hat{\mathbf{w}}_i^T(k)\Phi[\mathbf{x}(k-1)], \qquad (9.4)$$

and the estimation error (or innovations) for local model i is defined as

$$\epsilon_i(k) := y(k) - \hat{y}_i(k).$$

Since local model i is required to learn one particular mode, its parameters $\hat{\mathbf{w}}_i$ should only be adjusted during the times when that particular mode is active, using the information from the innovations ϵ_i. However, being a temporal multimodal problem, there is no direct information on which mode is active at any given time. Hence the particular local model that is set to identify the currently active mode must somehow be detected. In the case of a self-organized scheme, the system must also determine whether a mode is appearing for the first time so as to add a new local model to the set.

9.3 The Estimation Problem

In this section, for clarity of exposition, it is assumed that the number of plant modes H is known and that an equal number of local models is set up *a priori*. This assumption is eventually relaxed when the self-organized scheme for allocating local models in real time is described in Section 9.4.

9.3.1 Known Mode Case

To introduce the topic we will first consider how the parameter estimation problem is handled if the mode activity defined by $m(k)$ were known. Let us denote the sequence of modes from start till time k by $S(k)$, i.e.:

$$S(k) := \{m(1), m(2), \cdots, m(k)\}.$$

Since Equation (9.3) is linear in the parameters and noise $e(k)$ is Gaussian, a Kalman filter could be used to generate recursively the optimal conditional minimum mean square predictive estimate $\hat{\mathbf{w}}_i(k+1)$ of $\mathbf{w}_i^*$ and its covariance matrix $\mathbf{P}_i(k+1)$ *whenever* the mode corresponding to local model i is active, as indicated by $m(k)$. This is similar to the estimation procedure in Chapter 7, except that now H Kalman filters are required: one for every possible mode of operation. In addition, the use of a Kalman filter is subject to the following assumption being satisfied:

Assumption 9.3.1 *The optimal parameter vectors of the neural networks in all local models, $\mathbf{w}_i^*$ ($i = 1, \cdots, H$), are random variables whose initial value $\mathbf{w}_i^*(0)$ is Gaussian distributed with known mean $\mathbf{m}_i$ and covariance $\mathbf{R}_i$.*

The initial means $\mathbf{m}_i$ reflect any prior knowledge of $\mathbf{w}_i^*$, and the initial parameter covariance matrices $\mathbf{R}_i$ are chosen to be symmetric, positive definite and with large diagonal elements, because they reflect the uncertainty of the unknown parameter estimates which initially is large.

Therefore with H Kalman filters, one allocated to every local model, we are able to calculate the conditional minimum mean square estimate of the optimal network parameters characterizing every corresponding mode. Since *only* the Kalman filter of the model allocated to a presently active mode is updated at any one time, each model will "learn" a *distinct* mode. In the known mode case, the information on mode activity is provided directly by $m(k)$. This Kalman filter updating procedure is more conveniently handled by defining H *indicator variables*: $\gamma_1(k), \gamma_2(k) \cdots, \gamma_H(k)$, as follows:

$$\gamma_i(k) := \begin{cases} 1 \text{ if } m(k) = i \\ 0 \text{ otherwise} \end{cases} \tag{9.5}$$

leading to the following lemma on parameter estimation:

Lemma 9.3.1.

1. *Subject to Assumptions 9.2.1, 9.2.2, 9.2.5-9.3.1, and the known mode sequence $S(k)$; the distribution of $\mathbf{w}_i^*(k+1)$ conditioned on information state I^k and sequence $S(k)$ is Gaussian whose mean and covariance, defined and denoted as:*

$$\hat{\mathbf{w}}_i(k+1|S(k)) := E\{\mathbf{w}_i^*(k+1)|I^k\}$$
$$\mathbf{P}_i(k+1)|S(k)) := \text{cov}\{\mathbf{w}_i^*(k+1)|I^k\},$$

satisfy the following Kalman filter recursive equations:

$$\mathbf{K}_i(k) = \frac{\gamma_i(k)\mathbf{P}_i(k|S(k-1))\Phi[\mathbf{x}(k-1)]}{\sigma^2 + \Phi^T[\mathbf{x}(k-1)]\mathbf{P}_i(k|S(k-1))\Phi[\mathbf{x}(k-1)]}$$
$$\hat{\mathbf{w}}_i(k+1|S(k)) = \hat{\mathbf{w}}_i(k|S(k-1)) + \mathbf{K}_i(k)\epsilon_i(k) \tag{9.6}$$
$$\mathbf{P}_i(k+1|S(k)) = \{\mathbf{I} - \mathbf{K}_i(k)\Phi^T[\mathbf{x}(k-1)]\}\mathbf{P}_i(k|S(k-1))$$

with initial conditions $\hat{\mathbf{w}}_i(0) = \mathbf{m}_i$, $\mathbf{P}_i(0) = \mathbf{R}_i$ and indicator variables γ_i calculated by Equation (9.5).
   N.B. $S(k)$ *is included inside the arguments of $\hat{\mathbf{w}}_i$, $\mathbf{P}_i$ to explicitly denote the mode sequence that generated these estimates.*

2. *The probability density of $y(k)$ conditioned on $S(k)$ and I^{k-1}, denoted by $p(y(k)|S(k), I^{k-1})$, is also Gaussian. Specifically, its mean and variance are given by $\hat{\mathbf{w}}_{m(k)}^T(k|S(k-1))\Phi[\mathbf{x}(k-1)]$ and $\sigma^2 + \Phi^T[\mathbf{x}(k-1)]\mathbf{P}_{m(k)}(k|S(k-1))\Phi[\mathbf{x}(k-1)]$ respectively.*

Proof. The proof follows directly by applying H distinct predictive type Kalman filters to Equation (9.3), each updated when a particular mode is specified active by the indicator variables. This leads to model 1 being matched to mode 1, model 2 matched to mode 2 *etc.* The solution is optimal because the conditions for Kalman filtering are satisfied on a mode-by-mode basis. □

9.3.2 Unknown Mode Case

Since the mode index $m(k)$ is not actually known, an alternative solution for parameter estimation has to be found. One option is to extend the theory of MMAE for state estimation of jump systems described in Section 8.2.2, to this case of parameter estimation. This approach leads to the *optimal* estimate of the network parameters and also avoids direct estimation of $m(k)$. Essentially, at any time k, it involves consideration of the space of *all* possible mode sequences up to that time, of which there are H^k, and applying Equations (9.6) to *all* these possible outcomes.

Using the same notation of Chapter 8, if $M_i(k)$ denotes the event $\{m(k) = i\}$ and $S_j(k)$ denotes one specific sequence of such events from start till time k, the optimal predictive parameter estimates are given by

$$\hat{\mathbf{w}}_i(k+1) = \sum_{j=1}^{H^k} \Pr(S_j(k)|I^k)\hat{\mathbf{w}}_i(k+1|S_j(k)); \ \forall i = 1 \cdots H,$$

where $\Pr(S_j(k)|I^k)$ denotes the *a posteriori* probability of sequence $S_j(k)$ given the information state I^k. $\hat{\mathbf{w}}_i(k+1|S_j(k))$ denotes the parameter estimate from Equations (9.6) *but* subjected to sequence $S_j(k)$ rather than $S(k)$ by setting the indicator variables of Equation (9.5) according to the events $M_i(k)$ that make up sequence $S_j(k)$, instead of $m(k)$. Clearly this approach is impractical because at time k it requires H^{k+1} Kalman filters matched to all sequences $S_j(k)$ for all H modes, which is too memory intensive and increases exponentially in time. Therefore, in the following section, a suboptimal solution which avoids consideration of all possible mode sequences is proposed, by finding an estimate to the true but unknown mode sequence $S(k)$.

Mode Estimation. Taking some inspiration from the known mode case, the number of Kalman filters could be maintained equal to H by finding an estimate $\hat{S}(k)$ to the true, but unknown, mode sequence $S(k)$. The estimated sequence is represented by

$$\hat{S}(k) = \{\hat{m}(1), \hat{m}(2), \cdots, \hat{m}(k)\},$$

where $\hat{m}(k)$ denotes the index of the *model* estimated to be capturing the dynamics of the mode active at time k.

Index $\hat{m}(k)$ is calculated by introducing $\hat{M}_i(k)$ to denote the event $\{\hat{m}(k) = i\}$, i.e., the event that local model i represents the mode active at time k, and applying the following Maximum *a Posteriori* (MAP) criterion:

$$\hat{m}(k) = \arg\max_{i=1\cdots H} \Pr(\hat{M}_i(k)|\hat{S}(k-1), I^k), \tag{9.7}$$

where $\Pr(\hat{M}_i(k)|\hat{S}(k-1), I^k)$ is the *posterior probability* of event $\hat{M}_i(k)$ conditioned on the current information state and the previously estimated mode sequence. In effect, this MAP criterion is saying that the local model capturing the dynamics of the current active mode is the one whose posterior probability is largest. This requires calculation of all the model posterior probabilities, whose evaluation is given by the following lemma.

Lemma 9.3.2. *The posterior probability of event $\hat{M}_i(k)$ conditioned on $\hat{S}(k-1), I^k$ is given by:*

$$\Pr(\hat{M}_i(k)|\hat{S}(k-1), I^k) = \frac{p(y(k)|\hat{M}_i(k), \hat{S}(k-1), \tilde{I}^{k-1}) \Pr(\hat{M}_i(k)|\hat{S}(k-1), \tilde{I}^{k-1})}{\sum_{j=1}^{H} p(y(k)|\hat{M}_j(k), \hat{S}(k-1), \tilde{I}^{k-1}) \Pr(\hat{M}_j(k)|\hat{S}(k-1), \tilde{I}^{k-1})} \quad (9.8)$$

where $\tilde{I}^{k-1} := \{Y^{k-1}, U^{k-1}\}$ represents the set of outputs and inputs up to time step $(k-1)$. The term $p(y(k)|\hat{M}_i(k), \hat{S}(k-1), \tilde{I}^{k-1})$ denotes the probability density function of output $y(k)$ conditioned on the mode sequence $\{\hat{S}(k-1), \hat{M}_i(k)\}$ and $\tilde{I}^{k-1}$.

Proof.

$$\Pr(\hat{M}_i(k)|\hat{S}(k-1), I^k) = \Pr(\hat{M}_i(k)|\hat{S}(k-1), y(k), \tilde{I}^{k-1}) =$$
$$\frac{p(y(k)|\hat{M}_i(k), \hat{S}(k-1), \tilde{I}^{k-1}) \Pr(\hat{M}_i(k)|\hat{S}(k-1), \tilde{I}^{k-1})}{p(y(k)|\hat{S}(k-1), \tilde{I}^{k-1})},$$

where the last step follows by Bayes' rule. Since the events $\{\hat{M}_j(k)\}_{j=1}^H$ are required to be mutually exclusive and exhaustive, then marginalization leads to

$$p(y(k)|\hat{S}(k-1), \tilde{I}^{k-1}) =$$
$$\sum_{j=1}^{H} p(y(k)|\hat{M}_j(k), \hat{S}(k-1), \tilde{I}^{k-1}) \Pr(\hat{M}_j(k)|\hat{S}(k-1), \tilde{I}^{k-1}),$$

which proves the lemma. □

Remarks 9.3.1 For similar reasons as the known mode case (see Lemma 9.3.1), the density terms $p(y(k)|\hat{M}_i(k), \hat{S}(k-1), \tilde{I}^{k-1})$ appearing in Equation (9.8) are Gaussian distributed. Although the densities in Lemma 9.3.1 were conditioned on I^{k-1} rather than on $\tilde{I}^{k-1}$ (as in this case), the difference is of no consequence as far as the Gaussianity of the distribution is concerned. The reason is that $\tilde{I}^{k-1}$ differs from I^{k-1} by the inclusion of one additional term, $u(k-1)$, which is I^{k-1}-measurable anyway. Hence conditioning on I^{k-1} or $\tilde{I}^{k-1}$ is equivalent.

The mean m_{y_i} and variance r_{y_i} of $p(y(k)|\hat{M}_i(k), \hat{S}(k-1), \tilde{I}^{k-1})$ are therefore given by the equations

$$m_{y_i} = \hat{\mathbf{w}}_i^T(k|\hat{S}(k-1))\Phi[\mathbf{x}(k-1)] \quad (9.9)$$
$$r_{y_i} = \Phi^T[\mathbf{x}(k-1)]\mathbf{P}_i(k|\hat{S}(k-1))\Phi[\mathbf{x}(k-1)] + \sigma^2, \quad (9.10)$$

9.3 The Estimation Problem

where $\hat{\mathbf{w}}_i(k|\hat{S}(k-1))$, $\mathbf{P}_i(k|\hat{S}(k-1))$ are calculated using the same Kalman filter Equations (9.6) but conditioned on the sequence $\hat{S}(k-1)$. This means that instead of using Equation (9.5), the indicator variables $\gamma_i(k)$ are calculated as follows:

$$\gamma_i(k) = \begin{cases} 1 \text{ if } \hat{m}(k) = i \\ 0 \text{ otherwise} \end{cases}. \quad (9.11)$$

Note that at any time instant, only H Kalman filters are required to calculate the probability densities $p(y(k)|\hat{M}_i(k), \hat{S}(k-1), \tilde{I}^{k-1})$. The exact equation for the density of $y(k)$ is given by the standard Gaussian equation as

$$p(y(k)|\hat{M}_i(k), \hat{S}(k-1), \tilde{I}^{k-1}) = (2\pi r_{y_i})^{-\frac{1}{2}} \exp\left\{-\frac{|y(k) - m_{y_i}|^2}{2 r_{y_i}}\right\}. \quad (9.12)$$

To evaluate the prior probability terms $\Pr(\hat{M}_i(k)|\hat{S}(k-1), \tilde{I}^{k-1})$ that also appear in Equation (9.8), we assume that the mode transitions follow a Markov chain having a known Transition Probability Matrix $\Pi = [\pi_{i,j}]_{H \times H}$, where by definition

$$\pi_{i,j} = \Pr(M_j(k)|M_i(k-1)). \quad (9.13)$$

As far as the term $\Pr(\hat{M}_j(k)|\hat{S}(k-1), \tilde{I}^{k-1})$ is concerned, this Markov assumption means that

$$\Pr(\hat{M}_j(k)|\hat{S}(k-1), \tilde{I}^{k-1}) = \Pr(\hat{M}_j(k)|\hat{m}(k-1)),$$

where the right hand side represents the probability of transition from the mode captured by local model $\hat{m}(k-1)$ to that captured by model j. Unfortunately, Equation (9.13) still cannot be used to directly evaluate the term $\Pr(\hat{M}_j(k)|\hat{m}(k-1))$ because the former equation is indexed in terms of the *modes* and the latter in terms of the *models*, and the mapping between the index of the models and that of the modes is not necessarily known.

This problem is resolved if we assume that the transition probabilities $\pi_{i,j}$ could take only one of two distinct values, irrespective of which modes are actually concerned:

- the probability π^{ntr} (assumed known) that a transition does not take place
or,
- the probability π^{tr} that a transition takes place.

This is equivalent to saying that transition matrix Π has all diagonal terms equal to π^{ntr}, and all off-diagonal terms equal to π^{tr}. Since Π must satisfy the condition that $\sum_{j=1}^{H} \pi_{i,j} = 1$, the values of π^{tr} and π^{ntr} are not independent and the transition probabilities are given as:

$$\pi_{i,j} = \begin{cases} \pi^{ntr} & \text{if } i = j \\ \pi^{tr} = (1 - \pi^{ntr})/(H-1) & \text{otherwise} \end{cases}. \quad (9.14)$$

The transition probabilities are now not directly related to the actual modes concerned, but to whether a mode transition takes place. Hence the term $\Pr(\hat{M}_j(k)|\hat{m}(k-1))$ is evaluated as

$$\Pr(\hat{M}_j(k)|\hat{m}(k-1)) = \begin{cases} \pi^{ntr} & \text{if } \hat{m}(k-1) = j \\ \pi^{tr} & \text{otherwise} \end{cases}$$
$$= \pi_{\hat{m}(k-1),j}.$$

Equation (9.8) therefore reduces to

$$\Pr(\hat{M}_i(k)|\hat{S}(k-1), I^k) = \frac{p(y(k)|\hat{M}_i(k), \hat{S}(k-1), \tilde{I}^{k-1})\, \pi_{\hat{m}(k-1),i}}{\sum_{j=1}^{H} p(y(k)|\hat{M}_j(k), \hat{S}(k-1), \tilde{I}^{k-1})\, \pi_{\hat{m}(k-1),j}}. \tag{9.15}$$

To start off the evaluation of this equation from $k = 1$, the prior value of $\hat{m}$ (i.e., $\hat{m}(0)$) is set to any integer within the range $[1, H]$. π^{ntr} is a design parameter and its choice depends on the problem at hand. For example if mode transitions are not frequent, as would be the case for fault induced switches, $0.95 \leq \pi^{ntr} < 1$ was seen to give good performance in simulation experiments. Additionally, even when the mode sequence is not really a Markov chain, simulations have shown that this method works well.

Remarks 9.3.2 It is also possible to use the simpler, but cruder option for mode estimation as described in the previous chapter. This is based on ignoring the mode jumps and forcefully lower bounding the calculated posterior probabilities. Such an approach was investigated by the authors in [132].

Parameter Estimation. Having estimated the mode sequence, we now discuss the parameter estimation problem. At this point it is useful to distinguish clearly between the problem addressed in this chapter and classical MMAE methods. Rather than simply detecting "jumps" from a set of *known* modes as characterized by *pre-defined* local models, our problem involves the more complex task of on-line learning of H distinct mode dynamics, in addition to detection of mode transitions.

Hence, at any given time, we update the parameters of only one local model, according to the estimated mode sequence. Therefore during time k, having evaluated $\hat{m}(k)$ via Equation (9.7), the Kalman filter Equations (9.6) are used to generate the predictive parameter estimates $\hat{\mathbf{w}}_i(k+1|\hat{S}(k))$ and the covariance $\mathbf{P}_i(k+1|\hat{S}(k))$, with the indicator variables $\gamma_i(k)$ set according to Equation (9.11). This means that the only parameters to be updated are those of the local model estimated to be capturing the dynamics of the current mode. Note that these predictive estimates are also the same ones used to evaluate $p(y(k+1)|\hat{M}_i(k+1), \hat{S}(k), \tilde{I}^k)$ (via Equations (9.9), (9.10) and (9.12)) required in Equation (9.15) during the *next* time step.

The absence of prior knowledge on the mode dynamics means that initially none of the local models is in a position to capture accurately the dynamics of any of the modes. For this reason, the term

$$\max_{i=1\cdots H} \Pr(\hat{M}_i(k)|\hat{S}(k-1), I^k)$$

in Equation (9.7) might not always be unique, particularly during the initial phase of the process when all local models are equally valid contenders to start learning a mode. This is particularly true if all local models are configured to have the same initial priors $\mathbf{m}_i$, $\mathbf{R}_i$. The problem is solved by letting $\hat{m}(k)$ arbitrarily take the value of any one of these non-unique maxima whenever the condition occurs. This effectively *forces* the parameters of the selected model to match the dynamics of the presently active mode, leaving the rest of the models free to learn subsequent modes that have not yet been allocated a model. It also ensures that each local model will learn a distinct mode and that events $\left\{\hat{M}_j(k)\right\}_{j=1}^{H}$ are mutually exclusive.

Table 9.1 summarizes the proposed algorithm for mode and parameter estimation.

Table 9.1. The mode and parameter estimation algorithm at time k.

1.	Measure $y(k)$
2.	Calculate $p(y(k)|\hat{M}_i(k), \hat{S}(k-1), \tilde{I}^{k-1})$ using eqns. (9.9), (9.10), (9.12)
3.	Calculate the posteriors $\Pr(\hat{M}_i(k)|\hat{S}(k-1), I^k)$ using eqn.(9.15)
4.	Calculate $\hat{m}(k)$ from eqn. (9.7)
5.	Calculate $\{\gamma_i(k)\}$ from eqn. (9.11)
6.	Predict $\hat{\mathbf{w}}_i(k+1|\hat{S}(k))$, $\mathbf{P}_i(k+1|\hat{S}(k))$ from eqns. (9.6)

9.4 Self-organized Allocation of Local Models

In this section we describe a method for automatically allocating appropriate local models when the number of modes H is not known *a priori*. This takes place in an on-line manner, according to the mode activity detected during system operation. Hence, in addition to learning the various mode dynamics and detecting mode switches, the system will now automatically configure and grow the multiple model set itself. This technique will be referred to as *self-organized model allocation* [75].

To explain the logic behind the scheme, consider first the behaviour of the system when it is known that there is a total of H modes. As explained in Section 9.3, in this case H local models are set up at the outset. Index $\hat{m}$ of the local model estimated to be capturing the dynamics of the first active mode is determined by Equations (9.7) and (9.15). In case the maximum of Equation (9.7) is not unique, $\hat{m}$ is arbitrarily set to one of these non-unique solutions. This will force the selected model to learn the current mode by applying the Kalman filter algorithm to it *only*. Let us denote the index of this model as i. As long as the same mode remains active, this model's innovations will therefore start reducing and its posterior probability will increase and remain higher than the rest. Hence the MAP model selection criterion of Equation (9.7) ensures that its parameters continue to be adjusted.

Following a mode transition, the innovations of this model, $\epsilon_i(k) = y(k) - \hat{y}_i(k)$, increase appreciably because by now it would have captured the dynamics of the original mode quite accurately. Additionally, the conditional covariance $\mathbf{P}_i$ of model i's parameter estimates would have "reduced" substantially because its parameters have been subjected to learning via Equations (9.6) when its corresponding mode was previously active. Hence the variance of the associated Gaussian probability density $p(y(k)|\hat{M}_i(k), \hat{S}(k-1), \tilde{I}^{k-1})$ reduces as well (see Equation (9.10)) and the shape of this density function will be very sharp and narrow. This is shown as plot p_i in Figure 9.1. Consequently, during some time k_2 when the *new* mode is active, the probability density of model i evaluated at the measured value of $y(k_2)$ will be very small, as illustrated in Figure 9.1. Hence from Equation (9.15), the posterior probability of that model will quickly reduce towards zero as soon as the innovations start increasing as a result of the mode switch.

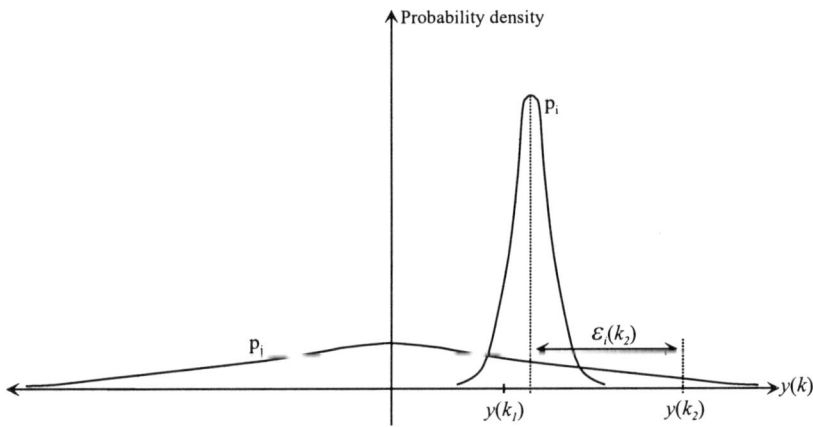

Fig. 9.1. The curves represent $p(y|\hat{M}, \hat{S}, \tilde{I})$, the probability density functions (pdf) for two models i and j. p_j represents the pdf for a model that has not yet been subjected to mode learning, and p_i is the pdf of model i at some time k after it was subjected to learning. The shape of the pdfs shows how the variance decreases as a consequence of learning. During activity of the mode captured by model i, the plant output will be well within the support of p_i, as shown by point $y(k_1)$. By contrast, if at some time k_2 a different mode is active, the innovations ϵ_i of model i will increase rapidly, leading to a small value of $p(y(k_2)|\hat{M}_i(k_2), \hat{S}(k_2 - 1), \tilde{I}^{k_2})$.

By contrast, due to the choice of a "large" initial covariance $\mathbf{R}_i$ and the fact that the rest of the models have not yet been subjected to learning, their associated Gaussian probability density $p(y(k)|\hat{M}_i, \hat{S}(k-1), \tilde{I}^{k-1})$ still has a large variance. Hence, even though their innovations are probably large (because they have not learned any mode dynamics yet), the posterior probability propagated by Equation (9.15) would be larger than the model that has learned the first mode, because their associated probability density is not so sharp. This is illustrated in Figure 9.1 by comparing the value of p_i with that of p_j at $y(k_2)$. Hence the MAP criterion would select one of these models that have not yet been selected to learn any mode. If it so happens that Equation (9.15) propagates the same probability for all these other $(H-1)$ models (reflecting that they are equally valid "fresh" contenders to start learning the new mode), one of these equiprobable contenders is chosen arbitrarily to learn the new mode by setting $\hat{m}$ equal to its index. The process of calculating the indicator variables and the parameter estimates is then repeated for the new model.

If after the first mode switch there were only one unallocated contender instead of $(H-1)$, by exactly the same arguments as before, Equation (9.15) would still propagate a high probability for this unallocated model. In this case there would be no ambiguity on the model to be selected for learning the new mode, simply because there is only one "fresh" contender. This point motivates the idea behind the self-organized scheme, which is detailed below:

1. We start off with only *two* local models and use the same algorithm of Table 9.1 with $H = 2$. This way, the first active mode is allocated to one of these two models and the second mode (once active) to the only remaining contender.
2. Just after the second model has been selected by the mode estimation algorithm as a result of a new mode being detected active, a third freshly-initialized local model is introduced by adding a newly initialized network parameter vector for this model. This fresh model is prepared to accept a new third mode, if and when it is active, by applying the usual mode estimation algorithm but with $H = 3$.
3. This procedure of adding a fresh local model once the previous one has been selected by the mode estimation algorithm, is repeated continuously. Hence, in effect there is always one *spare* local model ready to accept a new mode that has not appeared before.

Since models are being added in real time, Equations (9.7), (9.15) and (9.14) are respectively modified to:

$$\hat{m}(k) \doteq \arg\max_{i=1\cdots h(k)} \Pr(\hat{M}_i(k)|\hat{S}(k-1), I^k) \tag{9.16}$$

$$\Pr(\hat{M}_i(k)|\hat{S}(k-1), I^k) = \frac{p(y(k)|\hat{M}_i(k), \hat{S}(k-1), \tilde{I}^{k-1})\, \pi_{\hat{m}(k-1),i}}{\sum_{j=1}^{h(k)} p(y(k)|\hat{M}_j(k), \hat{S}(k-1), \tilde{I}^{k-1})\, \pi_{\hat{m}(k-1),j}} \tag{9.17}$$

and

$$\pi_{i,j} = \begin{cases} \pi^{ntr} & \text{if } i = j \\ (1-\pi^{ntr})/(h(k)-1) & \text{otherwise} \end{cases}, \tag{9.18}$$

where $h(k)$ denotes the number of local models created until time k.

9.5 The Control Law

The suboptimal dual control scheme described in Chapter 7 was shown to exhibit better performance over non-dual control laws. In this section we therefore extend the dual control methodology to the multiple model case of this chapter. As for the estimation problem, we will first assume that the number of plant modes H is known *a priori* and subsequently relax this assumption. Also, we start by considering the known mode case to motivate the algorithms subsequently proposed for the unknown mode case.

9.5.1 Known Mode Case

Recall that the performance index J_{inn} characterizing the IDC suboptimal dual scheme of Chapter 7 is given by

$$J_{inn} = E\left\{[y(k+1) - y_d(k+1)]^2 + qu^2(k) + r\epsilon^2(k+1)|I^k\right\}.$$

If the actual mode sequence $S(k)$ is assumed known then, as described in Section 9.3.1, H Kalman filters are enough to generate the optimal parameter estimates and covariance for all modes. In this case, the innovations term ϵ appearing in the above performance index is given by

$$\epsilon(k) := y(k) - \hat{y}(k), \tag{9.19}$$

where $\hat{y}(k) = \hat{\mathbf{w}}_{m(k)}^T(k|S(k-1))\Phi[\mathbf{x}(k-1)]$. This follows because the local model capturing the plant dynamics active at time k is the one indexed by $m(k)$, and so it only makes sense to use its innovations to induce caution and probing-like effects. Essentially, $\epsilon(k)$ captures the information regarding the uncertainty of the estimates of the *currently* active parameters $\mathbf{w}_{m(k)}^*$. The optimal solution for the known mode case is given by the following theorem.

Theorem 9.5.1. *The control law minimizing performance index J_{inn} subject to the system of Equation (9.1), all previous assumptions and knowledge of the mode sequence $S(k+1)$ is given by*

$$u^*(k) = \frac{(y_d(k+1) - \hat{f}_{m(k+1)})\hat{g}_{m(k+1)} - (1+r)\nu_{gf_{m(k+1)}}}{\hat{g}_{m(k+1)}^2 + q + (1+r)\nu_{gg_{m(k+1)}}} \tag{9.20}$$

where

$$\hat{f}_{m(k+1)} = \hat{\mathbf{w}}_{\mathbf{f}_{m(k+1)}}^T(k+1|S(k))\Phi_{\mathbf{f}}[\mathbf{x}(k)]$$
$$\hat{g}_{m(k+1)} = \hat{\mathbf{w}}_{\mathbf{g}_{m(k+1)}}^T(k+1|S(k))\Phi_{\mathbf{g}}[\mathbf{x}(k)]$$
$$\nu_{gf_{m(k+1)}} = \Phi_{\mathbf{g}}^T[\mathbf{x}(k)]\mathbf{P}_{\mathbf{gf}_{m(k+1)}}(k+1|S(k))\Phi_{\mathbf{f}}[\mathbf{x}(k)]$$
$$\nu_{gg_{m(k+1)}} = \Phi_{\mathbf{g}}^T[\mathbf{x}(k)]\mathbf{P}_{\mathbf{gg}_{m(k+1)}}(k+1|S(k))\Phi_{\mathbf{g}}[\mathbf{x}(k)].$$

$\hat{\mathbf{w}}_{\mathbf{f}_{m(k+1)}}(k+1|S(k))$ and $\hat{\mathbf{w}}_{\mathbf{g}_{m(k+1)}}(k+1|S(k))$ are the subvectors of $\hat{\mathbf{w}}_{m(k+1)}(k+1|S(k))$ from Lemma 9.3.1. Similarly $\mathbf{P}_{\mathbf{gf}_{m(k+1)}}(k+1|S(k))$, $\mathbf{P}_{\mathbf{gg}_{m(k+1)}}(k+$

$1|S(k))$ are submatrices of the covariance matrix $\mathbf{P}_{m(k+1)}(k+1|S(k))$ from the same lemma, where matrix partitioning is performed as described in Theorem 7.3.1 for the dual control law.

Proof. Since the mode sequence $S(k+1)$ is assumed known, this information must be included inside the performance index J_{inn} by conditioning upon it, giving rise to the performance index

$$J_{innk} = E\left\{[y(k+1) - y_d(k+1)]^2 + qu^2(k) + r\epsilon^2(k+1)|S(k+1), I^k\right\}.$$

The distribution of $y(k+1)$ conditioned on I^k and $S(k+1)$ is Gaussian, with mean and variance as specified in part 2 of Lemma 9.3.1. Hence it follows that

$$E\left\{[y(k+1) - y_d(k+1)]^2 | S(k+1), I^k\right\} =$$
$$\left\{\hat{\mathbf{w}}_{m(k+1)}^T(k+1|S(k))\Phi[\mathbf{x}(k)] - y_d(k+1)\right\}^2 +$$
$$\Phi^T[\mathbf{x}(k)]\mathbf{P}_{m(k+1)}(k+1|S(k))\Phi[\mathbf{x}(k)] + \sigma^2.$$

Additionally, the definition of ϵ given in Equation (9.19) and the conditional distribution of $y(k+1)$ yield,

$$E\left\{\epsilon^2(k+1)|S(k+1), I^k\right\} = \Phi^T[\mathbf{x}(k)]\mathbf{P}_{m(k+1)}(k+1|S(k))\Phi[\mathbf{x}(k)] + \sigma^2.$$

Re-expressing J_{innk} in the above terms and using the same minimization procedure as in Chapter 7, the proof is completed. Note that the optimal solution at time k requires knowledge of the mode during the next time step, $m(k+1)$. □

9.5.2 Unknown Mode Case

As in the estimation problem, the *optimal* solution to the unknown mode case involves considering the space of all possible mode sequences up to time $k+1$, of which there are H^{k+1}. Marginalizing the performance index J_{inn} across all these possible outcomes and using Lainiotis' Partition Theorem [150, 151, 262], leads to

$$J_{inn} = \sum_{j=1}^{H^{k+1}} E\{[y(k+1) - y_d(k+1)]^2 + qu^2(k) +$$
$$r\epsilon^2(k+1)|S_j(k+1), I^k\} \Pr(S_j(k+1)|I^k).$$

Note that when the mode sequence $S(k+1)$ is known,

$$\Pr(S_j(k+1)) = \begin{cases} 1 & \text{if } S_j(k+1) = S(k+1) \\ 0 & \text{otherwise} \end{cases},$$

and the marginalized performance index correctly reduces to J_{innk}. When $S(k+1)$ is unknown, the above performance index could be evaluated and minimized as in the known mode case via the estimates from Equations (9.6) conditioned on all sequences $S_j(k)$. However this would require H^{k+1} Kalman filters, which is clearly impractical and therefore demands the use of a suboptimal solution as in the estimation problem.

Following the basis of the parameter estimation scheme proposed in Section 9.3.2, the obvious approach to derive a suboptimal solution is to assume, at any time k, that the actual mode sequence is given by $\hat{S}(k)$. In mathematical terms, this means that J_{inn} will be conditioned on both the estimated mode sequence $\hat{S}(k)$ and the information set I^k. This leads to the following suboptimal version of performance index J_{inn}, denoted by J_{inns}:

$$J_{inns} = E\left\{[y(k+1) - y_d(k+1)]^2 + qu^2(k) + r\epsilon^2(k+1)|\hat{S}(k), I^k\right\}.$$

In addition, the ideal known mode solution requires that $m(k+1)$ be known as well. Since in this case $m(k+1)$ is neither known nor estimated, J_{inns} is marginalized across all possible events for $m(k+1)$, namely $\left\{\hat{M}_j(k+1)\right\}_{j=1}^{H}$, leading to the equation:

$$J_{inns} = \sum_{j=1}^{H} E\{[y(k+1) - y_d(k+1)]^2 + qu^2(k) +$$

$$r\epsilon^2(k+1)|\hat{M}_j(k+1), \hat{S}(k), I^k\} \Pr(\hat{M}_j(k+1)|\hat{S}(k), I^k). \quad (9.21)$$

Remarks 9.5.1

1. As in the estimation case, the term $\Pr(\hat{M}_j(k+1)|\hat{S}(k), I^k)$ is given by the transition probability $\pi_{\hat{m}(k),j}$.
2. The expectation terms are calculated via the predictive estimates from the parameter estimation scheme summarized in Table 9.1. Following the same reasoning as in the proof of Theorem 9.5.1, these terms are expressed as follows:

$$E\left\{[y(k+1) - y_d(k+1)]^2|\hat{M}_j(k+1), \hat{S}(k), I^k\right\} =$$

$$\left\{\hat{\mathbf{w}}_j^T(k+1|\hat{S}(k))\Phi[\mathbf{x}(k)] - y_d(k+1)\right\}^2 +$$

$$\Phi^T[\mathbf{x}(k)]\mathbf{P}_j(k+1|\hat{S}(k))\Phi[\mathbf{x}(k)] + \sigma^2.$$

$$E\left\{\epsilon^2(k+1)|\hat{M}_j(k+1), \hat{S}(k), I^k\right\} =$$

$$\Phi^T[\mathbf{x}(k)]\mathbf{P}_j(k+1|\hat{S}(k))\Phi[\mathbf{x}(k)] + \sigma^2.$$

We now describe two different ways of performing the minimization of performance index J_{inns}, one based around a simplifying assumption and the other exact.

9.5 The Control Law

(a) A MAP-based Approximation. The first control law is based on a simplification of Equation (9.21). Rather than finding $u(k)$ that minimizes the actual equation, only one term from within the summation is used; namely the one having the largest probability $\Pr(\hat{M}_j(k+1)|\hat{S}(k), I^k)$. Since this approximation is based on a maximum probability criterion, typical of MAP procedures, and the performance index is based around the IDC law, this controller shall be referred to as MAPIDC. In mathematical terms, the above means that the control law is given by

$$u_{map}(k) = \arg \min_{u(k)} E\{[y(k+1) - y_d(k+1)]^2 + qu^2(k)$$
$$+ r\epsilon^2(k+1)|\hat{M}_{j^*}(k+1), \hat{S}(k), I^k\}$$

where

$$j^* = \arg \max_{j=1,\cdots,H} \Pr(\hat{M}_j(k+1)|\hat{S}(k), I^k)$$
$$= \arg \max_{j=1,\cdots,H} \pi_{\hat{m}(k),j}. \tag{9.22}$$

This effectively means that j^* is taken to be the estimate of the index of the model that will capture the mode active at time $k+1$. Hence under this assumption, the optimal control law will be similar to the known mode case of Theorem 9.5.1 but with $m(k+1)$ replaced by j^* and $S(k)$ replaced by $\hat{S}(k)$, as shown below:

$$u_{map}(k) = \frac{(y_d(k+1) - \hat{f}_{j^*})\hat{g}_{j^*} - (1+r)\nu_{gf_{j^*}}}{\hat{g}_{j^*}^2 + q + (1+r)\nu_{gg_{j^*}}} \tag{9.23}$$

where

$$\hat{f}_{j^*} = \hat{\mathbf{w}}_{\mathbf{f}_{j^*}}^T(k+1|\hat{S}(k))\Phi_{\mathbf{f}}[\mathbf{x}(k)]$$
$$\hat{g}_{j^*} = \hat{\mathbf{w}}_{\mathbf{g}_{j^*}}^T(k+1|\hat{S}(k))\Phi_{\mathbf{g}}[\mathbf{x}(k)]$$
$$\nu_{gf_{j^*}} = \Phi_{\mathbf{g}}^T[\mathbf{x}(k)]\mathbf{P}_{\mathbf{gf}_{j^*}}(k+1|\hat{S}(k))\Phi_{\mathbf{f}}[\mathbf{x}(k)]$$
$$\nu_{gg_{j^*}} = \Phi_{\mathbf{g}}^T[\mathbf{x}(k)]\mathbf{P}_{\mathbf{gg}_{j^*}}(k+1|\hat{S}(k))\Phi_{\mathbf{g}}[\mathbf{x}(k)]$$

and the usual partitioning of the parameter vector and covariance matrix has been performed.

(b) Exact Minimization of J_{inns}. The second control law is based around minimization of J_{inns} as given in Equation (9.21), without any approximations. The resulting optimal control law is therefore an exact extension of the IDC performance index to the multiple model case and it shall be referred to as Multiple Model Innovations Dual Control (MMIDC). The MMIDC control law is precisely specified in the following proposition.

Proposition 9.5.1. *The control law minimizing performance index J_{inns} subject to the system of Equation (9.1) and all previous assumptions is given by*

$$u^*(k) = \frac{\sum_{j=1}^{H}[(y_d(k+1) - \hat{f}_j)\hat{g}_j - (1+r)\nu_{gf_j}]\pi_{\hat{m}(k),j}}{\sum_{j=1}^{H}[\hat{g}_j^2 + q + (1+r)\nu_{gg_j}]\pi_{\hat{m}(k),j}} \quad (9.24)$$

where

$$\hat{f}_j = \hat{\mathbf{w}}_{\mathbf{f}_j}^T(k+1|\hat{S}(k))\Phi_{\mathbf{f}}[\mathbf{x}(k)]$$
$$\hat{g}_j = \hat{\mathbf{w}}_{\mathbf{g}_j}^T(k+1|\hat{S}(k))\Phi_{\mathbf{g}}[\mathbf{x}(k)]$$
$$\nu_{gf_j} = \Phi_{\mathbf{g}}^T[\mathbf{x}(k)]\mathbf{P}_{\mathbf{gf}_j}(k+1|\hat{S}(k))\Phi_{\mathbf{f}}[\mathbf{x}(k)]$$
$$\nu_{gg_j} = \Phi_{\mathbf{g}}^T[\mathbf{x}(k)]\mathbf{P}_{\mathbf{gg}_j}(k+1|\hat{S}(k))\Phi_{\mathbf{g}}[\mathbf{x}(k)]$$

and $\hat{\mathbf{w}}_{\mathbf{f}_j}(k+1|\hat{S}(k))$, $\hat{\mathbf{w}}_{\mathbf{g}_j}(k+1|\hat{S}(k))$ *are the subvectors of* $\hat{\mathbf{w}}_i(k+1|\hat{S}(k))$. *The latter is obtained from Equations (9.6) with the indicator variables set according to criterion (9.11). Similarly* $\mathbf{P}_{\mathbf{gf}_j}(k+1|\hat{S}(k))$, $\mathbf{P}_{\mathbf{gg}_j}(k+1|\hat{S}(k))$ *are submatrices of the covariance matrix* $\mathbf{P}_i(k+1|\hat{S}(k))$ *from the same equations, with the partitioning performed as usual.*

Proof. Substituting the relations of Remarks 9.5.1 into Equation (9.21) yields

$$J_{inns} = \sum_{j=1}^{H}[(\hat{\mathbf{w}}_j^T(k+1|\hat{S}(k))\Phi[\mathbf{x}(k)] - y_d(k+1))^2 +$$
$$\Phi^T[\mathbf{x}(k)]\mathbf{P}_j(k+1|\hat{S}(k))\Phi[\mathbf{x}(k)] + \sigma^2 + qu^2(k) +$$
$$r(\Phi^T[\mathbf{x}(k)]\mathbf{P}_j(k+1|\hat{S}(k))\Phi[\mathbf{x}(k)] + \sigma^2)]\pi_{\hat{m}(k),j}.$$

The MMIDC control law follows after re-expressing $\hat{\mathbf{w}}_j$, $\Phi[\mathbf{x}(k)]$ and $\mathbf{P}_j$ in their usual subblocks related to terms of f and g, differentiating with respect to $u(k)$ and equating to zero. □

Remarks 9.5.2

1. Note that both the MAPIDC and the MMIDC laws rely on the predictive parameter estimates and covariance matrices $\hat{\mathbf{w}}_i$ and $\mathbf{P}_i$ from the Kalman filters.
2. To relax the assumption that H is known, thereby enabling the control laws to be used within the self-organized scheme, simply replace variable H by $h(k)$ in Equations (9.22) and (9.24).

9.6 Simulation Examples and Performance Evaluation

The simulation examples presented in this section concern a second order affine nonlinear plant having the form of Equation (9.1). The system is prone to exhibit three modes, each characterized by a pair of the following equations:

$$f_1 = \frac{-1.5y(k-1)y(k-2)}{1+y^2(k-1)+y^2(k-2)} + 0.35\sin[y(k-1)+y(k-2)], \quad g_1 = 5$$

$$f_2 = \frac{2.5y(k-1)y(k-2)}{1+y^2(k-1)+y^2(k-2)}, \quad g_2 = 1$$

$$f_3 = \frac{1.5y(k-1)y(k-2)}{1+y^2(k-1)+y^2(k-2)} + 0.35\cos[y(k-1)+y(k-2)), \quad g_3 = 3.$$

The nonlinear functions (f_i, g_i) and the number of modes $(H = 3)$ are considered unknown. The state vector $\mathbf{x}(k-1)$ is given by $[y(k-2)\ y(k-1)]^T$ and the zero mean, Gaussian additive noise e has known variance $\sigma^2 = 0.005$. The plant output $y(k)$ is required to track a reference input $y_d(k)$ obtained by sampling a low pass-filtered, unit amplitude, $0.1Hz$ square wave with a sampling period $T = 0.05$ seconds.

The $\hat{f}$ basis functions were placed on regular points of a mesh enclosing the two-dimensional region $[-2.0, 2.0] \times [-2.0, 2.0]$ along the $y(k-2), y(k-1)$ directions. The RBF centres were spaced out by a distance of 0.5 and the width parameter was set to 0.8 for all basis functions. The RBFs of the $\hat{g}$ network were placed on the one-dimensional region $[-2.0, 2.0]$ at a spacing of 1.5 and a width parameter of 3. The initial output layer weights were all set close to zero and the initial estimator covariance matrices $\mathbf{R}_i$ were set to $10\mathbf{I}$. The performance index weights were set at $r = -0.1$, $q = 0.001$, and the probability of no transition, π^{ntr}, was set to 0.999.

9.6.1 Example 1

In the first example, the three modes are activated during the time intervals shown in Table 9.2. To ensure a fair comparison, both the MAPIDC and MMIDC control laws are tested with the same noise sequence and initial parameter estimates. The results of the MAPIDC control law are shown in Figures 9.2 and 9.3.

Part (a) of Figure 9.2 shows the reference signal which the output is desired to track and part (b) shows the system output superimposed on this reference signal. Notice that despite the unknown mode transitions and mode dynamics, the output accurately tracks the reference signal, except for a short transient spike immediately following a mode transition. This spike is

Table 9.2. The mode activity for Experiment 1

Mode	Intervals of Activity /secs
1	[0, 20), (43, 57), (85, 100]
2	[20, 29), (71, 85]
3	[29, 43], [57, 71]

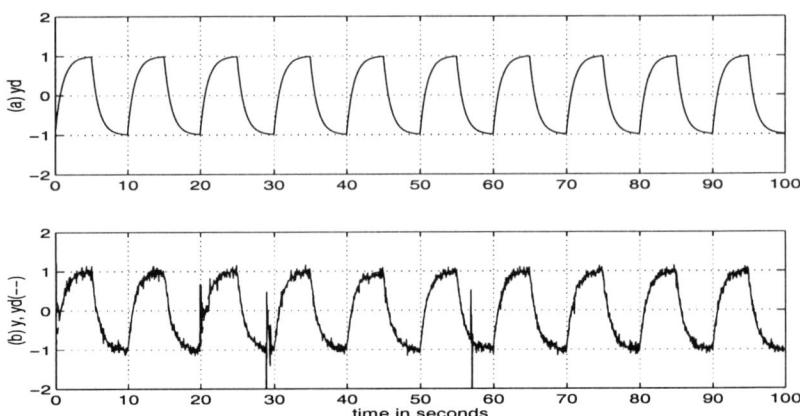

Fig. 9.2. MAPIDC control law; (a) reference input (b) plant output

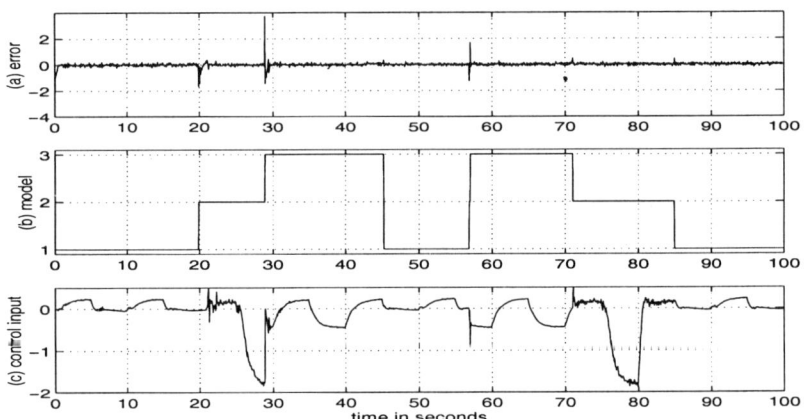

Fig. 9.3. MAPIDC control law; (a) tracking error (b) local model allocation (c) control

unavoidable for two reasons: (i) during those intervals that a particular mode becomes active for the very first time, the neural network parameters need some time to converge, as in all parameter estimation schemes (ii) when a mode makes a re-appearance, even though its model's parameters might have converged to their optimal value during the mode's previous presence, its new onset could only be detected *after* the transition actually takes place. This is due to the unavailability of any direct information on the mode activity, because of the temporal multimodal nature of the problem which implies random mode transitions.

Figure 9.3 (a) shows the tracking error $(y_d - y)$, which is a clearer representation of the deviation between the system output and reference input.

Part (b) of the same figure is a plot of $\hat{m}$ over time. This shows that the local models were allocated correctly: three local models were activated by the self-organizing subsystem, even though the number of modes $H = 3$ was not known *a priori*, and model switching is effected immediately after an actual mode change takes place. Part (c) of Figure 9.3 shows the control input. As expected, the quality of the control u changes according to the mode activity detected during the course of time so as to maintain good tracking in the face of mode transitions.

Figures 9.4 and 9.5 show the corresponding results of the MMIDC control law. These indicate that the performance of the system with the MMIDC law is practically identical to the MAPIDC case, both as far as tracking error and local model allocation are concerned. To confirm this, a Monte Carlo analysis of 100 trials with both control laws was performed. A fresh noise realization was used at every trial and the accumulated cost $\sum_{k=0}^{T}(y_d(k) - y(k))^2$ was calculated at the end of each trial. The average of the accumulated cost over 100 trials is practically the same, namely 80.7 and 79.0 for the MMIDC and MAPIDC laws respectively. This is expected in the present example because the probability of transition π^{tr} is very small (< 0.001).

It is interesting to analyse the behaviour of the model probability densities, particularly after the occurrence of a mode switch. Figures 9.6 to 9.8 show the distributions $p(y(k)|\hat{M}_i(k), \hat{S}(k-1), \tilde{I}^{k-1})$ for the local models, evaluated at 25, 29 and 40 seconds respectively. The first of these timings corresponds to an instant well into the first cycle of mode 2, the second one corresponds to the start of the first cycle of mode 3 and the third is an instant well into the first cycle of mode 3. Figure 9.6 shows that at the 25s instant the system has set up 3 local models, because up to this time two

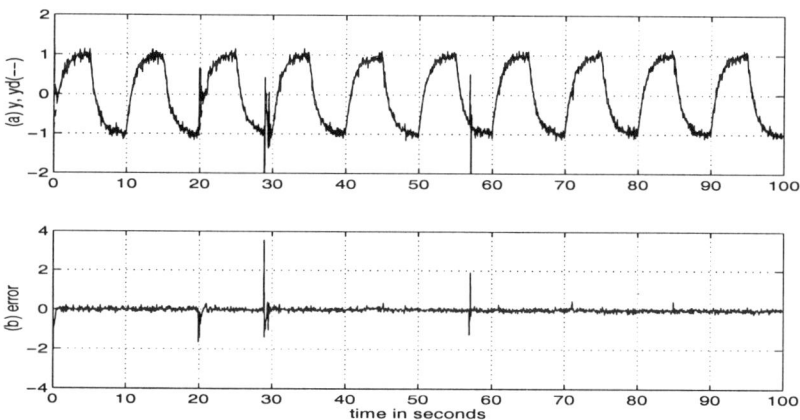

Fig. 9.4. MMIDC control law; (a) output and reference input (b) tracking error

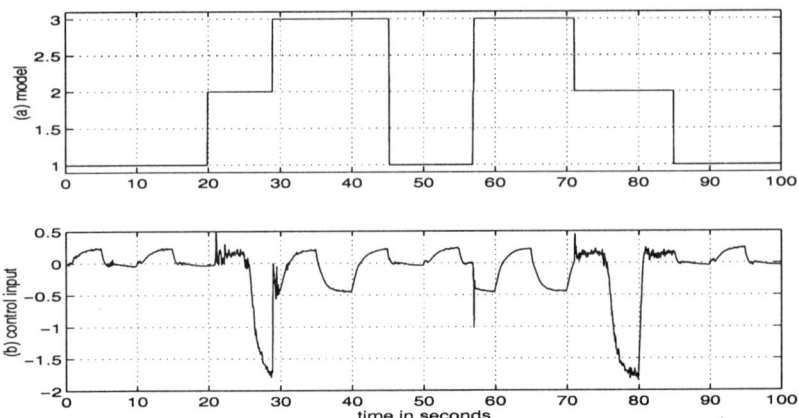

Fig. 9.5. MMIDC control law; (a) local model allocation (b) control

distinct modes were detected active and captured by models 1 and 2. Model 3 is a "spare", unallocated model, ready to learn a new mode when it appears. This explains why the spread of the density of model 3 is very large when compared to the other two. The actual value of y at this time instant is indicated by an asterisk (*) on the x-axis of the figure. It shows that the probability density for model 2 dominates at this value of y, because model 2 has captured the dynamics of the currently active mode and is effectively "tuned" to it.

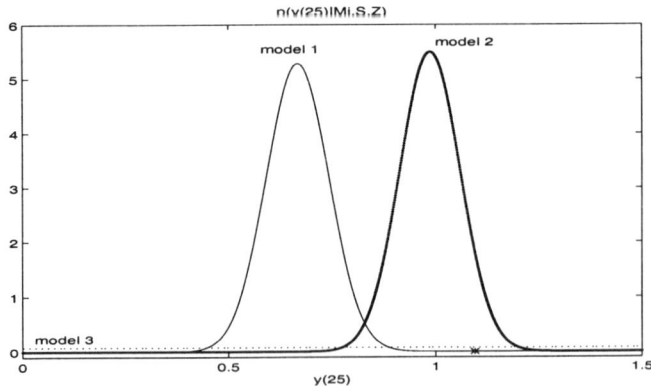

Fig. 9.6. Probability density function at $25s$ instant

Figure 9.7 coincides with the onset of a transition to a mode that had not existed before. As explained in Section 9.4, it clearly shows that at the current value of y (indicated by *), the probability density of the unallocated

9.6 Simulation Examples and Performance Evaluation

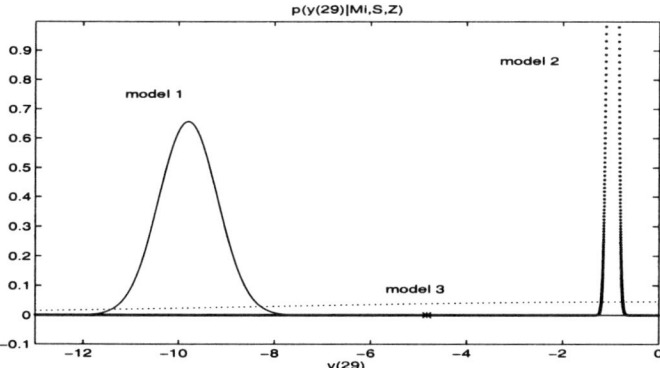

Fig. 9.7. Probability density function at $29s$ instant

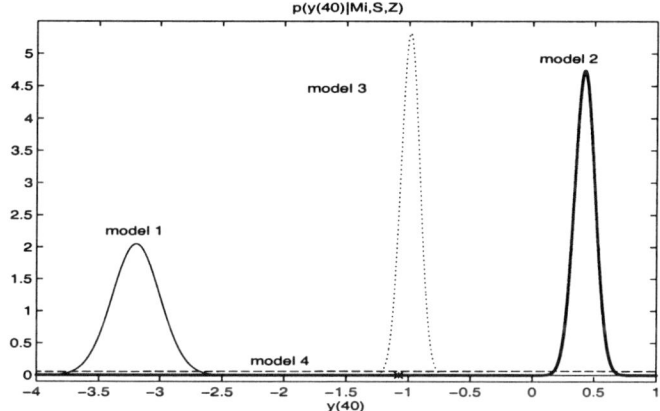

Fig. 9.8. Probability density function at $40s$ instant

model 3 now dominates. The reason is that although none of the models has as yet captured the dynamics of this new mode, models 1 and 2 propagate a negligibly low density because their variance is small and their density is tuned on to a different mode. This is reflected by the appreciably large distance separating their mean from the asterisk. By contrast, model 3 propagates a relatively much larger density because its spread is wide enough to enable it to capture any dynamics that fall outside the "range" of the other models. This favours model 3 to be selected for learning the new mode, as confirmed in Figure 9.8 which shows that $11s$ after the appearance of mode 3, the variance of the density conditioned on model 3 has already reduced appreciably. Additionally, its density dominates over the rest at the current value of y (indicated by *), because model 3 has now become tuned to this mode. Also, since model 3 is now committed, a new unallocated model (model 4) is initiated to accept any new future modes.

9.6.2 Example 2

In this example we compare the performance of the proposed multiple model system with a number of different indirect adaptive control schemes based on a single identification model, particularly those methods whose estimators are able to track time-varying parameters [164]. The mode sequence used in this comparative analysis is shown in Figure 9.9(a) and the reference input is the same as that of Experiment 1. Figure 9.9(b) shows the performance of the MMIDC control law. The plot shows the system output superimposed on the reference input. Clearly, despite the mode switches, the output is able to track adequately the reference input.

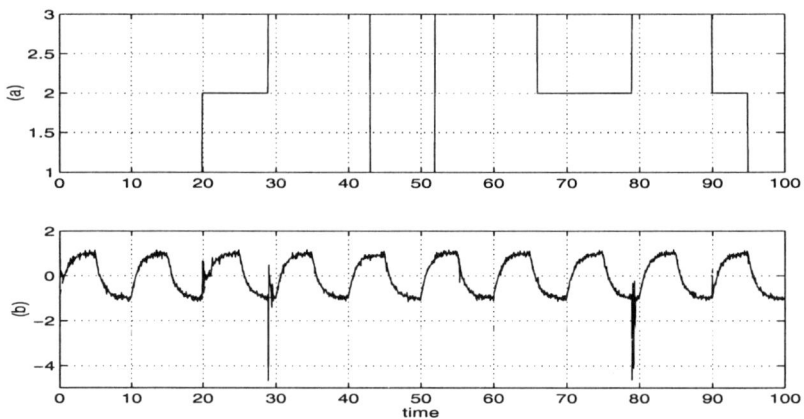

Fig. 9.9. (a) Mode sequence (b) MMIDC performance (y and y_d)

Figure 9.10 shows the output of the various single model adaptive schemes which, for fair comparison, were subjected to the same noise sequence and to the single model IDC control law developed in Chapter 7. In Figure 9.10(a), a standard Kalman filter was used as the estimator. As expected, the inability of this estimator to track time-varying parameters leads to a very bad control performance. In Figure 9.10(b), the estimator is modified to an *exponential-forgetting* type. It shows that although this method is normally suitable for tracking slow-varying parameters, it is inadequate for parameters that change abruptly. In Figure 9.10(c), a definitive improvement is obtained by assuming a *random walk model* for parameter variation. This assumes that the first of the set of Equations (9.3) includes an additional white noise term of constant covariance. Figure 9.10(d) shows a more or less similar improvement with a *covariance resetting* estimator, where the estimator covariance matrix is periodically reset to $\mathbf{R}_i$.

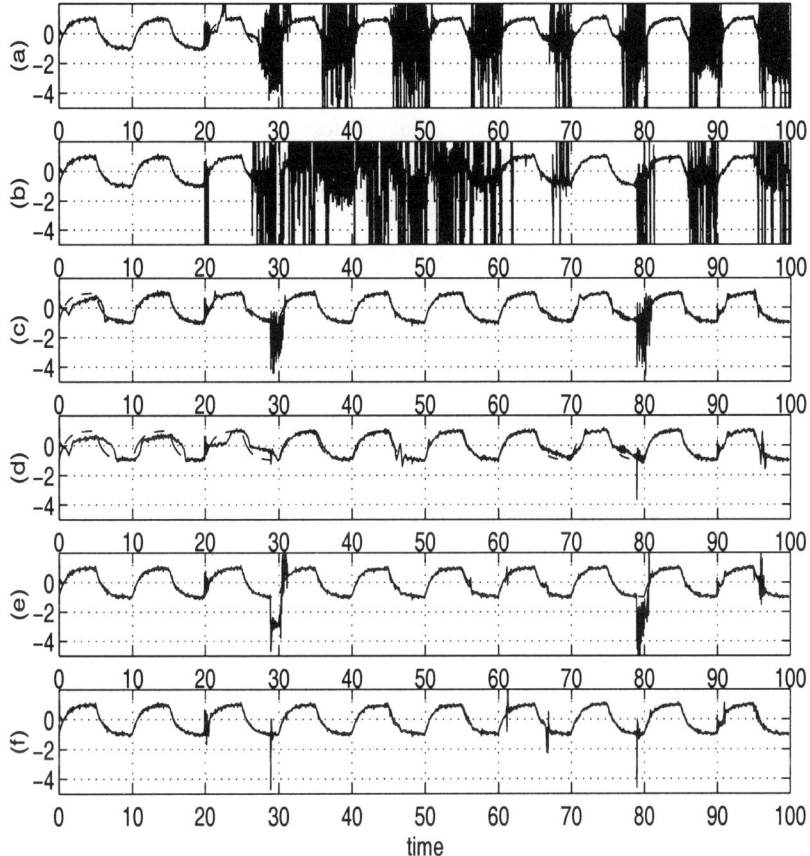

Fig. 9.10. Various single model adaptive controllers (a) Kalman filter estimator (b) Exponential forgetting (c) Random walk (d) Covariance resetting (e) Modified random walk (f) Modified covariance resetting

Finally Figures 9.10(e) and 9.10(f) respectively show the results of a modified version of the random walk and covariance resetting estimators based upon the detection of mode transitions. Such single model tracking estimator schemes have been previously proposed [35, 164], but here we utilize a novel mode transition sensing mechanism by calculating Equation (9.17) with $h(k)$ constantly set to 2. Since the parameters of only one model are actually being estimated,[1] the equation is able to sense the onset of a mode transition via a reduction of the model's posterior probability towards zero. Consequently a random walk model or a covariance matrix reset is effected *only* at the time instant that a mode transition is sensed. This method could be considered as a single model version of the multiple model scheme, with the difference that the previous mode dynamics are not memorized as separate models. Although the last two cases show relatively good performance, a comparison with Figure 9.9(b) shows that the multiple model scheme is superior, particularly because the local models preclude the need to re-estimate modes that had already been learned before. This capability, which amounts to *learning* rather than just *adaptation* [13], leads to a much faster reconfiguration and hence a better transient response when a previously-active mode makes a reappearance. This is clearly evident during the 66, 78, 90 and 95 second time instants in the simulation example.

9.7 Summary

A multiple model estimation and control scheme for a class of temporal multimodal, affine-nonlinear stochastic systems whose mode activity and mode dynamics are unknown, has been presented. The tasks of mode detection, mode estimation and control signal generation utilize the information provided by a set of Kalman filters. The multiple models consist of GaRBF neural networks that capture the nonlinear functions of the various modes. The proposed scheme includes a number of novel features when compared with the usual MMAC techniques. For example, it does not assume that the mode dynamics are identified prior to applying the control because they are estimated simultaneously with the mode activity, and it makes use of novel dual adaptive control laws. Additionally, when the number of modes that the plant may assume is unknown, a self-organizing scheme that autonomously creates and allocates models in real time is presented.

Simulation experiments have shown that the performance of the system is superior to that of single model schemes which utilize some form of tracking estimator, including those based on mode switching detection. On a mode transition, the multiple model technique allows for faster reconfiguration because the estimated mode dynamics are memorized. Hence the system possesses the important feature of learning, in addition to adaptation and it can handle temporal multimodality under conditions of mode dynamic uncertainty, which is also taken into consideration when generating the control via the MAPIDC or MMIDC dual control laws. All these are considered to be essential features of an intelligent control scheme that should be able to handle nonlinearity, uncertainty and multimodal complexity.

---

[1] the second model is a dummy model whose parameters are never adjusted and is introduced simply to enable discrimination between different modes

10. Multiple Model Dual Adaptive Control of Spatial Multimodal Systems

10.1 Introduction

The concept of multimodality suggests an interesting way of handling systems whose dynamics are characterized by nonlinear functions. The dynamics of such systems could be interpreted as a scheduled multimodal problem, where each mode captures the dynamics within a restricted (local) range of operating conditions and the scheduling is determined by the operating conditions themselves. This method is particularly appealing for spatially complex systems because they typically exhibit very different characteristics along different zones of the operating space. A multimodal interpretation is attractive because the local modes could be individually modelled by a less complex structure than would have been the case if one higher order model was chosen to capture the global nonlinear dynamics, such as a conventional neural network. This method of treating nonlinear systems naturally lends itself to multiple model based techniques, both for control and system identification. In control, it represents the fundamental principle behind the Gain Scheduling control technique [23, 217, 226, 227]. This scenario is distinct from the jump system case considered in the previous chapter because now the mode transitions are scheduled by some *measurable* operating conditions, instead of being arbitrary.

More specifically, in this chapter we consider control of a particular class of uncertain, spatial multimodal systems. The uncertainty refers to the fact that the local mode dynamics and scheduling are unknown. Hence the problem requires a different solution from the one provided by standard Gain Scheduling, because the latter relies on the availability of knowledge about both the mode dynamics and the scheduling. The uncertainty is handled by utilizing concepts from indirect adaptive control, in the sense that the plant is identified during control operation. However, whilst a conventional indirect adaptive scheme attempts to capture the global nonlinear plant dynamics by a *single* higher-order model, here we utilize a multiple model structure based on modular networks. This entails on-line estimation of the parameters of both the local candidate model set and the validity functions. The technique could therefore be interpreted as an adaptive form of Gain Scheduling. One obvious advantage of handling the problem in this manner is that the global nonlinear dynamics need not have a known functional form, *e.g.*, dynamics that are affine in the control. Only the functional form of the local mode dynamics needs to be assumed, and there is a certain degree of flexibility on the choice of this form. Indeed, most nonlinear functions could be approximated by a set of *linear* local dynamics, each valid over a particular region of the input/state space. The advantage of this follows because in general, knowledge of the functional form leads to simpler and more direct control design procedures, especially in the adaptive case.

Sequential identification of nonlinear systems by modular architectures has been documented before [125, 133, 131, 230], as well as the problem of controlling spatially complex systems by modular-type networks [45, 106, 118]. The local control schemes of [45] and [106] assume that the local model parameters and validity functions are known prior to controlling the system, possibly from a separate off-line identification phase, and so they do not represent an adaptive control methodology. The scheme of [118], which utilizes the Mixture of Experts architecture, is based on learning *feedforward* control laws. The feedforward controller consists of a modular network trained to learn the plant's inverse dynamics from the data generated by the system while operating with a fixed, feedback (PID) controller that runs in parallel with it. Hence the approach cannot really be classified as an adaptive control scheme, not only because of the feedforward nature of the control laws, but more importantly because of its reliance on a second, *known*, feedback controller.

The main novel results of this chapter, which represent an extensive development of the preliminary work presented by the authors in [73], hinge on the derivation of an indirect dual adaptive control scheme for a class of nonlinear systems whose identification model is of the modular type. A stochastic adaptive approach is taken in all aspects of the design: mode detection, local model and validity function estimation and control signal generation. A novel gating network for validity function estimation is also proposed and particular attention is given to the problem of generating an appropriate control signal when the control itself is one of the scheduling variables. The main contributions of this chapter could therefore be summarized as follows:

1. The design of a modular network, multiple model scheme to handle adaptive control of a nonlinear class of unknown stochastic plants, leading to an adaptive gain scheduling-like procedure.
2. The development of a mode detection, scheduling and local model estimation scheme within a probabilistic framework.
3. The development of a novel gating network that gives superior estimation to softmax gating and exhibits a better control performance when used in the proposed scheme.
4. The extension of the IDC suboptimal dual control law to the modular network case, with particular emphasis on the situation when the control itself is one of the scheduling variables.

10.2 Problem Formulation

The scheme addresses control of a class of discrete-time, nonlinear stochastic systems whose dynamics are characterized by a set of H linear equations, each of which is active only over a specific local region of the combined input and state space. These dynamics represent a piecewise-linear subclass of general nonlinear systems. In mathematical terms, the model is expressed by the following equation

$$y(k) = \mathbf{a}_{m(k)}^T \mathbf{x}(k-1) + b_{m(k)} u(k-1) + d_{m(k)} + e(k), \tag{10.1}$$

where $y(k)$ is the output, $u(k)$ is the control input, $\mathbf{x}(k) := [y(k) \cdots y(k-n+1)\, u(k-1) \cdots u(k-p)]^T$ is an $(n+p)$ dimensional state vector and $e(k)$ is additive noise.

The parameters $\mathbf{a}_i := [a_{1_i}\, a_{2_i} \cdots a_{(n+p)_i}]^T \in \Re^{n+p}$ and $b_i, d_i \in \Re$ take values from the following set of H constant elements, as indexed by the subscript $i \in \{1, \cdots, H\}$:

$$\{(\mathbf{a}_1, b_1, d_1), \cdots, (\mathbf{a}_H, b_H, d_H)\}.$$

Being a scheduled jump system, the parameter index in Equation (10.1) that is specified by the mode variable $m(k)$, depends on the operating conditions determined by the input $u(k-1)$ and the state $\mathbf{x}(k-1)$, according to the mapping:

$$m(k) = \Psi[u(k-1), \mathbf{x}(k-1)], \tag{10.2}$$

where $\Psi : \Re^{n+p} \times \Re \mapsto \{1, \cdots, H\}$. Hence the system is of the spatial multimodal type because the dynamics are scheduled through the state and the control input via mapping Ψ, that segments the space into a number of non-overlapping partitions, each of which corresponds to a particular mode.

In this chapter we consider the case where both the mapping Ψ and the parameters $(\mathbf{a}_i, b_i, d_i)$ for the different modes are unknown. The plant output is required to track a reference input y_d and the following conditions are assumed to hold:

Assumption 10.2.1 *Assumptions 9.2.1 - 9.2.3 of the previous chapter hold here as well.*

Assumption 10.2.2 *The number of modes H is known a priori.*

Assumption 10.2.3 *The dynamics of each mode are minimum phase (i.e., the roots of $b_i + a_{(n+1)_i} z^{-1} + \cdots + a_{(n+p)_i} z^{-p}$ lie within the unity circle) and $b_i \neq 0$ for all $i = 1, \cdots, H$.*

Assumption 10.2.4 *The deterministic part of Equation (10.1), represented by the function*

$$f[u(k-1), \mathbf{x}(k-1)] := \mathbf{a}_{m(k)}^T \mathbf{x}(k-1) + b_{m(k)} u(k-1) + d_{m(k)} \tag{10.3}$$

is globally invertible with respect to $u(k-1)$ and continuous.

The continuity in the last assumption implies that there always exist some $u(k-1), \mathbf{x}(k-1)$ that map to any desired value of f. The invertibility condition implies that any two distinct values of $u(k-1)$ map to two distinct values of f. This latter condition is satisfied if f is monotonic with respect to $u(k-1)$ [107].

To deal with the uncertainty in the system, an indirect adaptive Control solution is adopted. This requires the use of an identification model for the plant. The class of systems under consideration is naturally structured according to a finite set of linear dynamics, each characterized by parameters $(\mathbf{a}_i, b_i, d_i)$ that are active only over a localized region of the joint space of u and $\mathbf{x}$. A modular network is therefore ideally suited for setting up the identification model, because its features are directly compatible with the localized characteristics of the plant dynamics.

10.3 The Modular Network

To make the similarity between a modular network representation and the class of systems under consideration more transparent, Equation (10.1) is re-expressed in terms of indicator variables γ_i as follows,

$$y(k) = \sum_{i=1}^{H} \gamma_i[u(k-1), \mathbf{x}(k-1)] \mathbf{w}_i^{*T} \Phi(k-1) + e(k) \qquad (10.4)$$

where $\Phi(k) = [u(k)\ \mathbf{x}^T(k)\ 1]^T$, $\mathbf{w}_i^* = [b_i\ \mathbf{a}_i^T\ d_i]^T$ and

$$\gamma_i[u(k-1), \mathbf{x}(k-1)] = \begin{cases} 1 & \text{if } m(k) = i \\ 0 & \text{otherwise} \end{cases}. \qquad (10.5)$$

This equation represents an interconnection of H local models, each parameterized by vector $\mathbf{w}_i^*$, whose output is weighted by the indicator variables that specify which mode is active at any one time. Due to the plant uncertainty, the local model parameters $\{\mathbf{w}_i^*\}_{i=1}^{H}$ and the indicator variables $\{\gamma_i\}_{i=1}^{H}$ are unknown. Hence a plant identification model is set up. This consists of a modular network having:

1. H local candidate models parameterized by vectors $\{\hat{\mathbf{w}}_i\}_{i=1}^{H}$. The parameter vector of the ith local model, $\hat{\mathbf{w}}_i$ is an estimate of $\mathbf{w}_i^*$. In particular,

   $$\hat{\mathbf{w}}_i := [\hat{b}_i\ \hat{\mathbf{a}}_i^T\ \hat{d}_i]^T$$

   where $\hat{b}_i, \hat{\mathbf{a}}_i, \hat{d}_i$ are estimates of $b_i, \mathbf{a}_i, d_i$ respectively.
2. H validity functions $\hat{\gamma}_i[u(k-1), \mathbf{x}(k-1)]$; $i = 1, \cdots, H$, each of which is an estimate of indicator $\gamma_i[u(k=1), \mathbf{x}(k-1)]$, and which satisfy the conditions:

   $$0 \leq \hat{\gamma}_i \leq 1, \quad \sum_{i=1}^{H} \hat{\gamma}_i = 1. \qquad (10.6)$$

The output of the modular network identification model is hence given by

$$\hat{y}(k) = \sum_{i=1}^{H} \hat{\gamma}_i[u(k-1), \mathbf{x}(k-1)]\hat{y}_i(k) \tag{10.7}$$

where

$$\hat{y}_i(k) = \hat{\mathbf{w}}_i^T(k)\Phi(k-1).$$

Note that $\hat{y}_i$ represents the output of the ith local model, and $\hat{y}(k)$ consists of a weighted sum of the individual outputs from the H local models. The weights are provided by the validity functions.

Being part of an adaptive control scheme, the local model parameter estimates $\hat{\mathbf{w}}_i$ and the validity function estimates $\hat{\gamma}_i$ are recursively calculated in an on-line manner during control operation.

10.4 The Estimation Problem

10.4.1 Local Model Parameter Estimation

We will first treat the problem of local model parameter estimation under the assumption that the mode $m(k)$ is known at any time k. This will help to formulate the estimation algorithm for the unknown mode case.

Known Mode Case. During activity of some particular mode i, the mode variable $m(k)$ is equal to i and Equation (10.4) could be written in state-space form as:

$$\begin{aligned} \mathbf{w}_i^*(k+1) &= \mathbf{w}_i^*(k) \\ y(k) &= \mathbf{w}_i^{*T}(k)\Phi(k-1) + e(k). \end{aligned} \tag{10.8}$$

As in the previous chapter, since this equation is linear in the local model parameters and $e(k)$ is Gaussian, a Kalman filter could be used to generate recursively the optimal minimum mean square estimate of $\mathbf{w}_i^*$ and its covariance *whenever* mode i is active, subject to the following assumption,

Assumption 10.4.1 *The local model parameter vectors* $\{\mathbf{w}_i^*\}_{i=1}^{H}$ *are random variables with Gaussian distributed initial value* $\mathbf{w}_i^*(0)$ *of known mean* $\mathbf{m}_i$ *and covariance* $\mathbf{R}_i$.

As usual, the optimal predictive estimate at time k is given by the mean of $\mathbf{w}_i^*(k+1)$ conditioned on the information state I^k, according to the estimation procedure defined in the following lemma.

Lemma 10.4.1.

1. *Subject to the Gaussianity of the noise, Assumption 10.4.1 and assuming that the mode sequence* $S(k) := \{m(1), \cdots, m(k)\}$ *is known; the distribution of* $\mathbf{w}_i^*(k+1)$ *conditioned on information state* I^k *and sequence* $S(k)$ *is*

Gaussian with mean and covariance respectively denoted by $\hat{\mathbf{w}}_i(k+1|S(k))$ and $\mathbf{P}_i(k+1)|S(k))$ that satisfy the following Kalman filter recursive equations:

$$\mathbf{K}_i(k) = \frac{\gamma_i[u(k-1), \mathbf{x}(k-1)]\mathbf{P}_i(k|S(k-1))\Phi(k-1)}{\sigma^2 + \Phi^T(k-1)\mathbf{P}_i(k|S(k-1))\Phi(k-1)}$$
$$\hat{\mathbf{w}}_i(k+1|S(k)) = \hat{\mathbf{w}}_i(k|S(k-1)) + \mathbf{K}_i(k)(y(k) - \hat{y}_i(k)) \quad (10.9)$$
$$\mathbf{P}_i(k+1|S(k)) = \left\{\mathbf{I} - \mathbf{K}_i(k)\Phi^T(k-1)\right\}\mathbf{P}_i(k|S(k-1))$$

subject to initial conditions $\hat{\mathbf{w}}_i(0) = \mathbf{m}_i$, $\mathbf{P}_i(0) = \mathbf{R}_i$.

2. The probability density $p(y(k)|S(k), I^{k-1})$, is also Gaussian with mean and variance given by $\hat{\mathbf{w}}_{m(k)}^T(k|S(k-1))\Phi(k-1)$ and $\sigma^2 + \Phi^T(k-1)\mathbf{P}_{m(k)}(k|S(k-1))\Phi[\mathbf{x}(k-1)]$ respectively.

Proof. The proof is similar to that of Lemma 9.3.1 when applied on Equation (10.8). □

Unknown Mode Case. By similar reasoning to that of Chapter 9, when sequence $S(k)$ is unknown, the optimal solution to the parameter estimation problem involves considering the space of *all* possible mode sequences up to time k. However this solution is impractical because the number of Kalman filters increases exponentially and in addition, it ignores the fact that measurable variables $u(k-1), \mathbf{x}(k-1)$ convey information on $m(k)$ via the mapping Ψ of Equation (10.2), which is being learned indirectly through the gating function estimates $\hat{\gamma}_i$. Therefore we propose a MAP-based criterion to find an estimate $\hat{S}(k)$ of the true mode sequence $S(k)$. The estimated sequence $\hat{S}(k)$ is given by:

$$\hat{S}(k) = \{\hat{m}(1), \cdots, \hat{m}(k)\},$$

where $\hat{m}(k)$ denotes the index of the local model allocated to the dynamics of mode $m(k)$[1]. To evaluate $\hat{m}(k)$, denote the event $\{\hat{m}(k) = i\}$ by $M_i(k)$, i.e., the event that local model i represents the dynamics of mode $m(k)$. A similar MAP criterion as in the previous chapter is used to estimate $\hat{m}(k)$, namely

$$\hat{m}(k) = \arg\max_{i=1,\cdots,H} \Pr(\hat{M}_i(k)|\hat{S}(k-1), I^k). \quad (10.10)$$

In a similar way to Lemma 9.3.2, the posterior probability $\Pr(\hat{M}_i(k)|\hat{S}(k-1), I^k)$ is given by the equation

$$\Pr(\hat{M}_i(k)|\hat{S}(k-1), I^k) =$$
$$\frac{p(y(k)|\hat{M}_i(k), \hat{S}(k-1), \tilde{I}^{k-1})\ \Pr(\hat{M}_i(k)|\hat{S}(k-1), \tilde{I}^{k-1})}{\sum_{j=1}^{H} p(y(k)|\hat{M}_j(k), \hat{S}(k-1), \tilde{I}^{k-1})\ \Pr(\hat{M}_j(k)|\hat{S}(k-1), \tilde{I}^{k-1})} \quad (10.11)$$

---

[1] Note that we distinguish between a model $\hat{m}$ and a mode m because in general the local model index $\hat{m}$ need not be equal to the mode index m

where $p(y(k)|\hat{M}_i(k), \hat{S}(k-1), \tilde{I}^{k-1})$ denotes the probability density of output $y(k)$ conditioned on the mode sequence $\{\hat{S}(k-1), \hat{M}_i(k)\}$ and $\tilde{I}^{k-1}$, which was defined in Lemma 9.3.2.

Remarks 10.4.1

1. For similar reasons as in the known mode case (see Lemma 10.4.1), the density $p(y(k)|\hat{M}_i(k), \hat{S}(k-1), \tilde{I}^{k-1})$ is Gaussian with mean m_{y_i} and variance r_{y_i} given as

$$m_{y_i} = \hat{\mathbf{w}}_i^T(k|\hat{S}(k-1))\Phi(k-1) \tag{10.12}$$

$$r_{y_i} = \Phi^T(k-1)\mathbf{P}_i(k|\hat{S}(k-1))\Phi(k-1) + \sigma^2. \tag{10.13}$$

The terms $\hat{\mathbf{w}}_i(k|\hat{S}(k-1))$, $\mathbf{P}_i(k|\hat{S}(k-1))$ come from the H Kalman filter Equations (10.9), but subjected to sequence $\hat{S}$ instead of S. This requires replacing $\gamma_i[u(k-1), \mathbf{x}(k-1)]$ in (10.9) by $\gamma_i^{map}(k)$, defined as

$$\gamma_i^{map}(k) := \begin{cases} 1 & \text{if } \hat{m}(k) = i \\ 0 & \text{otherwise} \end{cases}. \tag{10.14}$$

2. The prior terms $\Pr(\hat{M}_i(k)|\hat{S}(k-1), \tilde{I}^{k-1})$ appearing in Equation (10.11) are unknown. However they are estimated by the validity functions $\hat{\gamma}_i[u(k-1), \mathbf{x}(k-1)]$. This interpretation of $\hat{\gamma}_i$ as a prior probability is sensible because (a) it does not depend on measurement $y(k)$, hence making it a *prior* estimate and (b) the restrictions on $\hat{\gamma}_i$ imposed by Equations (10.6) give it the characteristics of a probability measure. Hence Equation (10.11) becomes

$$\Pr(\hat{M}_i(k)|\hat{S}(k-1), I^k) = $$
$$\frac{p(y(k)|\hat{M}_i(k), \hat{S}(k-1), \tilde{I}^{k-1})\, \hat{\gamma}_i[u(k-1), \mathbf{x}(k-1)]}{\sum_{j=1}^{H} p(y(k)|\hat{M}_j(k), \hat{S}(k-1), \tilde{I}^{k-1})\, \hat{\gamma}_j[u(k-1), \mathbf{x}(k-1)]}.\tag{10.15}$$

3. The above point reflects the distinctive feature between the temporal multimodal case considered in Chapter 9 and the spatial multimodal case considered here. The former is characterized by *arbitrary* mode transitions, whilst the latter is subject to *scheduled* mode transitions. Hence, in the present scheme, it does not make sense to model the transitions as a Markov chain and estimate the priors by the transition probabilities of the chain. In this case, the transitions depend on the mapping Ψ which is a function of u and $\mathbf{x}$. Hence it is more reasonable to estimate the priors by the validity functions $\hat{\gamma}_i$ because they reflect the modes as a function of u and $\mathbf{x}$. This technique has been utilized in probabilistic approaches to modular network modelling such as [125] and [133].

In the absence of knowledge on the indicator variables γ_i, the local model parameter estimation procedure of Equations (10.9), which was valid for the known mode case, requires modification. In this case, the predictive parameter estimates at time k are calculated on the basis of the estimated mode sequence $\hat{S}(k)$ and are given by the same optimal algorithm of Equations (10.9), but with $\gamma_i[u(k-1), \mathbf{x}(k-1)]$ replaced by $\gamma_i^{map}(k)$ from Equation (10.14). Hence the prediction of $\mathbf{w}_i^*$ and its covariance, denoted by $\hat{\mathbf{w}}_i(k+1|\hat{S}(k))$ and $\mathbf{P}_i(k+1|\hat{S}(k))$ respectively, are given by the Kalman filter equations:

$$\mathbf{K}_i(k) = \frac{\gamma_i^{map}(k)\mathbf{P}_i(k|\hat{S}(k-1))\Phi(k-1)}{\sigma^2 + \Phi^T(k-1)\mathbf{P}_i(k|\hat{S}(k-1))\Phi(k-1)}$$
$$\hat{\mathbf{w}}_i(k+1|\hat{S}(k)) = \hat{\mathbf{w}}_i(k|\hat{S}(k-1)) + \mathbf{K}_i(k)(y(k) - \hat{y}_i(k)) \quad (10.16)$$
$$\mathbf{P}_i(k+1|\hat{S}(k)) = \{\mathbf{I} - \mathbf{K}_i(k)\Phi^T(k-1)\}\mathbf{P}_i(k|\hat{S}(k-1))$$

with initial conditions $\hat{\mathbf{w}}_i(0) = \mathbf{m}_i$, $\mathbf{P}_i(0) = \mathbf{R}_i$. Note that these are the same estimates used during the next time step to calculate the conditional probability densities $p(y(k+1)|\hat{M}_i(k+1), \hat{S}(k), \tilde{I}^k)$, as described in point (1) of Remarks 10.4.1.

Hence at any time instant, only H Kalman filters are required to calculate both the local model parameter estimates and the conditional probability densities of y. During the initial phase, not all models would have been subjected to parameter adjustment. Hence, for the same reasons explained in Chapter 8, the term $\max_{i=1,\ldots,H} \Pr(\hat{M}_i(k)|\hat{S}(k-1), I^k)$ required in Equation (10.10) might not be unique, especially if the Kalman filters of all local models are initialized with the same priors $\mathbf{m}_i$ and $\mathbf{R}_i$. In this case we adopt the solution of setting $\hat{m}(k)$ equal to any one of these non-unique maxima, thereby forcing the selected model to start learning the current mode.

10.4.2 Validity Function Estimation

The H validity functions $\hat{\gamma}_i$ are required for estimation of the prior terms $\Pr(\hat{M}_i(k)|\hat{S}(k-1), \tilde{I}^{k-1})$, $i = 1, \cdots, H$ of Equation (10.11). The validity functions map from the space of $u(k-1), \mathbf{x}(k-1)$ to the interval $[0, 1]$ subject to conditions (10.6). This represents a nonlinear mapping. Since this mapping is unknown, the validity functions will be learned on-line as part of the adaptation process by utilizing some kind of parameterized "neural" network that shall be called the *gating network*.

The Gating Network. Two types of gating network are considered in this chapter: Softmax gating and Gaussian Mixture Kernel (GMK) gating.

Softmax gating network. This gating network is the one used in the Mixture of Experts modular network architecture of Jacobs and Jordan [125]. Its parameterization is based around the *softmax function*. The ith validity function is given by

$$\hat{\gamma}_i[u(k-1), \mathbf{x}(k-1)] = \frac{\exp(\mathbf{h}_i^T \Phi(k-1))}{\sum_{j=1}^{H} \exp(\mathbf{h}_j^T \Phi(k-1))}, \quad (10.17)$$

where $\{\mathbf{h}_i\}_{i=1}^{H}$ are the H parameter vectors of the gating network that are recursively adjusted during the adaptation process. The softmax network satisfies the constraints represented by Equations (10.6). Note that the independent variable is not simply $u(k-1), \mathbf{x}(k-1)$, but $\Phi(k-1)$ because it includes the additional bias term of 1. Without this bias, irrespective of the gating parameters $\mathbf{h}_i$, all the validity functions would be equal to $1/H$ at the origin of the space of $(u(k-1), \mathbf{x}(k-1))$, which is generally not the case.

Gaussian mixture kernel gating network. In this section we propose a novel gating network structure for modular networks based upon mixture model density estimation techniques [39, 246]. The motivation for this approach in our case follows because the validity functions $\hat{\gamma}_i$ are approximating the prior probability terms $\Pr(\hat{M}_i(k)|\hat{S}(k-1), \bar{I}^{k-1})$. From Bayes' rule, this term could be expressed as

$$\Pr(\hat{M}_i(k)|\hat{S}(k-1), \bar{I}^{k-1}) = \Pr(\hat{M}_i(k)|\hat{S}(k-1), \tilde{\Phi}(k-1), \bar{I}^{k-2}) = \frac{p(\tilde{\Phi}(k-1)|\hat{M}_i, \hat{S}, \bar{I}^{k-2})\Pr(\hat{M}_i|\hat{S}, \bar{I}^{k-2})}{\sum_{j=1}^{H} p(\tilde{\Phi}(k-1)|\hat{M}_j, \hat{S}, \bar{I}^{k-2})\Pr(\hat{M}_j|\hat{S}, \bar{I}^{k-2})} \quad (10.18)$$

where $\bar{I}^{k-1}$ has been decomposed as $\bar{I}^{k-1} := \left\{\tilde{\Phi}(k-1), \bar{I}^{k-2}\right\}$, with $\tilde{\Phi}(k-1) := [u(k-1) \ \mathbf{x}^T(k-1)]^T$ representing the scheduling variables, and the time index of terms $\hat{M}_i$, $\hat{M}_j$ and $\hat{S}$ is k, k and $k-1$ respectively.

The probability density term $p(\tilde{\Phi}(k-1)|\hat{M}_i(k), \hat{S}(k-1), \bar{I}^{k-2})$ is not Gaussian in general. Hence it will be approximated by a Gaussian mixture model (also known as Gaussian Sum Representation [7]) as follows

$$p(\tilde{\Phi}(k-1)|\hat{M}_i(k), \hat{S}(k-1), \bar{I}^{k-2}) = \sum_{j=1}^{\zeta} g_j[\tilde{\Phi}(k-1), \mu_j, \sigma_j]\theta_{i_j} \quad (10.19)$$

where

$$g_j = \frac{\exp\left\{\left(-(\tilde{\Phi}(k-1) - \mu_j)^T(\tilde{\Phi}(k-1) - \mu_j)\right)/2\sigma_j^2\right\}}{(2\pi\sigma_j^2)^{0.5(n+p+1)}} \quad (10.20)$$

$$\theta_{i_j} = \frac{\exp(h_{i_j})}{\sum_{l=1}^{\zeta} \exp(h_{i_l})} \quad (10.21)$$

- The g_j terms of Equation (10.20) represent the ζ Gaussian components of the mixture model. They are parameterized in terms of the mean μ_j and variance σ_j, which are not adjustable but fixed *a priori* such that the ζ components cover appropriately the space where $\tilde{\Phi}$ is expected to be contained, much like the basis functions in a GaRBF neural network.
- The θ_{i_j} terms are the mixing coefficients. They are defined in terms of a softmax function parameterized by h_{i_j}; $i = 1, \cdots, H$, $j = 1, \cdots, \zeta$. This form is chosen because the mixing coefficients of a Gaussian sum need to satisfy the conditions

$$0 \leq \theta_{i_j} \leq 1, \ \sum_{j=1}^{\zeta} \theta_{i_j} = 1.$$

- Note that the Gaussian components $\{g_j\}_{j=1}^{\varsigma}$ are shared by all validity functions, whilst the mixing coefficients $\{\theta_{i_j}\}_{i=1}^{H}$ are model specific.

The other term appearing in Equation (10.18), $\Pr(\hat{M}_i(k)|\hat{S}(k-1), \bar{I}^{k-2})$ reflects the probability that, given $\hat{S}(k-1), \bar{I}^{k-2}$, the mode active at time k is captured by model i. In the class of systems under consideration, mode $m(k)$ is a direct function of $u(k-1), \mathbf{x}(k-1)$ and so it will be assumed that $m(k)$ is independent of $\hat{S}(k-1), \bar{I}^{k-2}$ and that under this conditioning, there is an equal chance that $m(k)$ takes any one of its H possible values. This assumption implies that $\Pr(\hat{M}_i(k)|\hat{S}(k-1), \bar{I}^{k-2})$ is independent and uniformly distributed and therefore equal to $1/H$. Substituting this and Equation (10.19) in Equation (10.18) leads to the following equation for validity function $\hat{\gamma}_i$,

$$\hat{\gamma}_i[u(k-1), \mathbf{x}(k-1)] = \frac{\mathbf{g}^T \Theta_i}{\sum_{j=1}^{H} \mathbf{g}^T \Theta_j} \qquad (10.22)$$

where $\mathbf{g} := [g_1 \ g_2 \cdots g_\varsigma]^T$ and $\Theta_i := [\theta_{i_1} \ \theta_{i_2} \cdots \theta_{i_\varsigma}]^T$.

Gate Parameter Estimation. The parameters of the gating network are adjusted sequentially together with the local model parameters. Applying on-line estimation techniques based on maximum likelihood methods to our case, the aim is to maximize the likelihood function $p(y(k)|\hat{S}(k-1), \bar{I}^{k-1})$. This means that the parameters should maximize the probability density of the output evaluated at the *actual* measurement $y(k)$. The likelihood is conditioned on $\hat{S}(k-1)$ in addition to $\bar{I}(k-1)$ because its calculation will inevitably depend on the parameter estimates of the local models, which are conditioned on $\hat{S}$ via the MAP mode estimation technique. This is clearly seen by partitioning the likelihood in terms of events $\hat{M}_j$ as follows

$$\begin{aligned} &p(y(k)|\hat{S}(k-1), \bar{I}^{k-1}) \\ &= \sum_{j=1}^{H} p(y(k)|\hat{M}_j(k), \hat{S}(k-1), \bar{I}^{k-1}) \Pr(\hat{M}_j(k)|\hat{S}(k-1), \bar{I}^{k-1}) \\ &= \sum_{j=1}^{H} p(y(k)|\hat{M}_j(k), \hat{S}(k-1), \bar{I}^{k-1}) \hat{\gamma}_j[u(k-1), \mathbf{x}(k-1)]. \end{aligned} \qquad (10.23)$$

The last equation follows because the terms $\Pr(\hat{M}_j(k)|\hat{S}(k-1), \bar{I}^{k-1})$ are being estimated via the validity functions. Note that the density terms $p(y(k)|\hat{M}_j(k), \hat{S}(k-1), \bar{I}^{k-1})$ are calculated from the parameter estimates of the local models (as explained in point (1) of Remarks 10.4.1) which are conditioned on sequence $\hat{S}$. This explains why the original likelihood function has also been conditioned on the same sequence.

To maximize the likelihood with respect to the gating network parameters, a gradient ascent algorithm is applied on the logarithm of $p(y(k)|\hat{S}(k-1), \bar{I}^{k-1})$, which will be denoted by $\mathcal{L}$. The use of logs is not compulsory, but it ultimately leads to more elegant parameter adjustment equations. In the following derivations, for notational convenience, we will write $p(y|\hat{M}_i)$ instead of $p(y(k)|\hat{M}_j(k), \hat{S}(k-1), \tilde{I}^{k-1})$ and $\hat{\gamma}_i$ instead of $\hat{\gamma}_i[u(k-1), \mathbf{x}(k-1)]$. Hence, using Equation (10.23), the log-likelihood function becomes

$$\mathcal{L} = \ln \sum_{j=1}^{H} p(y|\hat{M}_j)\hat{\gamma}_j. \tag{10.24}$$

To apply the gradient ascent procedure, we now require expressions for the partial derivative of $\mathcal{L}$ with respect to the gating parameters. Since two different gating network schemes have been proposed, the derivations are treated separately.

Softmax case.

For the softmax case, the gate parameters are given by the vectors $\mathbf{h}_i$. Using the chain rule,

$$\frac{\partial \mathcal{L}}{\partial \mathbf{h}_i} = \frac{\partial L}{\partial \hat{\gamma}_1} \frac{\partial \hat{\gamma}_1}{\partial \mathbf{h}_i} + \cdots + \frac{\partial L}{\partial \hat{\gamma}_H} \frac{\partial \hat{\gamma}_H}{\partial \mathbf{h}_i}. \tag{10.25}$$

From Equation (10.17), tedious but straightforward calculations give

$$\frac{\partial \hat{\gamma}_l}{\partial \mathbf{h}_i} = \begin{cases} -\hat{\gamma}_i \hat{\gamma}_l \Phi(k-1) & \text{if } i \neq l \\ \hat{\gamma}_i(1-\hat{\gamma}_i)\Phi(k-1) & \text{if } i = l \end{cases}. \tag{10.26}$$

Also, from Equation (10.24)

$$\frac{\partial \mathcal{L}}{\partial \hat{\gamma}_l} = \frac{p(y|\hat{M}_l)}{\sum_{j=1}^{H} p(y|\hat{M}_j)\hat{\gamma}_j}. \tag{10.27}$$

Substituting Equations (10.26) and (10.27) in Equation (10.25), and using Equation (10.15) yields,

$$\frac{\partial \mathcal{L}}{\partial \mathbf{h}_i} = [\Pr(M_i(k)|\hat{S}(k-1), I^k) - \hat{\gamma}_i]\Phi(k-1).$$

Hence the gradient ascent adjustment rule for parameter vector $\mathbf{h}_i$ is given by

$$\mathbf{h}_i(k+1) = \mathbf{h}_i(k) + \eta[\Pr(M_i(k)|\hat{S}(k-1), I^k) - \hat{\gamma}_i]\Phi(k-1), \tag{10.28}$$

where η is a positive constant often referred to as the *learning rate*.

Gaussian mixture kernel case.

In this case the adjustable parameters are h_{ij} and so we need to consider

$$\frac{\partial \mathcal{L}}{\partial h_{ij}} = \frac{\partial \mathcal{L}}{\partial \hat{\gamma}_1} \frac{\partial \hat{\gamma}_1}{\partial h_{ij}} + \cdots + \frac{\partial \mathcal{L}}{\partial \hat{\gamma}_H} \frac{\partial \hat{\gamma}_H}{\partial h_{ij}}. \tag{10.29}$$

Using the chain rule, Equations (10.21) and (10.22) and some algebraic manipulations, it follows that

$$\frac{\partial \hat{\gamma}_p}{\partial h_{ij}} = -\hat{\gamma}_p \left[\frac{\theta_{ij}}{\sum_{l=1}^{H} \mathbf{g}^T \Theta_l} \right] (g_j - \mathbf{g}^T \Theta_i) + \begin{cases} 0 & \text{if } p \neq i \\ \hat{\gamma}_i \left\{ \frac{g_j}{\mathbf{g}^T \Theta_i} - 1 \right\} \theta_{ij} & \text{if } p = i \end{cases}$$

Substituting into Equation (10.29), together with Equations (10.15), (10.27) and some algebraic manipulation yields,

$$\frac{\partial \mathcal{L}}{\partial h_{ij}} = \theta_{ij} \left(\frac{g_j}{\mathbf{g}^T \Theta_i} - 1 \right) \left[\Pr(\hat{M}_i(k)|\hat{S}(k-1), I^k) - \hat{\gamma}_i \right].$$

Hence the gradient ascent parameter adjustment rule for h_{ij} becomes

$$h_{ij}(k+1) = h_{ij}(k) +$$

$$\eta \left[\Pr(\hat{M}_i(k)|\hat{S}(k-1), I^k) - \hat{\gamma}_i \right] \theta_{ij} \left(\frac{g_j}{\mathbf{g}^T \Theta_i} - 1 \right), \quad (10.30)$$

where η is the constant and positive learning rate parameter.

Note that in both Equations (10.28) and (10.30), adjustment of the gating parameters depends directly on the error term $\left[\Pr(\hat{M}_i(k)|\hat{S}(k-1), I^k) - \hat{\gamma}_i \right]$, which is the difference between the prior and the posterior probability of event $\hat{M}_i(k)$. This is intuitively appealing because the posterior probability carries more recent information than the prior probability, and so the updating of the parameters is based on the error between them.

10.5 The Control Law

Before developing a control law for the adaptive case, we will first consider the situation when all the mode parameters (a_i, b_i, d_i) and the mapping Ψ are known. This will highlight some specific features of the class of systems in question, which are somewhat further complicated in the presence of uncertainty.

10.5.1 Known System Case

Lemma 10.5.1. *Given any desired value f_d for function $f[u(k), \mathbf{x}(k)]$ specified in Equation (10.3), there exists a function $\nu(\mathbf{x}, f)$ that generates a unique input $u_d(k) = \nu(\mathbf{x}(k), f_d)$ which satisfies $f[u_d(k), \mathbf{x}(k)] = f_d$.*

Proof. The existence and uniqueness of ν follow directly from the continuity and invertibility properties of function f specified in Assumption 10.2.4. $\square$

The existence of function $\nu(\cdot)$ enables the determination of an optimal control law as given in the following lemma.

Lemma 10.5.2. *The control law that minimizes the Performance Index*

$$J = E \left\{ \sum_{k=1}^{N} [y_d(k+1) - y(k+1)]^2 \right\}$$

is unique and given by

$$u^*(k) = \nu(\mathbf{x}(k), y_d(k+1)).$$

Proof. From the properties of function $\nu(\cdot)$ specified in Lemma 10.5.1, applying $u^*(k) = \nu(\mathbf{x}(k), y_d(k+1))$ as input to the system leads to,

$$f(u^*, \mathbf{x}(k)) = y_d(k+1).$$

Now Equations (10.1) and (10.3) yield that in general

$$y(k+1) = f[u(k), \mathbf{x}(k)] + e(k+1).$$

Hence when $u(k) = u^*(k)$,

$$y(k+1) = y_d(k+1) + e(k+1).$$

Substituting for $y(k+1)$ in Performance Index J yields

$$J = E\left\{\sum_{k=1}^{N} e^2(k+1)\right\},$$

which is optimal since the noise e is independent of all the other variables and unmeasurable, and so this cost cannot be minimized further. The uniqueness of $u^*(k)$ follows directly from Lemma 10.5.1. □

Lemma 10.5.2 shows that the optimal control law $u^*(k)$ should aim to force function $f[u(k), \mathbf{x}(k)]$ to be equal to $y_d(k+1)$, i.e., $f[u^*(k), \mathbf{x}(k)] = y_d(k+1)$. Substituting for f from Equation (10.3) leads to the functional form of $u^*(k)$ as follows:

$$u^*(k) = \frac{y_d(k+1) - \mathbf{a}_{m(k+1)}^T \mathbf{x}(k) - d_{m(k+1)}}{b_{m(k+1)}}. \tag{10.31}$$

Note that this equation depends on $m(k+1)$. But Equation (10.2) shows that $m(k+1)$ depends on $u(k)$, through function Ψ. Hence even if Ψ were known, the value of $m(k+1)$ could not be determined before $u(k)$ is known. On the other hand Equation (10.31) shows that the optimal value of $u(k)$ depends on $m(k+1)$, which means that $u^*(k)$ could not be determined before $m(k+1)$ is known! Hence Equation (10.31) does not really represent a closed form solution for $u^*(k)$. This situation arises because $u(k)$ not only affects $y(k+1)$, but is itself one of the scheduling variables that determine the mode active at time $(k+1)$. However, since $u^*(k)$ is unique, there can only be one value of $m(k+1)$ that satisfies Equation (10.31) for a given $\mathbf{x}(k)$. This leads us to derive a method for evaluating $u^*(k)$, shown in the following proposition.

Proposition 10.5.1. *The optimal control law $u^*(k)$ of Equation (10.31) is evaluated as*

$$u^*(k) = \sum_{i=1}^{H} \gamma_i[u_i(k), \mathbf{x}(k)] u_i(k) \tag{10.32}$$

where

$$u_i(k) := [y_d(k+1) - \mathbf{a}_i^T \mathbf{x}(k) - d_i]/b_i. \tag{10.33}$$

Proof. Define $m^*(k+1) := \Psi[u^*(k), \mathbf{x}(k)]$, i.e., the mode that results from applying the optimal control $u^*(k)$. From Equation (10.31) and the definition of $u_i(k)$ above, this means that

$$u^*(k) = u_{m^*(k+1)}(k). \tag{10.34}$$

We first show that the uniqueness of $u^*(k)$ implies that

$$\gamma_i[u_i(k), \mathbf{x}(k)] = \begin{cases} 1 \text{ for } & i = m^*(k+1) \\ 0 \text{ for } & i \neq m^*(k+1) \end{cases}. \tag{10.35}$$

Note that this relation is *not* an obvious consequence of the fact that the γ_i's are mutually exclusive, as reflected by the definition of γ_i given in Equation (10.5). The above equation differs from Equation (10.5) because the first argument of γ (i.e., $u_i(k)$) is changing with i and also, $m^*(k+1)$ is not a function of $u_i(k)$. This part of the proof follows by contradiction.

Define $f_i[u(k), \mathbf{x}(k)] := \mathbf{a}_i^T \mathbf{x}(k) + b_i u(k) + d_i$. Note from Equation (10.4) that this term is equivalent to $\mathbf{w}_i^{*T} \Phi(k)$. Substituting $u(k) = u_i(k)$ in this equation yields that $\forall i$:

$$f_i[u_i(k), \mathbf{x}(k)] = y_d(k+1). \tag{10.36}$$

Assume that

$$\gamma_i[u_i(k), \mathbf{x}(k)] = 1 \text{ for some } i = i' \neq m^*(k+1). \tag{10.37}$$

Setting $u(k) = u_{i'}(k)$ in Equation (10.4) $\Rightarrow$

$$y(k+1) = \sum_{i=1}^{H} \gamma_i[u_{i'}(k), \mathbf{x}(k)] \mathbf{w}_i^{*T} \Phi'(k) + e(k+1),$$

where $\Phi'(k) = [u_{i'}(k) \; \mathbf{x}^T(k) \; 1]^T$. Hence if condition (10.37) is true, and by the fact that for a specific input, the γ_i's are mutually exclusive, it follows that

$$\begin{aligned} y(k+1) &= \mathbf{w}_{i'}^{*T} \Phi'(k) + e(k+1) \\ &= f_{i'}[u_{i'}(k), \mathbf{x}(k)] + e(k+1) \\ &= y_d(k+1) + e(k+1) \text{ by (10.36).} \end{aligned}$$

This implies that setting $u(k) = u_{i'}(k)$ for some $i' \neq m^*(k+1)$ is an optimal solution. But this leads to a contradiction because the optimal solution is given by Equation (10.34), corresponding to $i = m^*(k+1)$, which is known to be *unique*. Hence the uniqueness of $u^*(k)$ implies that $\gamma_i[u_i(k), \mathbf{x}(k)]$ cannot be equal to 1 other than for $i = m^*(k+1)$, which proves (10.35). Clearly however, Equation (10.35) implies that

$$\sum_{i=1}^{H} \gamma_i[u_i(k), \mathbf{x}(k)] u_i(k) = u_{m^*(k+1)}(k),$$

which is indeed the required optimal solution. $\square$

Note that unlike Equation (10.31), it is now possible to evaluate Equation (10.32) because it does not depend on $m(k+1)$. Every term $u_i(k)$ reflects the input that would be applied if $m(k+1)$ were equal to i. A helpful way of interpreting control law (10.32) is that $u^*(k)$ is the particular member of the set of individual control laws $\{u_i(k)\}_{i=1}^{H}$ that "satisfies" its own partition, i.e., the one whose indicator $\gamma_i[u_i(k),\mathbf{x}(k)] = 1$. The continuity and invertibility of f ensure the *existence* and *uniqueness* of this condition.

The logic behind the proof could be clearly seen in Figure 10.1 which, for simplicity, shows a 3 mode system whose partitioning is independent of $\mathbf{x}$ and depends only on u. The figure shows the example when $m^*(k+1) = 3$ because $u^*(k)$ falls in the partition of mode $m = 3$. Note that $\gamma_1(u_1) = \gamma_2(u_2) = 0$, whilst *only* $\gamma_3(u_3) = 1$. This allows Equation (10.32) to deduce that $u^*(k) = u_3(k)$.

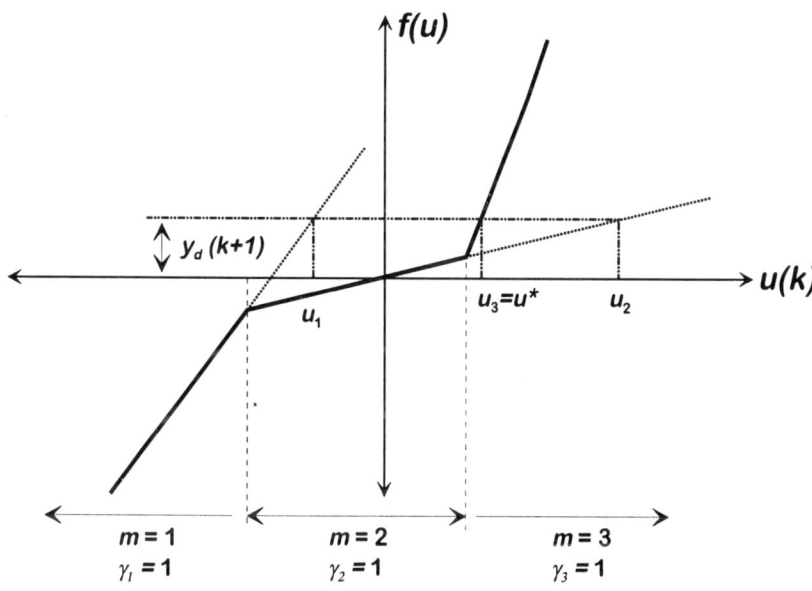

Fig. 10.1. Illustration of the proof of Proposition 10.5.1 for a simple three mode system. Note that only u_3 "satisfies" its own partition.

10.5.2 Unknown System Case

In the unknown mode case, neither γ_i nor $(\mathbf{a}_i, b_i, d_i)$ are known. One possible way of obtaining an adaptive control law is to replace these values by their estimates $\hat{\gamma}_i, \hat{\mathbf{a}}_i, \hat{b}_i, \hat{d}_i$ and use them directly in the known case control law of equations (10.32) and (10.33). However this simple replacement of parameters $\hat{\mathbf{a}}_i, \hat{b}_i, \hat{d}_i$ in Equation (10.33) amounts to a heuristic certainty equivalent approach. It is better if, instead, a suboptimal dual control law were used to introduce the desirable effects of caution and probing. Since each individual mode is a linear system, any suboptimal dual scheme developed for linear IO systems could be applied. For continuity with the previous work documented in this book, we select Milito et al.'s IDC law [176] which for mode i takes the form

$$\hat{u}_i(k) = \frac{\{y_d(k+1) - \hat{\mathbf{w}}_i^T(\cdot)\breve{\Phi}(k)\}\hat{b}_i(\cdot) - (1+r)[p_{b_i}\ \mathbf{p}_{1_i}^T]\breve{\Phi}(k)}{q + \hat{b}_i^2(\cdot) + (1+r)p_{b_i}}, \quad (10.38)$$

where the argument $(\cdot)$ of terms $\hat{\mathbf{w}}_i^T$ and $\hat{b}_i$ is $(k+1|\hat{S}(k))$, parameters $-1 \leq r \leq 0, q > 0$ are the performance index weights determining the amount of caution and the penalty on the control as in Chapter 7, $\breve{\Phi} := [0\ \mathbf{x}^T\ 1]^T$ and $p_{b_i}, \mathbf{p}_{1_i}$ are partitions of the covariance matrix $\mathbf{P}_i(k+1|\hat{S}(k))$ from the Kalman filter equations (10.16) as shown below:

$$\mathbf{P}_i(k+1|\hat{S}(k)) = \begin{vmatrix} p_{b_i} & \vdots & \mathbf{p}_{1_i}^T \\ \cdots & \cdots & \cdots \\ \mathbf{p}_{1_i} & \vdots & \mathbf{P}_2 \end{vmatrix},$$

with p_{b_i} denoting the $(1,1)$ element of matrix $\mathbf{P}_i$. This way, following the pattern of Equation (10.32), the adaptive control law becomes

$$u(k) = \sum_{i=1}^{H} \hat{\gamma}_i[\hat{u}_i(k), \mathbf{x}(k)]\hat{u}_i(k). \quad (10.39)$$

In effect, this control law represents a mixture of all local dual control signals weighted by the corresponding validity functions.

Note that although the known system control law of Equation (10.32) has a similar form, it is not actually *mixing* the local control signals because in Equation (10.32) the indicator variables $\gamma_i[u_i(k), \mathbf{x}(k)]$ are binary and satisfy Equation (10.35). Hence only one of the indicators, corresponding to $i = m^*(k+1)$, is actually equal to one. In fact the known system control law could be interpreted equivalently as follows:

$$u^*(k) = u_{m^*(k+1)}(k) \quad (10.40)$$

where $m^*(k+1) = \{i|\gamma_i[u_i(k), \mathbf{x}(k)] = 1;\ i = 1, \cdots, H\}. \quad (10.41)$

In this form, we are determining which index i satisfies the condition $\gamma_i[u_i(k), \mathbf{x}(k)] = 1$ and then applying its corresponding control as input. Equations (10.40), (10.41) therefore suggest another possibility for the adaptive control law. Unlike the indicator variables, the validity functions $\hat{\gamma}_i$ are not binary and can take any value between 0 and 1. However the condition $\gamma_i = 1$ in Equation (10.41) could be reinterpreted to mean that $m^*(k+1)$ is the value of i for which γ_i is a *maximum* (rather than unity), because it is known that the rest are 0. This way, its extension in terms of $\hat{\gamma}$ becomes more sensible because it is not based on specific values of 0 or 1, but on relative values. Hence the corresponding alternative adaptive control law would be

$$u(k) = \hat{u}_{\hat{m}^*(k+1)}(k)$$
$$\text{where } \hat{m}^*(k+1) = \arg\max_{i=1,\cdots,H} \{\hat{\gamma}_i[\hat{u}_i(k), \mathbf{x}(k)]\}. \tag{10.42}$$

Unlike adaptive control law (10.39), this one selects only one of the H local dual control signals; namely the one that is most likely to satisfy its own partition as indicated by the value of $\hat{\gamma}_i$. It is possible that the maximum term in Equation (10.42) may not always be unique. In this case, $\hat{m}^*(k+1)$ is set equal to any one of these multiple maxima.

Remarks 10.5.1

1. In simulation experiments, control law (10.42) was found to give much better performance than control law (10.39). There are two main reasons for this:
    - Since the gating is being learned during system operation, there is the possibility that initially all the validity functions $\hat{\gamma}_i[\hat{u}_i(k), \mathbf{x}(k)]; i = 1, \cdots, H$ are very small. Consequently $u(k)$ calculated from Equation (10.39) is also small and the excitation of the system remains weak. This could have the effect of inhibiting further the gate learning, especially when using the GMK network because of the local support of the Gaussian components. Hence the validity functions and control signal remain small, leading to a virtual "turn-off" of the system.
    - Even if this type of turn-off does not occur, unless the inputs are persistently exciting there is no guarantee that the mapping estimated by the gating network will converge substantially close to the actual segmentation. This was found to be particularly true for the softmax network because it typically does not pinpoint accurately the boundary between partitions. Hence if a validity function does not accurately characterize its actual partition (*e.g.*, it gives a value of 0.3 when in fact it should have been closer to 0), a particularly erroneous control signal might result from using Equation (10.39) because it mixes all validity functions and local controls. Under such a situation it would be better not to mix all local controls at all and apply *only* the control from the model whose validity function dominates over the rest, as in control law (10.42).
2. To avoid the possibility of numerical errors, the following additional precautions are introduced:
    - A lower bound $\underline{b}$ of the order of 10^{-5} is imposed when calculating the denominator terms of the validity function Equations (10.17) or (10.22) and the posterior probability Equation (10.15).
    - In the GMK gating, since the kernels have local approximation properties, parameter adjustment is halted during those instants that $\tilde{\Phi}$ drifts outside the overall approximation region covered by the kernels.

3. Simulations indicated that occasionally the parameter estimates of some local model might diverge, *i.e.*, they converge to values that do not reflect any one of the actual modes. This situation becomes particularly critical at the point when all local models would have been subjected to mode learning, because then no more models are available to learn the mode that was erratically captured by the divergent model. Ideally this condition should be detected and the divergent model reset so as to restart learning of this mode. Hence the following procedure is introduced:

- When no model has been detected to be capturing the dynamics of the current mode and all models have already been previously allocated to learn some mode, the terms $p(y(k)|\hat{M}_i(k), \hat{S}(k-1), \tilde{I}^{k-1})$ will be negligibly small (≈ 0) for all i. Hence the above-mentioned critical condition could be detected by monitoring the value of the denominator term of Equation (10.15), *i.e.*,

$$\delta(k) := \sum_{i=1}^{H} p(y(k)|\hat{M}_i(k), \hat{S}(k-1), \tilde{I}^{k-1}) \hat{\gamma}_i[u(k-1), \mathbf{x}(k-1)], \quad (10.43)$$

and checking if $\delta(k)$ has reached its lower bound $\underline{b}$.

- If this condition is detected, the divergent model needs to be found. This is performed by accumulating the posterior probabilities for each model, *i.e.*,

$$P_{a_i}(k) := \sum_{t=1}^{k} \Pr(\hat{M}_i(t)|\hat{S}(t-1), I^t); i = 1, \cdots, H.$$

Note that not to avoid storing the full history of the posterior probabilities, in practice P_{a_i} are calculated recursively as follows

$$P_{a_i}(k) = P_{a_i}(k-1) + \Pr(\hat{M}_i(k)|\hat{S}(k-1), I^k). \quad (10.44)$$

Since the mode captured by the divergent model never really exists, the value of P_{a_i} for the divergent model is bound to be the smallest because that model was never found to be active for any appreciable amount of time. Hence the index of the divergent model is given by

$$m_d = \arg \min_{i=1,\cdots,H} \{P_{a_i}(k)\}. \quad (10.45)$$

- Parameter learning of model m_d is then restarted by applying the Kalman filter Equations (10.16) for that model only, *after* increasing the covariance matrix $\mathbf{P}_{m_d}(k|\hat{S}(k-1))$ by $\rho \mathbf{I}$, where ρ is a positive constant. This entails resetting $\gamma_i^{map}(k) = 1$ for model $i = m_d$ only. This step therefore "opens up" once again the covariance and forces the filter of model m_d to forget previous estimates, just like "random walk" tracking estimators. Note however that in our case this is only done instantaneously to bring the divergent model back in line whenever the critical condition is detected, and not continuously.

Table 10.1 gives an outline of the adaptive control algorithm described in this chapter.

Table 10.1. The adaptive control algorithm using modular networks

1.	Measure $y(k)$
2.	Calculate validity functions $\hat{\gamma}_i[u(k-1), \mathbf{x}(k-1)]$; eqns. (10.17) or (10.22)
3.	Calculate $p(y(k)|\hat{M}_i(k), \hat{S}(k-1), \tilde{I}^{k-1})$; eqns. (10.12), (10.13)
4.	Calculate $\Pr(\hat{M}_i(k)|\hat{S}(k-1), I^{k-1}), P_{a_i}(k), \delta(k)$; eqns. (10.15), (10.44), (10.43)
5.	IF $\delta(k) \neq \underline{b}$: Calculate $\hat{m}(k)$; eqn.(10.10) Calculate γ_i^{map}; eqn. (10.14) Predict $\hat{\mathbf{w}}_i$ and $\mathbf{P}_i$; eqns. (10.16) Update gating network parameters; eqns. (10.28) or (10.30) ELSE : Calculate m_d; eqn. (10.45) Apply equations (10.16) on m_d only, after increasing its covariance
6.	Calculate $u(k)$; eqns. (10.39) or (10.42)

10.6 Simulation Examples and Performance Evaluation

10.6.1 Example 1

The plant considered in this example is partitioned into three distinct linear local regions whose scheduling depends on the input $u(k-1)$ as shown below:

$$y(k) = 3u(k-1) - 5 + e(k) \qquad \text{if } u(k-1) \geq 2.5$$
$$y(k) = u(k-1) + e(k) \qquad \text{if } -1 < u(k-1) < 2.5$$
$$y(k) = 2u(k-1) + 1 + e(k) \qquad \text{if } u(k-1) \leq -1.$$

The actual parameters for each mode, $\mathbf{w}_i^* = [3\ -5]^T, [1\ 0]^T, [2\ 1]^T$ are unknown, as well as the scheduling boundaries -1 and 2.5 for $u(k-1)$. The zero-mean Gaussian noise $e(k)$ has variance of 0.01. The reference input $y_d(k)$ is obtained by low-pass filtering a $0.1Hz$ square wave and discretizing the output with a sampling frequency of $20Hz$. Two simulation experiments were carried out, one using Softmax gating and another using the GMK gating network. In both cases, the initial covariance matrices $\mathbf{R}_i$ of all the local Kalman filter estimators were set to $5\mathbf{I}$ and the initial parameter estimates $\mathbf{m}_i$ were selected at random. The performance index weights were set to $r = -0.9, q = 0.001$.

Experiment 1: Softmax Gating Network. The softmax network parameter vectors $\mathbf{h}_i$ were all initialized at a value of $[0.1\ 0.1]^T$ and the gradient ascent learning rate parameter η was set to 0.1. Figure 10.2 shows the results of the simulation when using adaptive control law (10.42). Part (a) of the figure shows that after a short initial transient, the plant output tracks the reference input y_d. This is confirmed in part (b) of the figure which shows that the tracking error $(y - y_d)$ converges to a region around zero.

Figure 10.3 shows how the model estimates are progressing in time. Note that the estimates of each local model converge to the true model parameters

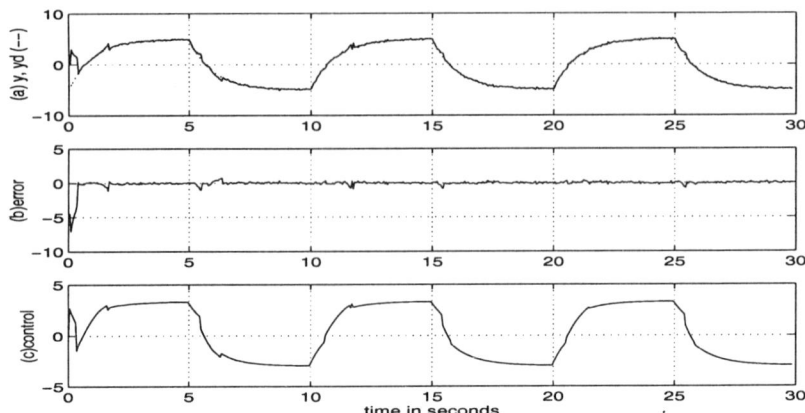

Fig. 10.2. Example 1 with softmax gating and control law (10.42).

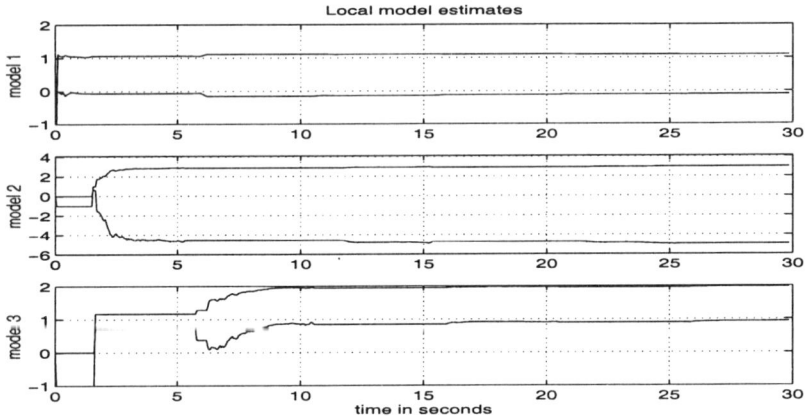

Fig. 10.3. Local model parameter estimation for Example 1.

of every mode, as desired. Figure 10.4(a) shows the three validity functions plotted against $u(k-1)$ as estimated by the softmax gating network when utilizing the parameters obtained after 30 seconds of simulation. Note that the segmentation calculated from the validity functions is close to, but not really a very accurate representation, of the actual partitioning, which specifies two boundaries at $u = -1$ and $u = 2.5$. In addition, the rising and falling edges of the validity functions are not sharp, with the consequence that a validity function has the tendency to remain appreciably high even within the partition of its neighbouring mode. The effect of this is reflected further in part (b) of the same figure, which shows the plot of $y(k)$ against $u(k-1)$ as defined by the actual plant equations (in bold), and also as estimated by the modular network when using Equation (10.7) (dashed). Note that,

10.6 Simulation Examples and Performance Evaluation   233

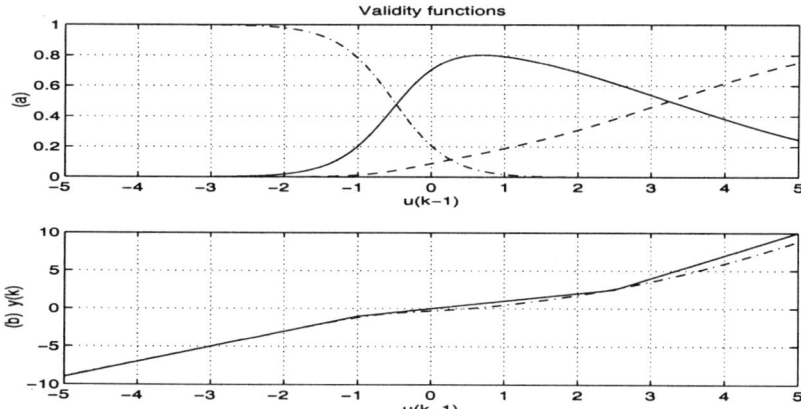

Fig. 10.4. Example 1 with softmax gate; (a) Validity functions (b)Actual and estimated (dashed) plant dynamics

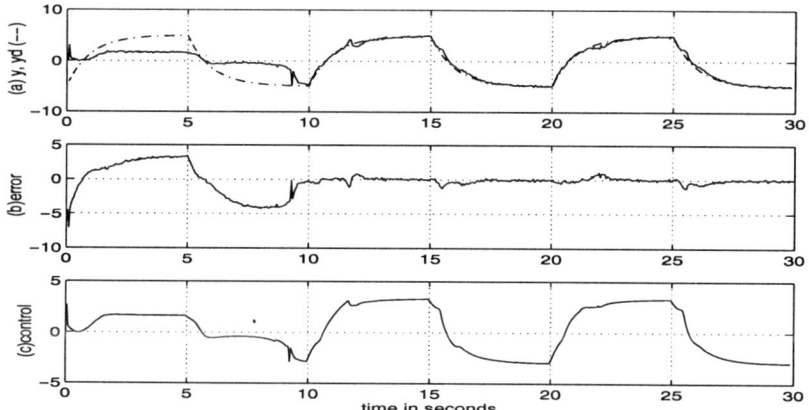

Fig. 10.5. Example 1 with softmax gating and control law (10.39)

particularly for $u(k-1) > 0$, there is a significant discrepancy between the two plots. Clearly this is due to the low quality gate estimation rather than a bad local model estimation, because the model parameters were seen to converge very close to the actual values in Figure 10.3. This suggests that an inferior performance would be expected if the alternative control law of Equation (10.39) were to be applied because it mixes all the local controls according to the validity functions, which are not very accurate. This is confirmed in Figure 10.5. A comparison with Figure 10.2 shows that the inferior performance of control law (10.39) is not limited only to the initial transient but also appears during the steady state.

234   10. Multiple Model Dual Adaptive Control of Spatial Multimodal Systems

Experiment 2: Gaussian Mixture Kernel Gating Network. Figures 10.6 to 10.8 show the performance of the system with control law (10.42) when the softmax network is replaced by the GMK gating network. The means μ_j of the Gaussian components were chosen so that they uniformly cover the range $[-5, 5]$ with a separation of 0.5, and their variance was set to 0.2. The gate learning rate η was set to 1.

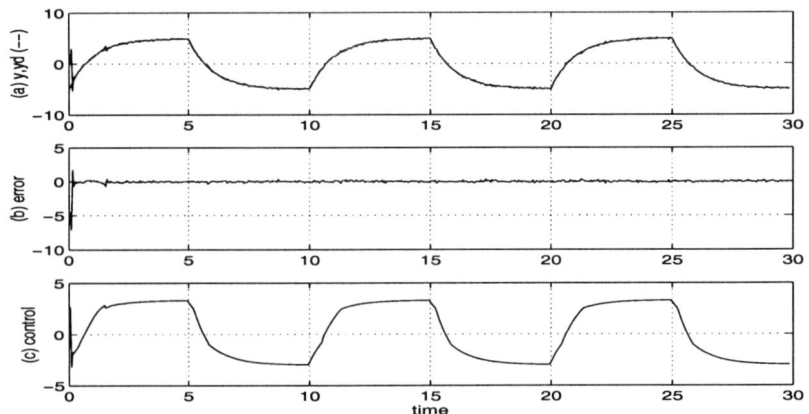

Fig. 10.6. Example 1 with GMK gating and control law (10.42)

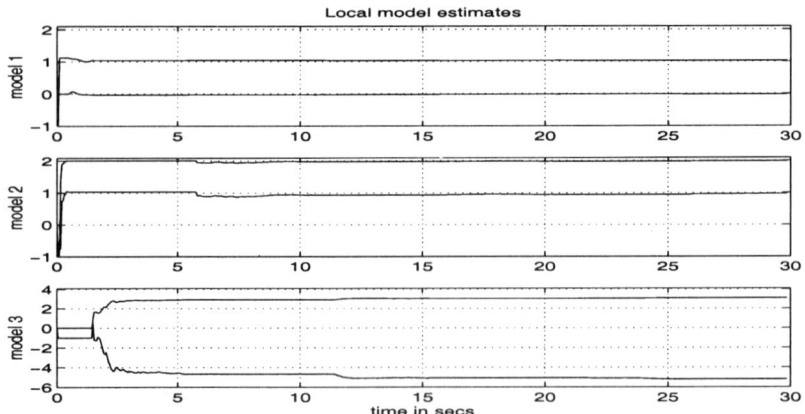

Fig. 10.7. Local model parameter estimates for example 1 with GMK gate

Figure 10.6 shows that the output tracks the reference input with a better transient response than for the softmax network, and Figure 10.7 shows that the local model estimation has similar characteristics. The validity functions of the GMK gating represent far more accurately the actual segmentation; both in terms of boundary location and the definition of their rising and falling edges. This is reflected in Figure 10.8(b) which shows that between the range $-3.5 \leq u(k-1) \leq 3.5$, the modular network provides a very accurate representation of the actual plant dynamics when using the gating parameters obtained at the end of the simulation. The validity function approximation is restricted to within this range of $u(k-1)$ because the localized support of the Gaussian components gives rise to a significant gate parameter adjustment only over the region where u has been active. Part (a) of the same figure shows that the three validity functions (solid curves) capture much better the actual partition boundaries, especially the one at $u = 2.5$, and that their rising and falling edges are sharper. The dashed curves show the Gaussian mixture approximation to the probability density given by Equation (10.19), for all three models. As expected, the plots confirm that these densities are not normally distributed.

Finally Figure 10.9 show once again that the mixing control law (10.39) yields bad performance. In fact, now it is worse than for the softmax case because the system practically remains turned-off. As explained previously in Comments 10.5.1, this is not surprising because gate learning critically depends on sufficient excitation of u, especially when localized Gaussian components are used.

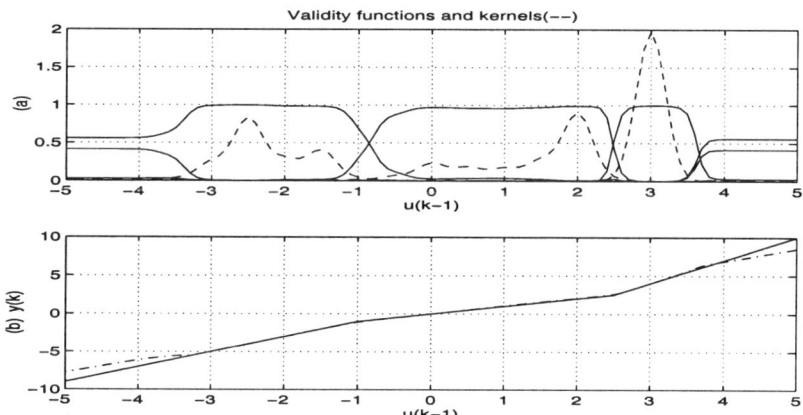

Fig. 10.8. Example 1 with GMK gate; (a) Validity functions and Kernels (dashed) (b)Actual and estimated (dashed) dynamics

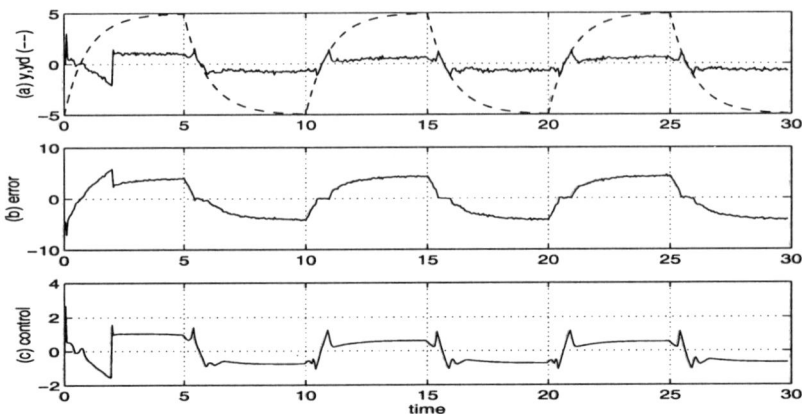

Fig. 10.9. Example 1 with GMK gate and control law (10.39)

10.6.2 Example 2

In this second example, the plant scheduling is a function of both the output and the input. There are 3 modes that exhibit the following dynamics:

$y(k) = 0.7y(k-1) + 2u(k-1) - 1.8 + e(k)$ for Mode 1
$y(k) = -y(k-1) + 0.2u(k-1) + e(k)$ for Mode 2
$y(k) = 0.7y(k-1) + 2u(k-1) + 1.8 + e(k)$ for Mode 3

The noise variance is 0.01 and the mode scheduling is given as follows:

Mode 1 is active when:
$1.7y(k-1) + 1.8u(k-1) + 1.8 \geq 0$ and
$1.7y(k-1) + 1.8u(k-1) - 1.8 > 0$

Mode 2 is active when:
$1.7y(k-1) + 1.8u(k-1) + 1.8 < 0$ and
$1.7y(k-1) + 1.8u(k-1) - 1.8 > 0$

Mode 3 is active when:
$1.7y(k-1) + 1.8u(k-1) + 1.8 < 0$ and
$1.7y(k-1) + 1.8u(k-1) - 1.8 \leq 0$.

The reference input consists of a sinusoidal wave of frequency $0.3Hz$ sampled at a rate of $20Hz$. The control performance index weights are chosen as $q = 0.01, r = -0.5$ and control law (10.42) is used. The initial Kalman filter covariance matrices are set to $300\mathbf{I}$ and the initial parameter estimates are selected at random. For the softmax gating network, the parameter learning rate is set to $\eta = 0.5$. The GMK network consists of Gaussian components having variance 0.4, located on a regular mesh covering the region $[-6, 6] \times [-6, 6]$ with a spacing of 1.0, and the gate parameter learning rate $\eta = 2$.

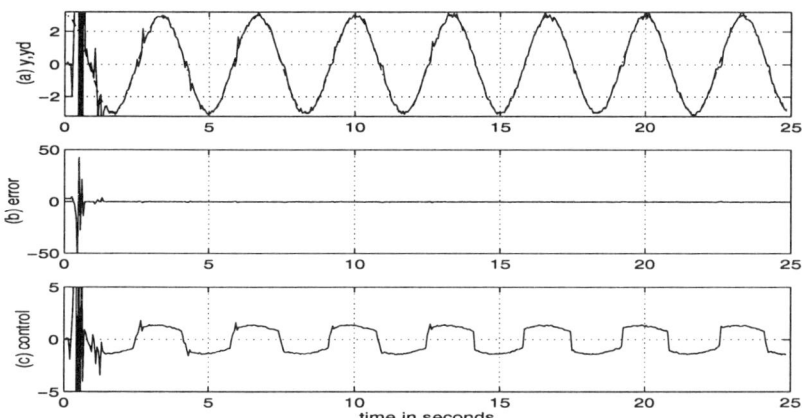

Fig. 10.10. Example 2 with softmax gate

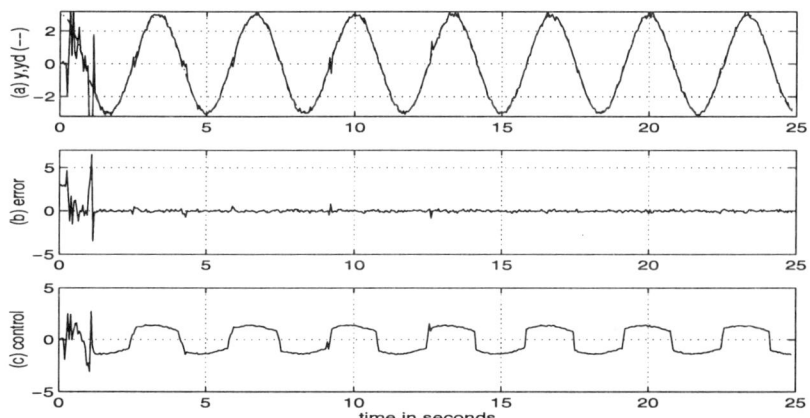

Fig. 10.11. Example 2 with GMK gate

Figures 10.10 and 10.11 show the results of the simulation with the softmax and GMK gating networks respectively. In both cases the output tracks the reference input after an initial transient, during which the mode parameters and the gating are learned. The transient response of the GMK case is much better than for the softmax case, particularly in terms of overshoot. Figures 10.12 and 10.13 show the local model parameter estimation process.

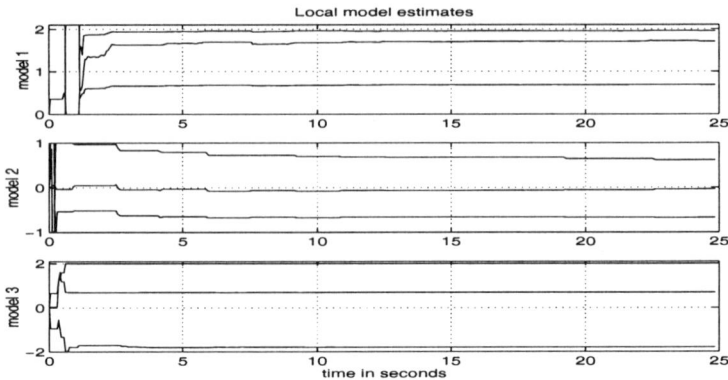

Fig. 10.12. Local model estimates for Example 2 with softmax gate

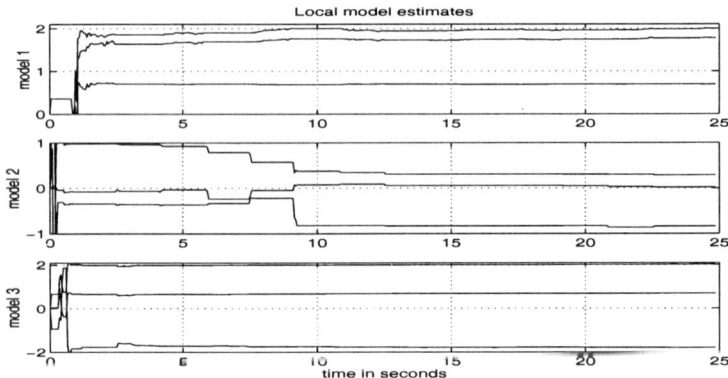

Fig. 10.13. Local model estimates for Example 2 with GMK gate

Note that for the GMK gating case, the parameter estimates converge close to the actual parameters of each particular mode. For the softmax gating case however, the estimates of local model 2 do not capture very accurately the parameters of mode 2. Note that in general, convergence of the parameters to the actual values is not guaranteed unless the signals in the system are persistently exciting.

Figures 10.14 and 10.15 show how the mode segmentation has been estimated by the softmax and GMK networks respectively, when using the gate parameters obtained at the end of the simulation. The plots show the 0.5 level contour of the validity functions on the (u, y) plane (dashed curves), which gives a good indication of the how the partition boundaries have been estimated. The true boundaries are also shown superimposed on the same diagrams as solid curves. Note that within the space where u and y have been active, the validity functions of the GMK gating network discriminate reasonably well the 3 different mode regions. By contrast, the softmax gating is far from being an accurate representation of the actual mode segmentation.

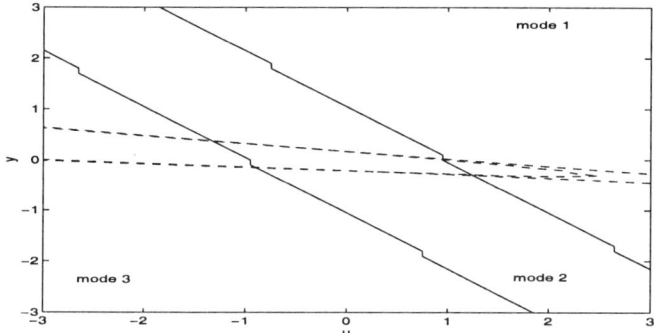

Fig. 10.14. Partition boundaries with softmax: actual (solid) and estimated (dashed)

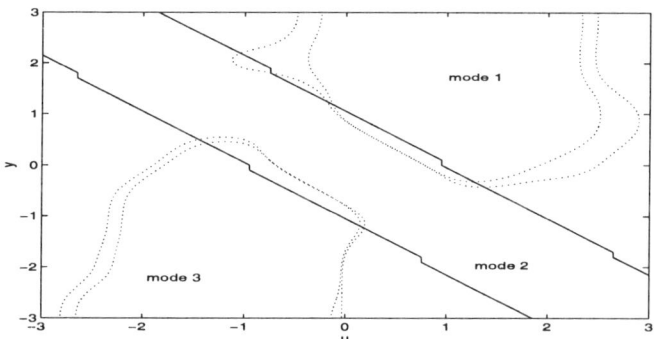

Fig. 10.15. Partition boundaries with GMK; actual (solid) and estimated (dashed)

Finally Figure 10.16 show the degradation in performance when the mixing control law of Equation (10.39) is used, both for the softmax and GMK gates. The figure shows the system output (solid) superimposed on the reference input (dashed). Note that in the softmax case the output is unable to track the reference input, and the GMK case takes a long time to track.

10.6.3 Performance Evaluation

Examples 1 and 2 show that the adaptive controller with GMK gating gives better general performance than softmax gating, particularly as far as tran-

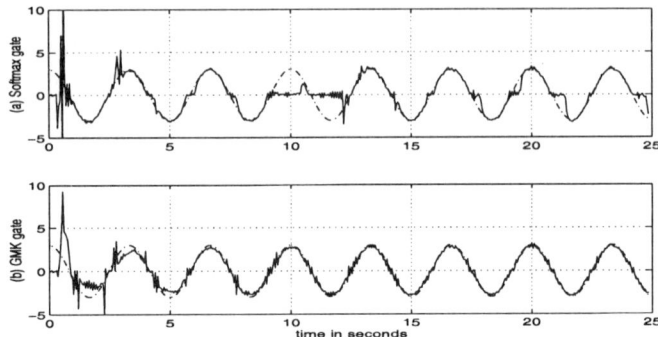

Fig. 10.16. System output of Example 2 with control law (10.39); (a) Softmax gate (b) GMK gate

sient overshoot is concerned. In addition, the partitioning estimated by GMK is much more accurate and better defined than for the softmax case. This improvement is attributed to the compact support of the individual Gaussian components used in the GMK network, which help to approximate better the characteristics of the mode segmentation on a localized basis. However there is a price to pay for this improvement, namely a greater number of adjustable parameters and *a priori* determination of the Gaussian components' means and variance.

As far as control is concerned, the examples confirm that the law of Equation (10.42) gives better performance than the mixing law of Equation (10.39). Since the gating is generally not learned perfectly well, very bad control could be calculated from Equation (10.39) because it places too much emphasis on the actual value of the validity functions, rather than on the one that dominates over the rest. In the GMK case, the mixing law could lead to turn-off because it attenuates the input excitation, as shown in Example 1. In the softmax case, since the partition boundaries are typically not estimated very accurately, the mixing law often leads to bad output tracking, as seen in Example 2.

10.7 Summary

In this chapter we have presented a multiple model adaptive estimation and control scheme for the piecewise linear class of spatial multimodal, stochastic systems that are subject to uncertainty. The uncertainty concerns *both* the mode dynamics and the function that maps the scheduling variables onto the modes. The problem therefore represents an adaptive form of gain scheduling where the system dynamics are learned in an on-line manner on the same philosophy as indirect adaptive control schemes. In contrast to standard indirect

10.7 Summary

adaptive schemes however, the identification model used here is a modular network that comprises multiple linear local models and a gating network, instead of one higher-order homogeneous global model. This facilitates the determination of the control law because it is based on the simpler linear functional form of the local models.

Mode activity is estimated by a Bayesian method that utilizes the information from a set of Kalman filters. The same filters are used for local model parameter estimation. The parameters of the gating network, which estimates the unknown scheduling function, are adjusted by a maximum likelihood technique. A gating network based on Gaussian Mixtures has been proposed. This gating network exhibits superior performance to the conventional softmax gating when applied to the scheme. Two different control laws have been presented, one of which was found to be superior to the other. They control laws are based on the IDC suboptimal dual scheme, but extended to this multiple model case. In particular, special attention has been given to the conditions that the system must satisfy to ensure the existence of a unique optimal control when this signal itself is one of the scheduling variables.

The proposed scheme represents a solution for handling spatial multimodality. It could be interpreted as an alternative way of tackling nonlinear functional uncertainty, by combining traditional gain scheduling with adaptive control principles. In this manner, complex nonlinear functions are partitioned into a number of simpler local problems so as to facilitate the design of a global adaptive control scheme.

Part IV

Conclusions

11. Conclusions

Modern plants and processes are often characterized by highly complex structures and dynamic behaviour typified by nonlinearities, time-varying dynamics and the influence of unpredictable disturbances. Under these circumstances, knowledge about the system is usually incomplete and subject to high levels of uncertainty. The task of controlling this kind of system is therefore a considerable challenge, especially when expected to operate safely, reliably and efficiently within a wide range of operating conditions and with as little human intervention as possible. This combination of complexity coupled with strict performance specifications is not uncommon in modern aircraft, marine and aerospace systems or even industrial processes that are highly sensitive to disturbances in the environment and in the quality of the process inputs.

The modern field of *intelligent control* emerged as a response to the challenge of controlling such systems by combining concepts from control theory and artificial intelligence. AI techniques, for example neural networks and fuzzy logic, are a powerful way of representing nonlinear dynamics, whilst control theoretic techniques offer solutions for handling parametric uncertainty and nonlinearity. It is generally accepted that an ideal intelligent controller should be able to cope with functional uncertainty, nonlinear dynamics and complexity such as component failure, unpredictable disturbances, multimodality and high dimensional spaces. To achieve such levels of performance, the intelligent controller must embody adaptation, learning and autonomy as its key features.

Intelligent control is still an active research area with scope for much further development, both theoretical and practical. Techniques that help to enhance any of the key features of intelligent control, as well as designs that improve the performance of the system in the face of increasingly complex or uncertain conditions, are all valid contributions to the field. The work reported in the book is a step in this direction. Various situations involving nonlinear functional uncertainty have been considered. These include deterministic and stochastic conditions, continuous and discrete-time dynamics and temporal and spatial multimodality. Nonlinearities were represented by neural networks of different types: GaRBF, sigmoidal MLP and modular architectures. In the deterministic case, stability and convergence issues have been addressed from the perspective of Lyapunov theory. Stochastic systems were handled by using concepts from stochastic adaptive optimal control theory, and multiple model techniques were used to address the complexity arising from both temporal and spatial multimodality. All these techniques were shown to enhance the performance of the control system, thereby leading to higher levels of "intelligence".

The first part of the book was devoted to deterministic systems. Chapter 3 introduced a control scheme based on dynamic structure GaRBF neural networks for stable and robust adaptive control of an affine class of functionally uncertain nonlinear systems in continuous time. The objective behind this scheme is to mitigate the problem of the curse of dimensionality associated with the use of network basis functions that have local representation. By introducing basis functions in an on-line manner while the system dynamics are evolving in time, it was shown that for situations where the state spans a relatively small subset of space, great savings in terms of network size and memory requirements are obtained. This technique has important practical implications regarding the implementation of neural-adaptive control schemes. In addition, the adaptation and control laws are derived in a way so as to ensure system stability in a Lyapunov sense and robustness to uncertain dynamics.

Chapter 4 introduced the use of composite adaptation for functional adaptive control of affine nonlinear systems in continuous time by GaRBF neural networks. The aim is to improve the transient response; an issue that has not been tackled previously for this class of controllers. The neural control scheme of [255] and the composite adaptation law of [233] were combined together and appropriately modified to yield a control scheme that maintains closed-loop stability and offers an improved transient response. The modifications represent an extension of the original concepts found in both the neural control and the composite adaptation schemes of [255] and [233]. The former is rendered more "intelligent" by the composite adaptation law which utilizes richer information for adjustment of the network parameters, while the latter is enhanced by making it robust against the de-stabilizing effect of the additive disturbance arising from the network's inherent approximation error. The beneficial effects of composite adaptation on the transient response were clearly demonstrated in the simulation experiments.

Adaptive control of uncertain, discrete-time, nonlinear systems was addressed in Chapter 5. The discrete-time problem merits particular attention because most modern controllers are implemented as a computer program. Moreover from a theoretical perspective, when addressing stable adaptive control of discrete-time systems, one has to face specific problems that are not present in the continuous-time counterpart. In particular, adaptive control methods based on Lyapunov stability are harder to formulate and very few results abound. Chapter 5 aims to enrich these results by deriving a novel scheme for stable adaptive control of a class of functional uncertain, nonlinear, discrete-time systems. The approach uses augmented error techniques and, in contrast to previous work, it does not impose any restriction on the neural network initial parameter error, nor demand prior knowledge on the value of the unknown optimal network parameters. The influence of the network approximation error on stability is kept in check by dead-zone adaptation, and an original analysis technique is introduced to prove signal boundedness, convergence and stability. Additionally, the system is modified so as to develop a new, stable adaptive sliding mode control scheme for the same class of nonlinear systems. The stability results obtained in this case are more complete and general than other attempts at adaptive sliding mode control of uncertain nonlinear systems by neural networks.

The second part of the book was devoted to stochastic systems. A stochastic perspective was introduced not only to handle external random disturbances, but also as a powerful method of dealing with uncertainty. Chapter 7 introduced the concept of dual adaptive control within the functional adaptive methodology. Both GaRBF and MLP networks were considered for control of a class of functional uncertain nonlinear systems. The method vastly improves the transient response of the system, especially during those periods when the uncertainty is high, by taking it into consideration inside the control law. This permits control and estimation to be performed simultaneously from the outset, eliminating the need of performing off-line system identification prior to applying the control. The ideas presented in this chapter represent a generalization of the IDC suboptimal dual control scheme originally developed for linear ARX systems [176]. The generalizations introduced are twofold: the extension to functional uncertain nonlinear systems and the extension to models that are not linearly parameterized, as in the case of MLP networks. This work has important implications for intelligent control because of the way that the control law utilizes the rich information from the estimation subsystem so as to encourage caution and probing-like effects, leading to a scheme that inherently provides a control which takes into consideration the uncertainty of the estimates.

The problem of adaptive control under multimodal complexity was treated in Chapters 9 and 10. The temporal multimodality problem was addressed in Chapter 9 for a class of nonlinear systems whose dynamics are both unknown and prone to jump arbitrarily in time. Although the solution to the problem was based upon the Bayesian methodology of classical multiple model systems, the approach was extensively modified because the case under consideration involves functional and not state uncertainty. This complicates the detection of mode jumps because the local models have to learn the nonlinear mode dynamics at the same as time as the modes are being estimated. The resulting scheme is a novel strategy that integrates the processes of mode detection, mode dynamics estimation and control signal generation within one unified stochastic framework. The dual control ideas of Chapter 7 were extended to this case, leading to new dual control laws for multiple model schemes. Another novelty concerns the development of a self-organizing algorithm for autonomously creating and assigning local models in real time when the number of modes is also unknown. This technique augments the intelligence of the control scheme presented in Chapter 7 by enabling it to handle not only uncertainty, but also temporal complexity. The local models also introduce learning, as opposed to mere adaptation, which leads to fast control reconfiguration; an important issue when controlling jump systems.

Chapter 10 addressed the spatial multimodal complexity problem for a piecewise linear class of nonlinear systems having unknown dynamics and subject to stochastic disturbances. In this approach, nonlinear systems are treated as a multimodal problem whose mode activity is not arbitrary, but scheduled according to the value of the state and/or the input. Although the idea is reminiscent of Gain Scheduling, a new dimension has been introduced because it is assumed that the local mode dynamics and the mapping that specifies the scheduling, are both unknown. This concept leads to a novel control scheme that could be interpreted as an adaptive form of Gain Scheduling. A multiple model modular network architecture is used as a plant identification model. The network's local models capture the different mode dynamics, and the scheduling is captured by a gating subsystem that estimates the mapping between the scheduling variables and the active mode in terms of a set of validity functions. As opposed to those techniques that attempt to capture the nonlinearity by one global complex model, modularization leads to an easier and more direct way of determining the control law because it is based on the relatively simpler functional form of the local mode dynamics. As in Chapter 9, the design is rooted in multiple model probabilistic and Bayesian concepts. This includes the mode and local model estimation algorithms, the validity function estimation and also the control law, which is based on dual control techniques. Special attention is given to the situation when the control itself is one of the scheduling variables. In particular, when this situation occurs, the conditions that the plant must satisfy to ensure the existence of a unique optimal control signal are clearly specified. An additional novelty is the development of a gating network based on Gaussian Mixtures. This was found to give a better estimation of the actual mode partitions and leads to an improved control performance. The approach presented in this chapter therefore offers a solution for handling control of complex nonlinear systems by subdivision into simpler local dynamics. The originality of the scheme centres on its ability to control functional uncertain systems by sequential estimation of both the local dynamics and the scheduling in an on-line manner during control operation, as well as the extension of suboptimal dual control schemes to this case.

In conclusion, this book has presented a wide range of techniques that lead to novel strategies for effecting intelligent control of complex systems. Various conditions were taken into consideration, including functional uncertainty, nonlinearity and multimodal complexity. The novel designs, which are based on adaptation, learning, multiple models, stability theory, stochastic adaptive control and neural networks, were shown to be effective for handling such stringent system conditions. These techniques should provide a good basis for future extensions aimed at increasing a control system's capacity for handling even higher levels of uncertainty and complexity, with less prior knowledge and increased autonomy.

References

1. G. A. Ackerson and K. S. Fu. On state estimation in switching environments. *IEEE Transactions on Automatic Control*, AC-15(1):10–17, February 1970.
2. H. Akashi and H. Kumamoto. Random Sampling Approach to state estimation in switching environments. *Automatica*, 13:429–434, 1977.
3. H. Akashi, H. Kumamoto, and K. Nose. Application of Monte Carlo method to optimal control for linear systems under measurement noise with Markov dependent statistical property. *International Journal of Control*, 22(6):821–836, 1975.
4. J. S. Albus. A new approach to manipulator control: the cerebellar model articulation controller. *Trans ASME, G, J. Dyn. Syst., Meas. Control*, 97:220–227, 1975.
5. B. J. Allison, J. E. Ciarniello, J. C. Tessier, and G. A. Dumont. Dual adaptive control of chip refiner motor load. *Automatica*, 31(8):1169–1184, 1995.
6. D. L. Alspach. Dual control based on approximate a posteriori density functions. *IEEE Transactions on Automatic Control*, AC-17:689–693, October 1972.
7. D. L. Alspach and H. W. Sorenson. Nonlinear Bayesian estimation using Gaussian Sum approximations. *IEEE Transactions on Automatic Control*, AC-17(4):439–448, August 1972.
8. J. Alster and P. R. Bélanger. A technique for dual adaptive control. *Automatica*, 10:627–634, 1974.
9. B.O. Anderson and J.B. Moore. *Optimal Filtering*. Prentice-Hall, U.S.A, 1979.
10. C. W. Anderson. Q-learning with hidden-unit restarting. In *Advances in Neural Information Processing Systems 5*, San Mateo, CA, 1993. Morgan Kaufmann.
11. P. Andersson. Adaptive forgetting in recursive identification through multiple models. *International Journal of Control*, 42(5):1175–1193, 1985.
12. P. J. Antsaklis. Defining intelligent control: Report of the Task Force on intelligent control. *IEEE Control Systems Magazine*, pages 4–66, June 1994.
13. P. J. Antsaklis. Intelligent learning control. *IEEE Control Systems Magazine*, pages 5–7, June 1995.
14. M. Aoki. *Optimization of stochastic systems: Topics in discrete-time systems*. Number 32 in Mathematics in Science and Engineering. Academic Press, Inc., New York, 1967.
15. M. Aoki. *Optimization of stochastic systems: Topics in discrete-time dynamics*. Economic Theory, Econometrics, and Mathematical Economics. Academic Press, Inc., San Diego, CA, 2nd edition, 1989.
16. K.-E. Arzen. An architecture for expert system based feedback control. *Automatica*, 25(6):813–827, 1989.
17. K. J. Åström. *Introduction to Stochastic Control Theory*. Academic Press, New York, 1970.
18. K. J. Åström, U. Borrison, L. Ljung, and B. Wittenmark. Theory and applications of self-tuning regulators. *Automatica*, 17:457–476, 1977.

19. K. J. Åström and A. Helmersson. Dual control of an integrator with unknown gain. *Comp. & Maths. with Applications*, 12A(6):653–662, 1986.
20. K. J. Åström and T. J. McAvoy. Intelligent control: An overview and evaluation. In D. A. White and D. A. Sofge, editors, *Handbook of intelligent control: Neural, fuzzy and adaptive approaches*, chapter 1, pages 3–34. Van Nostrand Reinhold, New York, 1992.
21. K. J. Åström and B. Wittenmark. Problems of identification and control. *Journal of Mathematical Analysis and Applications*, 34:90–113, 1971.
22. K. J. Åström and B. Wittenmark. On self-tuning regulators. *Automatica*, 9:185–199, 1973.
23. K. J. Åström and B. Wittenmark. *Adaptive Control*. Addison-Wesley, Reading, MA, U.S.A, 1989.
24. M. Athans, D. Castanon, K. P. Dunn, C. S. Greene, W. H. Lee, N. R. Sandell, and A. S. Willsky. The stochastic control of the F-8C aircraft using a multiple model adaptive control (MMAC) Method-Part 1: Equilibrium flight. *IEEE Transactions on Automatic Control*, AC-22(5):768–780, October 1977.
25. W. L. Baker and J. A. Farrell. An introduction to connectionist learning control systems. In D. A. White and D. A. Sofge, editors, *Handbook of Intelligent Control: Neural, fuzzy and adaptive approaches*, chapter 2, pages 35–63. Van Nostrand Reinhold, New York, 1992.
26. S. P. Banks and S. A. Khathur. Structure and control of piecewise-linear systems. *International Journal of Control*, 50(2):667–686, 1989.
27. Y. Bar-Shalom. Stochastic dynamic programming: Caution and probing. *IEEE Transactions on Automatic Control*, AC-26(5):1184–1195, October 1981.
28. Y. Bar-Shalom and T. E. Fortmann. *Tracking and Data Association*, volume 179 of *Mathematics in Science and Engineering*. Academic Press, Inc, San Diego, 1988.
29. Y. Bar-Shalom and E. Tse. Dual effect, certainty equivalence and separation in stochastic control. *IEEE Transactions on Automatic Control*, AC-19:494–500, October 1974.
30. Y. Bar-Shalom and E. Tse. Concepts and methods in stochastic control. In O. T. Leondes, editor, *Control and Dynamic Systems*, volume 12, pages 99–172. Academic Press Inc., New York, 1976.
31. Y. Bar-Shalom and K. D. Wall. Dual adaptive control and uncertainty effects in macroeconomic systems optimization. *Automatica*, 16:147–156, 1980.
32. A. G. Barto. Reinforcement learning and adaptive critic methods. In D. A. White and D. A. Sofge, editors, *Handbook of intelligent control: Neural, fuzzy and adaptive approaches*, chapter 12, pages 469–491. Van Nostrand Reinhold, New York, 1992.
33. A. G. Barto. Reinforcement learning. In O. Omidvar and D. L. Elliott, editors, *Neural Systems for Control*, chapter 2, pages 7–30. Academic Press, San Diego, CA, 1997.
34. G. Bartolini, A. Ferrara, and V. I. Utkin. Adaptive sliding mode control in discrete-time systems. *Automatica*, 31(5):769–773, 1995.
35. M. Basseville and A. Benveniste, editors. *Detection of Abrupt Changes in Signals and Dynamical Systems*. Number 77 in Lecture Notes in Control and Information Sciences. Springer-Verlag, Berlin, 1986.
36. R. Bellman. *Adaptive Processes: A Guided Tour*. Princeton University Press, Princeton, N.J., 1961.
37. D. P. Bertsekas. *Dynamic Programming: Deterministic and Stochastic Models*. Prentice-Hall, Englewood Cliffs, N. J., 1987.
38. S. A. Billings and W. S. F. Voon. Piecewise linear identification of non-linear systems. *International Journal of Control*, 46(1):215–235, 1987.

39. C. M. Bishop. *Neural Networks for Pattern Recognition.* Oxford University Press, New York, 1995.
40. H. A. P. Blom and Y. Bar-Shalom. The interacting multiple model algorithm for systems with Markovian switching coefficients. *IEEE Transactions on Automatic Control*, 33(8):780–784, August 1988.
41. J. D. Boskovic. Stable adaptive control of a class of first-order nonlinearly parameterized plants. *IEEE Transactions on Automatic Control*, AC-40:347–350, 1995.
42. M. S. Branicky. Hybrid dynamical systems, or HDS: The ultimate switching experience. In A. S. Morse, editor, *Control using Logic-Based Switching*, Lecture Notes in Control and Information Sciences, chapter 1, pages 1–12. Springer-Verlag, London, 1997.
43. L. Breiman. Hinging hyperplanes for regression, classification, and function approximation. *IEEE Transactions on Information Theory*, 39(3):999–1013, 1993.
44. L. Breiman, J. H. Friedman, R. A. Olshen, and C. J. Stone. *Classification and Regression Trees.* Wadsworth International Group, Belmont, CA, 1984.
45. M. D. Brown, G. Lightbody, and G. W. Irwin. Nonlinear internal model control using local model networks. *IEE Proceedings- Control Theory and Applications*, 144(6):505–514, November 1997.
46. J. B. D. Cabrera and K. S. Narendra. The general tracking problem for discrete-time dynamical systems. In *Proceedings of the 36th IEEE Conference on Decision and Control*, pages 1451–1456, San Diego, California, December 1997.
47. L. Campo, Y. Bar-Shalom, and X. Rong Li. Control of discrete-time hybrid stochastic systems. *Control and Dynamic Systems*, 76:341–361, 1996.
48. B. Castillo and S. Di Gennaro. Asymptotic output tracking for SISO nonlinear discrete systems. In *Proceedings of the 30th IEEE Conference on Decision and Control*, pages 1802–1806, Brighton, England, December 1991.
49. C. Y. Chan. Discrete adaptive sliding-mode tracking controller. *Automatica*, 33(5):999–1002, 1997.
50. C. Y. Chan. Discrete adaptive sliding mode control of a state-space system with a bounded disturbance. *Automatica*, 34(12):1631–1635, 1998.
51. S. S. Chan and M. B. Zarrop. A suboptimal dual controller for stochastic systems with unknown parameters. *International Journal of Control*, 41(2):507–524, 1985.
52. C. B. Chang and M. Athans. State estimation for discrete systems with switching parameters. *IEEE Transactions on Aerospace and Electronic Systems*, AES-14(3):418–425, May 1978.
53. F. C. Chen and H. K. Khalil. Adaptive control of a class of nonlinear discrete-time systems. *IEEE Transactions on Automatic Control*, 40(5), May 1995.
54. F.-C. Chen and C.-C. Liu. Adaptively controlling nonlinear continuous-time systems using multilayer neural networks. *IEEE Transactions on Automatic Control*, 39(6):1306–1310, June 1994.
55. D. W. Clarke and P. J. Gawthrop. A self-tuning controller. *IEE Proceedings*, 122:929–934, 1975.
56. D. W. Clarke and P. J. Gawthrop. Self-tuning control. *IEE Proceedings*, 126:633–640, 1979.
57. D. W. Clarke, C. Mohtadi, and P. S. Tuffs. Generalized predictive control - Part I. The basic algorithm. *Automatica*, 23:137–148, 1987.
58. D. W. Clarke, C. Mohtadi, and P. S. Tuffs. Generalized predictive control - Part II. Extensions and interpretations. *Automatica*, 23:149–160, 1987.

59. M. L. Corradini and G. Orlando. A discrete adaptive variable-structure controller for MIMO systems, and its application to an underwater ROV. *IEEE Transactions on Control Systems Technology*, 5(3):349–359, May 1997.
60. J. J. Craig. *Adaptive Control of Mechanical Manipulators*. Addison-Wesley, Reading, MA, 1988.
61. G. Cybenko. Approximations by superpositions of a sigmoidal function. *Mathematics of Control, Signals and Systems*, 2:303–314, 1989.
62. A. Datta and P. A. Ioannou. Performance analysis and improvement in model reference adaptive control. *IEEE Transactions on Automatic Control*, 39(12):2370–2387, December 1994.
63. A. P. Dempster, N. M. Laird, and D. B. Rubin. Maximum likelihood from incomplete data via the EM algorithm. *Journal of the Royal Statistical Society, B*, 39(1):1–38, 1977.
64. P. L. Dersin, M. Athans, and D. A. Kendrick. Some properties of the dual adaptive stochastic control algorithm. *IEEE Transactions on Automatic Control*, AC-26(5):1001–1008, October 1981.
65. J. G. Deshpande, T. N. Upadhyay, and D. G. Lainiotis. Adaptive control of linear stochastic systems. *Automatica*, 9:107–115, 1973.
66. M. A. Duarte and K. S. Narendra. Combined direct and indirect approach to adaptive control. *IEEE Transactions on Automatic Control*, 34(10):1071–1075, October 1989.
67. C. G. Economou and M. Morari. Internal model control, 5. Extension to nonlinear systems. *Ind. Eng. Chem. Process Des. Dev.*, 25(2):403–411, 1986.
68. B. Egardt. *Stability of Adaptive Controllers*. Springer-Verlag, Berlin, 1979.
69. B. Egardt. Stability analysis of discrete-time adaptive control schemes. *IEEE Transactions on Automatic Control*, AC-25(4):710–716, August 1980.
70. S. Fabri and V. Kadirkamanathan. Dynamic structure neural networks for stable adaptive control of nonlinear systems. *IEEE Transactions on Neural Networks*, 7(5):1151–1167, September 1996.
71. S. Fabri and V. Kadirkamanathan. Neural control of nonlinear systems with composite adaptation for improved convergence of Gaussian networks. In *4th European Control Conference ECC97*, Brussels, Belgium, July 1997.
72. S. Fabri and V. Kadirkamanathan. Dual adaptive control of nonlinear stochastic systems using neural networks. *Automatica*, 34(2):245–253, February 1998.
73. S. G. Fabri and V. Kadirkamanathan. Adaptive gain scheduling with modular models. In *Proceedings of the UKACC International Conference on CONTROL'98*, volume 1, pages 44–48, Swansea, Wales, UK, September 1998.
74. S. G. Fabri and V. Kadirkamanathan. Discrete-time adaptive control of nonlinear systems using neural networks. In *Preprints of the 1998 IFAC Workshop on AdaptiveSystems in Control and Signal Processing*, pages 153–158, Glasgow, Scotland, UK, August 1998.
75. S. G. Fabri and V. Kadirkamanathan. A self-organized multiple model approach for neural-adaptive control of jump nonlinear systems. In *Preprints of the 1998 IFAC Workshop on Adaptive Systems in Control and Signal Processing*, pages 147–152, Glasgow, Scotland, UK, August 1998.
76. A. A. Fel'dbaum. Dual control theory I-II. *Automation and Remote Control*, 21:874–880, 1033–1039, 1960.
77. A. A. Fel'dbaum. Dual control theory III-IV. *Automation and Remote Control*, 22:1–12, 109–121, 1961.
78. A. A. Fel'dbaum. *Optimal Control Systems*. Academic Press, New York, 1965.
79. N. M. Filatov, U. Keuchel, and H. Unbehauen. Dual control for an unstable mechanical plant. *IEEE Control Systems Magazine*, August 1996.

80. N. M. Filatov, H. Unbehauen, and U. Keuchel. Dual version of direct adaptive pole placement controller. In *Preprints of the 5th IFAC Symposium on Adaptive Systems in Control and Signal Processing*, pages 449–454, Hungary, June 1995.
81. J. H. Friedman. Multivariate adaptive regression splines. *Annals of Statistics*, 19:1–141, 1991.
82. K. S. Fu. Learning control systems- Review and outlook. *IEEE Transactions on Automatic Control*, AC-15:210–221, April 1970.
83. S. Fujita and T. Fukao. Optimal stochastic control for discrete-time linear system with interrupted observations. *Automatica*, 8:425–432, 1972.
84. K. Funahashi. On the approximate realization of continuous mappings by neural networks. *Neural Networks*, 2:183–192, 1989.
85. K. Furuta. Sliding mode control of a discrete system. *Systems and Control Letters*, (14):145–152, 1990.
86. K. Furuta. VSS type self-tuning control. *IEEE Transactions on Industrial Electronics*, 40(1):37–44, 1993.
87. W. Gao, Y. Wang, and A. Homaifa. Discrete-time variable structure control systems. *IEEE Transactions on Ind. Electronics*, 42(2):117–122, 1995.
88. S. S. Ge, C. C. Hang, and T. Zhang. A direct adaptive controller for dynamics systems with a class of nonlinear parameterizations. *Automatica*, 35:741–747, 1999.
89. C. J. Goh. Model reference control of non-linear systems via implicit function emulation. *International Journal of Control*, 60(1):91–115, 1994.
90. H. Gollee and K. J. Hunt. Nonlinear modelling and control of electrically stimulated muscle: a local model network approach. *International Journal of Control*, 68(6):1258–1288, 1997.
91. G. C. Goodwin, P. J. Caines, and P. E. Caines. Discrete time stochastic adaptive control. *SIAM Journal on Control and Optimization*, 19(6):829–853, 1981.
92. G. C. Goodwin and R. L. Payne. *Dynamic System Identification: Experiment Design and Data Analysis*. Academic Press, New York, 1977.
93. G. C. Goodwin, P. J. Ramadge, and P. E. Caines. Discrete-time multivariable adaptive control. *IEEE Transactions on Automatic Control*, AC-25(3):449–456, June 1980.
94. G. C. Goodwin and K. S. Sin. *Adaptive Filtering, Prediction and Control*. Prentice-Hall, Englewood Cliffs, New Jersey, 1984.
95. B. E. Griffiths and K. A. Loparo. Optimal control of jump linear Gaussian systems. *International Journal of Control*, 42(4):791–819, 1985.
96. J. W. Grizzle and P. V. Kokotović. Feedback linearization of sampled-data systems. *IEEE Transactions on Automatic Control*, 33(9):857–859, September 1988.
97. L. Guo. Further results on least squares based adaptive minimum variance control. *SIAM Journal on Control and Optimization*, 32(1):187–212, January 1994.
98. L. Guo and H-F. Chen. The Åström-Wittenmark self-tuning regulator revisited and ELS-based adaptive trackers. *IEEE Transactions on Automatic Control*, 36(7):802–812, 1991.
99. J. A. Gustafson and P. S. Maybeck. Flexible spacestructure control via moving-bank multiple model algorithms. *IEEE Transactions on Aerospace and Electronic Systems*, 30(3):750–757, July 1994.
100. C. J. Harris, C. G. Moore, and M. Brown. *Intelligent Control: Aspects of fuzzy logic and neural nets*, volume 6 of *World Scientific Series in Robotics and Automated Systems*. World Scientific Publishing Co. Pte. Ltd., Singapore, 1993.
101. S. Haykin. *Neural Networks: A comprehensive foundation*. Prentice-Hall, Upper Saddle River, NJ, 2nd edition, 1999.

102. J. H. Holland. *Adaptation in Natural and Artificial Systems*. University of Michigan Press, Ann Arbor, Michigan, 1975.
103. K. Hornik, M. Stinchcombe, and H. White. Multilayer feedforward networks are universal approximators. *Neural Networks*, 2:359–366, 1989.
104. J. Hu, D. M. Dawson, and Y. Qian. Position tracking control of an induction motor via partial state feedback. *Automatica*, 31(7):989–1000, July 1995.
105. D. J. Hughes and O. L. R. Jacobs. Turn-off, escape and probing in non-linear stochastic control. In *Preprints of the IFAC Symposium on Stochastic Control*, Budapest., 1974.
106. K. J. Hunt and T. A. Johansen. Design and analysis of gain-scheduled control using local controller networks. *International Journal of Control*, 66(5):619–651, 1997.
107. K. J. Hunt and D. Sbarbaro. Neural networks for nonlinear internal model control. *IEE Proceedings-D*, 138(5):431–438, September 1991.
108. K. J. Hunt, D. Sbarbaro, R. Zbikowski, and P. J. Gawthrop. Neural networks for control systems- A survey. *Automatica*, 28(6):1083–1112, 1992.
109. P. A. Ioannou and J. Sun. *Stable and Robust Adaptive Control*. Prentice-Hall, Englewood Cliffs, NJ, 1995.
110. P. A. Ioannou and G. Tao. Dominant richness and improvement of performance of robust adaptive control. *Automatica*, 25(2):287–291, March 1989.
111. T. Ionescu and R. Monopoli. Discrete model reference adaptive control with an augmented error signal. *Automatica*, 13:507–517, 1977.
112. T. Ishihara, K.-I. Abe, and H. Takeda. Extensions of innovations dual control. *International Journal of Systems Science*, 19(4):653–667, 1988.
113. A. Isidori. *Nonlinear Control Systems: An Introduction*. Springer-Verlag, Berlin, 1989.
114. O. L. R. Jacobs. *Introduction to Control Theory*. Oxford University Press, New York, 2nd edition, 1993.
115. O. L. R. Jacobs and J. W. Patchell. Caution and probing in stochastic control. *International Journal of Control*, 16:189–199, 1972.
116. O. L. R. Jacobs and R. V. Potter. Optimal control of a stochastic bilinear system. In M. J. Gregson, editor, *Recent Theoretical Developments in Control*, chapter 22, pages 403–419. Academic Press, USA, 1978.
117. R. A . Jacobs, M. I. Jordan, S. J. Nowlan, and G. E. Hinton. Adaptive mixtures of local experts. *Neural Computation*, 3:79–87, 1991.
118. R. A. Jacobs and M. I. Jordan. Learning piecewise control strategies in a modular neural network architecture. *IEEE Transactions on Systems, Man and Cybernetics*, 23(2):337–345, March/April 1993.
119. A. G. Jaffer and S. C. Gupta. On estimation of discrete processes under multiplicative and additive noise conditions. *Information Sciences*, 3:267–276, 1971.
120. S. Jagannathan and F. L. Lewis. Multilayer discrete-time neural-net controller with guaranteed performance. *IEEE Transactions on Neural Networks*, 7(1):107–130, January 1996.
121. A. H. Jazwinski. *Stochastic Processes and Filtering Theory*. Academic Press, New York, 1970.
122. T. A. Johansen and B. A. Foss. Constructing NARMAX models using ARMAX models. *International Journal of Control*, 58(5):1125–1153, 1993.
123. T. A. Johansen and B. A. Foss. Identification of non-linear system structure and parameters using regime decomposition. *Automatica*, 31(2):321–326, 1995.
124. T. A. Johansen and R. Murray-Smith. The operating regime approach to nonlinear modelling and control. In R. Murray-Smith and T.A. Johansen, editors,

Multiple Model Approaches to Modelling and Control, chapter 1, pages 3–72. Taylor and Francis, U. K., 1997.
125. M. I. Jordan and R. A. Jacobs. Hierarchical mixtures of experts and the EM algorithm. *Neural Computation*, (6):181–214, 1994.
126. M. I. Jordan and L. Xu. Convergence results for the EM approach to Mixtures of Experts architectures. *Neural Networks*, 8(9):1409–1431, 1995.
127. V. Kadirkamanathan. *Sequential learning in artificial neural networks.* PhD Thesis, Cambridge University Engineering Department, Cambridge, UK, September 1991.
128. V. Kadirkamanathan. A statistical inference based growth criterion for the RBF network. In *Proceedings of the IEEE Workshop on Neural Networks for Signal Processing IV*, pages 12–21, 1994.
129. V. Kadirkamanathan. Recursive nonlinear identification using multiple model algorithm. In *Proceedings of the IEEE Workshop on Neural Networks for Signal Processing V*, pages 171–180, 1995.
130. V. Kadirkamanathan and S. G. Fabri. Discrete-time adaptive sliding mode control of nonlinear systems using neural networks. In A. Tornambe, G. Conte, and A. M. Perdon, editors, *Theory and Practice of Control and Systems: Proceedings of the 6th IEEE Mediterranean Conference on Control and Systems*, pages 361–366, Alghero, Italy, June 1998. World Scientific Publishing, Singapore.
131. V. Kadirkamanathan and S. G. Fabri. Recursive structure estimation for nonlinear identification with modular networks. In T. Constantinides, S. Y. Kung, M. Niranjan, and E. Wilson, editors, *Proceedings of the 1998 IEEE Signal Processing Society Workshop on Neural Networks for Signal Processing VIII*, pages 343–350, New York, August 1998. IEEE.
132. V. Kadirkamanathan and S. G. Fabri. A stochastic method for neural-adaptive control of multi-modal nonlinear systems. In *Proceedings of the UKACC International Conference on CONTROL'98*, volume 1, pages 49–53, Swansea, Wales, UK, September 1998.
133. V. Kadirkamanathan and M. Kadirkamanathan. Kalman filter based estimation of dynamic modular networks. In *Proceedings of the IEEE Workshop on Neural Networks for Signal Processing VI*, pages 180–189, September 1996.
134. V. Kadirkamanathan and M. Kadirkamanathan. Recursive estimation of dynamic modular RBF networks. In D. S. Touretzky, editor, *Advances in Neural Information Processing Systems 8*, pages 239–245. MIT Press, 1996.
135. V. Kadirkamanathan and M. Niranjan. Application of an architecturally dynamic network for speech pattern recognition. *Proceedings of the Institute of Acoustics*, 14(6):343–350, 1992.
136. V. Kadirkamanathan and M. Niranjan. A function estimation approach to sequential learning with neural networks. *Neural Computation 5*, 6:854–975, 1993.
137. I. Kanellakopoulos, P. V. Kokotović, and R. Marino. An extended direct scheme for robust adaptive nonlinear control. *Automatica*, 27:247–255, 1991.
138. I. Kannellakopoulos. A discrete-time adaptive nonlinear system. *IEEE Transactions on Automatic Control*, 39(11):2362–2365, November 1994.
139. H. K. Khalil. *Nonlinear Systems.* Prentice-Hall, Upper Saddle River, NJ, 2nd edition, 1996.
140. A. Kimura, I. Arizono, and H. Ohta. An improvement of a back propagation algorithm by the extended Kalman filter and demand forecasting by layered neural networks. *International Journal of Systems Science*, 27(5):473–482, 1996.
141. P. V. Kokotović, editor. *Foundations of Adaptive Control.* Springer-Verlag, Berlin, 1991.

142. P. V. Kokotović, I. Kanellakopoulos, and A. S. Morse. Adaptive feedback linearization of nonlinear systems. In P. V. Kokotović, editor, *Foundations of Adaptive Control*, Lecture notes in Control and Information Sciences, pages 311-346. Springer-Verlag, Berlin, 1991.
143. U. Kotta. Comments on the stability of discrete-time sliding mode control systems. *IEEE Transactions on Automatic Control*, AC-34:1021-1022, 1989.
144. M. Krstić, I. Kanellakopoulos, and P. Kokotović. *Nonlinear and Adaptive Control Design*. Wiley Series on Adaptive and Learning Systems for Signal Processing, Communications, and Control. Wiley-Interscience, U.S.A, 1995.
145. M. Krstić, I. Kanellakopoulos, and P. V. Kokotović. Adaptive nonlinear control without overparametrization. *Systems and Control Letters*, 19:177-185, 1992.
146. M. Krstić and P. V. Kokotović. Modular approach to adaptive nonlinear stabilization. *Automatica*, 32(4):625-629, April 1996.
147. G. J. Kulawski and M. A. Brdyś. Stable adaptive control with recurrent networks. In *4th European Control Conference ECC97*, Brussels, Belgium, July 1997.
148. C. Kulcsár, L. Pronzato, and E. Walter. A dual control policy for linearly parameterized systems. In *Proceedings of the 3rd European Control Conference*, volume 1, pages 55-60, Rome, September 1995.
149. P. R. Kumar. A survey of some results in stochastic adaptive control. *SIAM Journal on Control and Optimization*, 23(3):329-380, May 1985.
150. D. G. Lainiotis. Optimal adaptive estimation: Structure and parameter adaptation. *IEEE Transactions on Automatic Control*, AC-16(2):160-170, 1971.
151. D. G. Lainiotis. Partitioning: A unifying framework for adaptive systems, Parts I, II. *Proceedings of the IEEE*, 64(8):1126-1143, 1182-1198, August 1976.
152. R. Langari and H. R. Berenji. Fuzzy logic in control engineering. In D. A. White and D. A. Sofge, editors, *Handbook of Intelligent Control: Neural, fuzzy and adaptive approaches*, chapter 4, pages 93-140. Van Nostrand Reinhold, New York, 1992.
153. A. Y. Lee and C. S. Sims. Adaptive estimation and stochastic control for uncertain models. *International Journal of Control*, 19(3):625-639, 1974.
154. I. J. Leontaritis and S. A. Billings. Input-output parametric models for nonlinear systems, Part I: Deterministic nonlinear systems. *International Journal of Control*, 41(2):303-328, 1985.
155. A. U. Levin and K. S. Narendra. Control of nonlinear dynamic systems using neural networks- Part II: Observability, Identification and Control. *IEEE Transactions on Neural Networks*, 7(1):30-42, January 1996.
156. F. L. Lewis, A. Yeşildirek, and K. Liu. Multilayer neural-net robot controller with guaranteed tracking performance. *IEEE Transactions on Neural Networks*, 7(2):388-399, March 1996.
157. X.-R. Li and Y. Bar-Shalom. Multiple-model estimation with variable structure. *IEEE Transactions on Automatic Control*, 41(4):478-493, April 1996.
158. G. Lightbody and G. W. Irwin. Nonlinear control structures based on embedded neural system models. *IEEE Transactions on Neural Networks*, 8(3):553-567, May 1997.
159. D. A. Linkens and H. O. Nyongesa. Learning systems in intelligent control: an appraisal of fuzzy, neural and genetic algorithm control applications. *IEE Proc.-Control Theory Applications*, 143(4):367-386, 1996.
160. G. P. Liu, V. Kadirkamanathan, and S. A. Billings. Predictive control of nonlinear systems using neural networks. *International Journal of Control*, 67(11):1645-1657, November 1998.

161. G. P. Liu, V. Kadirkamanathan, and S. A. Billings. Neural network based variable structure system nonlinear control. *International Journal of System Science*, 30(10):1153-1160, 1999.
162. G. P. Liu, V. Kadirkamanathan, and S. A. Billings. Variable neural networks for adaptive control of nonlinear systems. *IEEE Transactions in Systems, Man and Cybernetics, Part C*, 29(1):34-43, 1999.
163. L. Ljung. Asymptotic behaviour of the extended Kalman filter as a parameter estimator for linear systems. *IEEE Transactions on Automatic Control*, AC-24:36-50, February 1979.
164. L. Ljung and S. Gunnarsson. Adaptation and tracking in system identification: A survey. *Automatica*, 26(1):7-21, 1990.
165. A. Madani, S. Monaco, and D. Normand-Cyrot. Adaptive control of discrete-time dynamics in parametric strict-fedback form. In *Proceedings of the 35th Conference on Decision and Control*, pages 2659-2664, Kobe, Japan, December 1996. IEEE.
166. D. T. Magill. Optimal adaptive estimation of sampled processes. *IEEE Transactions on Automatic Control*, AC-10:434-439, October 1965.
167. A. L. Maitelli and T. Yoneyama. A two-stage suboptimal controller for stochastic systems using approximate moments. *Automatica*, 30(12):1949-1954, 1994.
168. R. Marino and P. Tomei. Global adaptive output-feedback control of nonlinear systems, Part II: Nonlinear parameterization. *IEEE Transactions on Automatic Control*, AC-38:17-48, 1993.
169. J. Mason and P. Parks. Selection of neural network structures: Some approximation theory guidelines. In D. A. White and D. A. Sofge, editors, *Neural Networks for Control and Systems*, IEE Control Engineering Series 42, pages 151-180. Reading, MA, 1992.
170. P. S. Maybeck and P. D. Hanlon. Perfromance enhancement of a multiple model adaptive estimator. *IEEE Transactions on Aerospace and Electronic Systems*, 31(4):1240-1253, October 1995.
171. P. S. Maybeck and D. L. Pogoda. Multiple model adaptive controller for the STOL F-15 with sensor/actuator failures. In *28th IEEE Conference on Decision and Control*, pages 1566-1572, Tampa, FL, December 1989.
172. P. S. Maybeck and R. D. Stevens. Reconfigurable flight control via multiple model adaptive control methods. *IEEE Transactions on Aerospace and Electronic Systems*, 27(3):470-479, May 1991.
173. E. Mazor, A. Averbuch, Y. Bar-Shalom, and J. Dayan. Interacting Multiple Model methods in target tracking: A survey. *IEEE Transactions on Aerospace and Electronic Systems*, 34(1):103-123, January 1998.
174. J. L. McClelland and D. E. Rumelhart, editors. *Parallel Distributed Processing: Explorations in the Microstructure of Cognition.* Bradford Books/ MIT Press, Cambridge, MA, U. S. A, 1986.
175. G. McLachlan and K. E. Basford. *Mixture Models: Inference and Applications to Clustering.* Marcel Dekker, 1988.
176. R. Milito, C. S. Padilla, R. A. Padilla, and D. Cadorin. An innovations approach to dual control. *IEEE Transactions on Automatic Control*, AC-27(1):133-137, February 1982.
177. M. Millnert. Identification of ARX models with Markovian parameters. *International Journal of Control*, 45(6):2045-2058, 1987.
178. S. Monaco and D. Normand-Cyrot. On the linearizing feedback in nonlinear sampled data control schemes. In *Proceedings of the 25th IEEE Conference on Decision and Control*, pages 2056-2060, Athens, 1986.

179. S. Monaco and D. Normand-Cyrot. Minimum phase nonlinear discrete-time systems and feedback stabilization. In *Proceedings of the 26th IEEE Conference on Decision and Control*, pages 979–986, 1987.
180. A. S. Morse. High-order parameter tuners for the adaptive control of linear and nonlinear systems. In *Proceedings of the US-Italy joint seminar "Systems, models and feedback: theory and computation"*, Capri, Italy, June 1992.
181. D. Q. Muñoz and D. H. Sbarbaro. Discrete robust adaptive controller based on artificial neural networks. In *4th European Control Conference*, Brussels, Belgium., July 1997.
182. K. S. Narendra. The maturing of adaptive control. In P. V. Kokotović, editor, *Founations of Adaptive Control*, Lecture notes in Control and Information Sciences, pages 3–36. Spinger-Verlag, Berlin, 1991.
183. K. S. Narendra. Neural networks for control: Theory and practice. *Proceedings of the IEEE*, 84(10):1385–1406, October 1996.
184. K. S. Narendra and A. M. Annaswamy. Robust adaptive control. In K. S. Narendra, editor, *Adaptive and Learning Systems: Theory and Applications*, pages 3–31. Plenum, New York, 1986.
185. K. S. Narendra and A. M. Annaswamy. A new adaptive law for robust adaptation without persistent excitation. *IEEE Transactions on Automatic Control*, AC-32(2):134–157, February 1987.
186. K. S. Narendra and A. M. Annaswamy. *Stable Adaptive Systems*. Prentice-Hall, Englewood Cliffs, New Jersey, 1989.
187. K. S. Narendra and J. Balakrishnan. Adaptive control using multiple models. *IEEE Transactions on Automatic Control*, 42(2):171–187, February 1997.
188. K. S. Narendra, J. Balakrishnan, and M. Kemal Ciliz. Adaptation and learning using multiple models, switching, and tuning. *IEEE Control Systems Magazine*, pages 37–51, June 1995.
189. K. S. Narendra and Y. H. Lin. Stable discrete adaptive control. *IEEE Transactions on Automatic Control*, AC-25(3):456–461, June 1980.
190. K. S. Narendra and S. Mukhopadhyay. Intelligent control using neural networks. *IEEE Control Systems Magazine*, pages 11–18, April 1992.
191. K. S. Narendra and S. Mukhopadhyay. Adaptive control using neural networks and approximate models. *IEEE Transactions on Neural Networks*, 8(3):475–485, May 1997.
192. K. S. Narendra, R. Ortega, and P. Dorato, editors. *Advances in Adaptive Control*. IEEE Press, U. S. A, 1991.
193. K. S. Narendra and K. Parthasarathy. Identification and control of dynamical systems using neural networks. *IEEE Transactions on Neural Networks*, 1(1):4–27, 1990.
194. R. Ortega. On Morse's new adaptive controller: Parameter convergence and transient performance. *IEEE Transactions on Automatic Control*, 38(8):1191–1202, August 1993.
195. C. S. Padilla, J. B. Cruz, and R. A. Padilla. A simple algorithm for SAFER control. *International Journal of Control*, 32(6):1111–1118, 1980.
196. J. Park and I. W. Sandberg. Universal approximation using radial basis function networks. *Neural Computation*, 3(2):246–257, 1991.
197. J. Park and I. W. Sandberg. Approximation and radial basis function networks. *Neural Computation*, 5(2):305–316, 1993.
198. K. M. Passino. Intelligent control for autonomous systems. *IEEE Spectrum*, pages 55–62, June 1995.
199. W. Pedrycz. *Fuzzy Control and Fuzzy Systems*. Wiley, New York, U. S. A, 2nd edition, 1993.

200. B. B. Peterson and K. S. Narendra. Bounded error adaptive control. *IEEE Transactions on Automatic Control*, AC-27(6):1161–1168, December 1982.
201. N. B. O. L. Pettit and P. E. Wellstead. Analyzing piecewise-linear dynamic systems. *IEEE Control Systems Magazine*, 15(5):43–50, 1995.
202. J. C. Platt. A resource allocating network for function interpolation. *Neural Computation*, 3(2):213–225, 1991.
203. T. Poggio and F. Girosi. Networks for approximation and learning. *Proceedings of the IEEE*, 78(9):1481–1497, September 1990.
204. M. Polycarpou and P. Ioannou. Identification and control of nonlinear systems using neural network models: Design and stability analysis. Technical Report 91-09-01, Department of Electrical Engineering Systems, University of Southern California, Los Angeles, USA, September 1991.
205. M. M. Polycarpou. Stable adaptive neural control scheme for nonlinear systems. *IEEE Transaction on Automatic Control*, 41(3):447–451, March 1996.
206. J. B. Pomet and L. Praly. Adaptive nonlinear regulation: estimation from the Lyapunov equation. *IEEE Transactions on Automatic Control*, 37:729–740, 1992.
207. M. J. D. Powell. Radial basis functions for multivariable interpolation: A review. In J. C. Mason and M. G. Cox, editors, *Algorithms for Approximation*, pages 143–167. Oxford University Press, Oxford, UK, 1987.
208. T. J. Procyk and E. H. Mamdani. A linguistic self-organising process controller. *Automatica*, 15:15–30, 1977.
209. L. Pronzato, C. Kulcsár, and E. Walter. An actively adaptive control policy for linear models. *IEEE Transactions on Automatic Control*, 41(6):855–858, June 1996.
210. D. Psaltis, A. Saridis, and A. A. Yamamura. A multilayered neural network controller. *IEEE Control Systems Magazine*, 8:17–21, 1988.
211. M. S. Radenković. Convergence of the generalized dual control algorithm. *International Journal of Control*, 47(5):1419–1441, 1988.
212. A. Rantzer and M. Johansson. Piecewise linear quadratic optimal control. In *Proceedings of the 1997 American Control Conference*, volume 3, pages 1749–1753, Albuquerque, NM, 1997.
213. H. E. Rauch. Autonomous control reconfiguration. *IEEE Control Systems Magazine*, pages 37–48, December 1995.
214. T. RayChaudhuri, L. G. C. Hamey, and R. D. Bell. From conventional control to autonomous intelligent methods. *IEEE Control Systems Magazine*, pages 78–84, October 1996.
215. G. A. Rovithakis and M. A. Christodoulou. Adaptive control of unknown plants using dynamical neural networks. *IEEE Transactions on Systems, Man and Cybernetics*, 24(3), March 1994.
216. G. A. Rovithakis and M. A. Christodoulou. Direct adaptive regulation of unknown nonlinear dynamical systems via dynamic neural networks. *IEEE Transactions on Systems, Man and Cybernetics*, 25(12):1578–1594, December 1995.
217. W. J. Rugh. Analytical framework for Gain Scheduling. *IEEE Control Systems Magazine*, pages 79–84, January 1991.
218. D. E. Rumelhart, G. E. Hinton, and Williams. Learning internal representations by error propagation. In J. L. McClelland and D. E. Rumelhart, editors, *Parallel Distributed Processing: Explorations in the Microstructure of Cognition*, volume 1, pages 318–361. Bradford Books/MIT Press, Cambridge, MA, 1986.
219. R. M. Sanner and J.-J. E. Slotine. Gaussian networks for direct adaptive control. *IEEE Transactions on Neural Networks*, 3(6), November 1992.

220. R. M. Sanner and J.-J. E. Slotine. Stable adaptive control of robot manipulators using "neural" networks. *Neural Computation*, 7:753–790, 1995.
221. G. N. Saridis and T. K. Dao. A learning approach to the parameter-adaptive self-organizing control problem. *Automatica*, 8:589–597, 1972.
222. S. Z. Sarptürk, Y. Istefanopulos, and O. Kaynak. On the stability of discrete-time sliding mode control systems. *IEEE Transactions on Automatic Control*, 32(10):930–932, 1987.
223. S. Sastry and M. Bodson. *Adaptive Control: Stability, Convergence and Robustness*. Prentice-Hall International, Englewood Cliffs, New Jersey, 1989.
224. S. S. Sastry and A. Isidori. Adaptive control of linearizable systems. *IEEE Transactions on Automatic Control*, 34:1123–1131, 1989.
225. F. C. Schweppe. *Uncertain Dynamic Systems*. Prentice-Hall, Inc., Englewood Cliffs, New Jersey, 1973.
226. J. S. Shamma and M. Athans. Analysis of gain scheduled control for nonlinear plants. *IEEE Transactions on Automatic Control*, 35:898–907, 1990.
227. J. S. Shamma and M. Athans. Gain Scheduling: Potential hazards and possible remedies. *IEEE Control Systems Magazine*, pages 101–107, June 1992.
228. K. S. Sin and G. C. Goodwin. Stochastic adaptive control using a modified least squares algorithm. *Automatica*, 18(3):315–321, 1982.
229. H. Sira-Ramirez. Non-linear discrete variable structure systems in quasi-sliding mode. *International Journal of Control*, 54(5):1171–1187, 1991.
230. A. Skeppstedt, L. Ljung, and M. Millnert. Construction of composite models from observed data. *International Journal of Control*, 55(1):141–152, 1992.
231. J.-J. E. Slotine and J. A. Coetsee. Adaptive sliding controller synthesis for nonlinear systems. *International Journal of Control*, 43:1631–1651, 1986.
232. J.-J. E. Slotine and W. Li. On the adaptive control of robot manipulators. *International Journal of Robotics Research*, 6:49–59, 1987.
233. J.-J. E. Slotine and W. Li. Composite adaptive control of robot manipulators. *Automatica*, 25(4):508–519, 1989.
234. J.-J. E. Slotine and W. Li. *Applied Nonlinear Control*. Prentice-Hall International, Englewood Cliffs, New Jersey, 1991.
235. J.-J. E. Slotine and S. S. Sastry. Tracking control of nonlinear systems using sliding surfaces, with application to robot manipulators. *International Journal of Control*, 38(2):465–492, 1983.
236. T. Söderström. *Discrete-time Stochastic Systems: Estimation and Control*. Prentice-Hall International, U.K., 1994.
237. E. D. Sontag. Nonlinear regulation: The piecewise linear approach. *IEEE Transactions on Automatic Control*, AC-26(2):346–357, April 1981.
238. H. W. Sorenson. An overview of filtering and stochastic control in dynamic systems. In C. T. Leondes, editor, *Control and Dynamic Systems*, volume 12, pages 1–61. Academic Press Inc., New York, 1976.
239. J. T. Spooner and K. M. Passino. Stable adaptive control using fuzzy systems and neural networks. *IEEE Transactions on Fuzzy Systems*, 4(3):339–359, August 1996.
240. F. Sun, Z. Sun, and P.-Y. Woo. Stable neural-network-based adaptive control for sampled-data nonlinear systems. *IEEE Transactions on Neural Networks*, 9(5):956–968, September 1998.
241. J. Sun. A modified model reference adaptive control scheme for improved transient performance. *IEEE Transactions on Automatic Control*, 38(8):1255–1259, August 1993.
242. T. Takagi and M. Sugeno. Fuzzy identification of systems and its application for modelling and control. *IEEE Transactions on Systems, Man and Cybernetics*, 15(1):116–132, 1985.

243. S. Tan, C.-C. Hang, and J.-S. Chai. Gain Scheduling: from conventional to neuro-fuzzy. *Automatica*, 33(3):411–419, 1997.
244. D. Taylor, P. V. Kokotović, R. Marino, and I. Kanellakopoulos. Adaptive regulation of nonlinear systems with unmodelled dynamics. *IEEE Transactions on Automatic Control*, 34:405–412, 1989.
245. A. R. Teel, R. R. Kadiyala, P. V. Kokotović, and S. S. Sastry. Indirect techniques for adaptive input-output linearization of non-linear systems. *International Journal of Control*, 53(1):193–222, 1991.
246. D. Titterington, A. F. M. Smith, and U. E. Makov. *Statistical Analysis of Finite Mixture Distributions*. John Wiley and Sons, Chichester, 1985.
247. E. Tse and M. Athans. Adaptive stochastic control for a class of linear systems. *IEEE Transactions on Automatic Control*, 17:38, 1972.
248. E. Tse and Y. Bar-Shalom. An actively adaptive control for linear systems with random parameters via the dual control approach. *IEEE Transactions on Automatic Control*, AC-18(2):109–117, April 1973.
249. E. Tse and Y. Bar-Shalom. Actively adaptive control for nonlinear stochastic systems. *Proceedings of the IEEE*, 64(8):1172–1181, August 1976.
250. E. Tse, Y. Bar-Shalom, and L. Meier. Wide-sense adaptive dual control for nonlinear stochastic systems. *IEEE Transactions on Automatic Control*, AC-18(2):98–108, April 1973.
251. Y. Tsypkin. *Adaptation and Learning in Automatic Systems*. Academic Press, New York, U. S. A, 1971.
252. J. K. Tugnait. Control of stochastic systems with Markov interrupted observations. *IEEE Transactions on Aerospace and Electronic Systems*, AES-19:232–239, 1983.
253. J. K. Tugnait and A. H. Haddad. State estimation under uncertain observations with unknown statistics. *IEEE Transactions on Automatic Control*, AC-24(2):201–210, 1979.
254. E. Tzirkel-Hancock. *Stable control of nonlinear systems using neural networks*. PhD Thesis, Cambridge University Engineering Department, Cambridge University, Cambridge, UK, June 1992.
255. E. Tzirkel-Hancock and F. Fallside. Stable control of nonlinear systems using neural networks. *International Journal of Robust and Nonlinear Control*, 2:63–86, May 1992.
256. V. I. Utkin. Sliding mode and its applications to variable structure systems. *IEEE Transactions on Automatic Control*, 22(2):212–222, 1977.
257. V. I. Utkin. *Sliding Modes and their Application in Variable Structure Systems*. MIR Publishers, Moscow, 1978.
258. V. I. Utkin. *Sliding Modes in Control and Optimization*. Springer, Berlin, 1992.
259. V. I. Utkin and S. V. Drakunov. On discrete-time sliding mode control. In *Preprints IFAC Conference on Nonlinear Control*, pages 484–489, Capri, Italy, 1989.
260. L. X. Wang. *Adaptive Fuzzy Systems and Control: Design and Stability Analysis*. Prentice-Hall, Englewood Cliffs, U. S. A., 1994.
261. K. Watanabe. A passive type multiple-model adaptive control (MMAC) of linear discrete-time stochastic systems with uncertain observation subsystems. *International Journal of Systems Science*, 15(6):647–659, 1984.
262. K. Watanabe. *Adaptive Estimation and Control: Partitioning Approach*. Prentice-Hall International, U.K., 1991.
263. K. Watanabe, T. Fukuda, and S. Tzafestas. Learning algorithms of layered neural networks via extended Kalman filters. *International Journal of Systems Science*, 22(4):753–768, 1991.

264. K. Watanabe and S. G. Tzafestas. Multiple-model adaptive control for jump-linear stochastic systems. *International Journal of Control*, 50(5):1603–1617, 1989.
265. K. Watanabe and S. G. Tzafestas. A hierarchical multiple model adaptive control of discrete-time stochastic systems for sensor and actuator uncertainties. *Automatica*, 26(5):875–886, 1990.
266. C. J. Wenk and Y. Bar-Shalom. A multiple model adaptive dual control algorithm for stochastic systems with unknown parameters. *IEEE Transactions on Automatic Control*, AC-25(4):703–710, 1980.
267. P. J. Werbos. Neurocontrol and supervised learning: An overview and evaluation. In D. A. White and D. A. Sofge, editors, *Handbook of intelligent control: Neural, fuzzy and adaptive approaches*, chapter 3, pages 65–89. Van Nostrand Reinhold, New York, 1992.
268. D. A. White and D. A. Sofge, editors. *Handbook of intelligent control:Neural, fuzzy and adaptive approaches*. Van Nostrand Reinhold, New York, 1992.
269. J. Wieslander and B. Wittenmark. An approach to adaptive control using real-time identification. *Automatica*, 7:211–217, 1971.
270. B. Wittenmark. Stochastic adaptive control methods: a survey. *International Journal of Control*, 21:705–730, 1975.
271. B. Wittenmark. Adaptive dual control methods: An overview. In *Preprints of the 5th IFAC Symposium on Adaptive Systems in Control and Signal Processing*, pages 67–72, Hungary, June 1995.
272. B. Wittenmark and C. Elevitch. An adaptive control algorithm with dual features. In *Proceedings of the 7th IFAC/IFORS Symposium on Identification and System Parameter Estimation*, pages 587–592, York, U.K., 1985.
273. L. L. Xie and L. Guo. Adaptive control of a class of discrete-time affine nonlinear systems. *Systems and Control Letters*, 35:201–206, 1998.
274. A. Yeşildirek and F. L. Lewis. Feedback linearization using neural networks. *Automatica*, 31(11):1659–1664, 1995.
275. P. C. Yeh and P. V. Kokotović. Adaptive output-feedback design for a class of nonlinear discrete-time systems. *IEEE Transactions on Automatic Control*, 40:1663–1668, 1995.
276. S.-H. Yu and A. M. Annaswamy. Adaptive control of nonlinear systems using θ-adaptive neural networks. *Automatica*, 33(11):1975–1995, 1997.
277. L. A. Zadeh. Fuzzy sets. *Information and control*, (8):338–352, 1965.
278. J. Zhang and A. J. Morris. Fuzzy neural networks for nonlinear systems modelling. *IEE Proc.-Control Theory Appl.*, 142(6):551–561, November 1995.
279. T. Zhang, S. S. Ge, and C. C. Hang. Neural-based direct adaptive control for a class of general nonlinear systems. *International Journal of Systems Science*, 28(10):1011–1020, 1997.
280. T. Zhang, S. S. Ge, and C. C. Hang. Direct adaptive control of non-affine nonlinear systems using multilayer neural networks. In *Proceedings of the 1998 American Control Conference*, volume 1, pages 515–519, Philadelphia, Pennsylvania, June 1998.
281. J. Zhao and I. Kanellakopoulos. Adaptive control of discrete-time output-feedback nonlinear systems. In *Proceedings of the 5th IEEE Mediterranean Conference on Control and Systems*, Cyprus, July 1997. Focus Interactive Technology Inc.
282. J. Zhao and I. Kanellakopoulos. Adaptive control of discrete-time strict-feedback nonlinear systems. In *Proceedings of the 1997 American Control Conference*, volume 1, pages 828–832, Albuquerque, NM, 1997.

Index

Φ-based activation, 68

AAC, 143
adaptation, 4
adaptive control, 29
– functional, 31, 44, 50, 81, 101, 103, 147, 149
– nonlinear, 29, 30, 42
ASOD, 143
augmented error control, 101, 104

backstepping control, 27
basis functions, 32
Bellman Equation, 134
Bellman's Principle of Optimality, 133
bicriterial approach, 144
block-strict-feedback systems, 28

caution, 136
cautious control, 139
certainty equivalence, 137
closed-loop policy, 140
combined adaptive control, 80
composite adaptive control, 79, 80, 84
curse of dimensionality, 34, 47, 134, 153

dead-zone adaptation, 48, 86, 105
detection-estimation algorithm, 173
dual control, 14, 136, 147
DUL algorithm, 168
dynamic programming, 133
dynamic structure network, 48, 54

expert systems, 6
extended Kalman filter, 154

feedback linearization, 24
functional adaptive systems, 8, 29
fuzzy logic, 6

gain scheduling, 13, 180, 213
gating network, 220

– Gaussian mixture kernel, 221
– softmax, 220
generalized minimum variance control, 139
generalized pseudo-Bayes, 173, 174

hard partitioning, 181
heuristic certainty equivalent control, 138

IDC, 144, 147, 228
indicator variable, 180, 192
information state, 133
intelligent control, 4, 5, 8, 245
interacting multiple model, 173, 176
inverse control, 8

jump systems, 170, 187

Kalman filter, 139, 151, 167, 173, 191, 217

learning, 4, 12, 187
Lie derivative, 24
Linear Quadratic Gaussian, 135, 137, 169
Lyapunov stability, 35, 59, 88, 101, 105, 123

MAPIDC, 203
minimum variance control, 138
mixture modelling, 183
mixture of experts, 183
MMIDC, 203
modular networks, 13, 182, 183, 216
multimodality
– spatial, 11, 12, 180, 213
– temporal, 10, 11, 170, 187
multiple model systems, 10, 165
– adaptive control, 168
– dual adaptive control, 187, 189, 213, 216

– state estimation, 167

NARMA representation, 40
– NARMA-L1, 42
– NARMA-L2, 42, 101
neural control, 34
neural networks, 7, 31
– Gaussian radial basis function, 32
– multi-layer perceptron, 32
neutrality, 137
nonlinear systems
– continuous-time, 23
– discrete-time, 36
normal form, 26, 39

OLOF policy, 140
optimal control, 132
– policy, 132

Partitioning Theory, 11
predictive control, 139
probing, 136
pure-feedback systems, 28

random sampling algorithm, 173
reinforcement learning, 15, 48

self-organized model allocation, 197
self-tuning control, 138
separability, 137
sliding control, 48, 88
– adaptive, 120
– boundary layer, 65
– discrete-time, 117
– discrete-time condition, 120
– sliding sector, 45, 120
soft partitioning, 181
stochastic adaptive control, 14
stochastic control, 131
strict-feedback systems, 27
suboptimal-dual control, 141
supervised control, 6

turn-off, 140

uncertainty
– functional, 6, 29
– parametric, 5, 29
Universal Approximation Property, 33, 52, 103, 150, 190

validity functions, 181, 217, 220

zero dynamics, 26, 39